W0107080

THE BRIGHTEST BINARIES

ASTROPHYSICS AND SPACE SCIENCE LIBRARY

VOLUME 232

Executive Committee

W. B. BURTON, *Sterrewacht, Leiden, The Netherlands*
J. M. E. KUIJPERS, *Faculty of Science, Nijmegen, The Netherlands*
E. P. J. VAN DEN HEUVEL, *Astronomical Institute, University of Amsterdam, The Netherlands*
H. VAN DER LAAN, *Astronomical Institute, University of Utrecht, The Netherlands*

Editorial Board

I. APPENZELLER, *Landessternwarte Heidelberg-Königstuhl, Germany*
J. N. BAHCALL, *The Institute for Advanced Study, Princeton, U.S.A.*
F. BERTOLA, *Università di Padova, Italy*
W. B. BURTON, *Sterrewacht, Leiden, The Netherlands*
J. P. CASSINELLI, *University of Wisconsin, Madison, U.S.A.*
C. J. CESARSKY, *Centre d'Etudes de Saclay, Gif-sur-Yvette Cedex, France*
O. ENGVOLD, *Institute of Theoretical Astrophysics, University of Oslo, Norway*
J. M. E. KUIJPERS, *Faculty of Science, Nijmegen, The Netherlands*
R. McCRAY, *University of Colorado, JILA, Boulder, U.S.A.*
P. G. MURDIN, *Royal Greenwich Observatory, Cambridge, U.K.*
F. PACINI, *Istituto Astronomia Arcetri, Firenze, Italy*
V. RADHAKRISHNAN, *Raman Research Institute, Bangalore, India*
K. SATO, *School of Science, The University of Tokyo, Japan*
F. H. SHU, *University of California, Berkeley, U.S.A.*
B. V. SOMOV, *Astronomical Institute, Moscow State University, Russia*
R. A. SUNYAEV, *Space Research Institute, Moscow, Russia*
Y. TANAKA, *Institute of Space & Astronautical Science, Kanagawa, Japan*
S. TREMAINE, *CITA, University of Toronto, Canada*
E. P. J. VAN DEN HEUVEL, *Astronomical Institute, University of Amsterdam, The Netherlands*
H. VAN DER LAAN, *Astronomical Institute, University of Utrecht, The Netherlands*
N. O. WEISS, *University of Cambridge, U.K.*

THE BRIGHTEST BINARIES

by

D. VANBEVEREN

Astrophysical Institute, Vrije Universiteit Brussel, Belgium
Institute of Technology Louvain, Leuven, Belgium

W. VAN RENSBERGEN

Astrophysical Institute, Vrije Universiteit Brussel, Belgium

and

C. DE LOORE

Astrophysical Institute, Vrije Universiteit Brussel, Belgium

KLUWER ACADEMIC PUBLISHERS

DORDRECHT / BOSTON / LONDON

A C.I.P. Catalogue record for this book is available from the Library of Congress.

ISBN-13: 978-1-4020-0376-9 e-ISBN-13: 978-94-010-9831-1
DOI: 10.1007/978-94-010-9831-1

Published by Kluwer Academic Publishers,
P.O. Box 17, 3300 AA Dordrecht, The Netherlands.

Sold and distributed in North, Central and South America
by Kluwer Academic Publishers,
101 Philip Drive, Norwell, MA 02061, U.S.A.

In all other countries, sold and distributed
by Kluwer Academic Publishers,
P.O. Box 322, 3300 AH Dordrecht, The Netherlands.

Printed on acid-free paper

All Rights Reserved
©1998 Kluwer Academic Publishers
Softcover reprint of the hardcover 1st edition 1998
No part of the material protected by this copyright notice may be reproduced or
utilized in any form or by any means, electronic or mechanical,
including photocopying, recording or by any information storage and
retrieval system, without written permission from the copyright owner

Dedication:

To those who have taught us the physics of the sun and the stars: Marcel Minnaert, Cees de Jager and Ed van den Heuvel

Table of contents

Table of contents 1
Introduction 8
List of abbreviations 10
Constants 12

Chapter I: MASSIVE STARS 13

1. Definition 13
2. General observations 13
   2.1. The HR diagram of massive stars with spectral type O3 - M5 15
      2.1.1. The effective temperature and bolometric correction of massive stars 15
      2.1.2. The absolute visual magnitude M_v of massive stars 18
      2.1.3. The spectral type - luminosity class - M_{bol} - T_{eff} - (B-V) calibration of massive stars 19
   2.2. The stellar wind of massive OB type stars 26
   2.3. The chemical abundances in massive OBA type stars 31
   2.4. Rotational properties of OB type stars 32
      2.4.1. 'Normal' OB type stars 34
      2.4.2. OBe type stars 35
   2.5. LBVs 36
   2.6. YSGs, RSGs and hypergiants 38
   2.7. The WR stars 39
   2.8. Circumstellar shells around massive stars 42
   2.9. The SN explosion of a massive star 43
   2.10. Summary 47

Chapter II: MASSIVE SINGLE STARS 49

3. The equations of stellar structure of a non-magnetic, non-rotating single star 49
   3.1. The continuity of mass 50
   3.2. The physical state of the plasma 50
      3.2.1. Hydrostatic equilibrium 50

3.2.2. Non-degeneracy and degeneracy: definition 51
3.2.3. The equation of state of a non-degenerate plasma 51
3.2.4. The equation of state of a degenerate plasma 52
3.2.5. The Chandrasekhar mass limit of degenerate stars 53
3.2.6. Neutron star masses 54
3.3. The energy equation 55
3.3.1. The energy generation rate by nuclear reactions ε 55
3.3.2. The gravitational energy generation 59
3.4. Energy transport 60
3.4.1. Radiative energy transport 60
3.4.2 Convective energy transport and mixing 61
3.4.3. The boundary between convective and radiative layers ... 63
3.5. The time-variation of mass in evolutionary computations 65
3.6. Summary 66
3.7. Surface layers - Boundary conditions 67
4. Evolutionary computations of non-rotating massive single stars 70
4.1. Small convective core overshooting 71
4.1.1. The evolution of single stars with $5\ M_o \leq M \leq 12\ M_o$ 71
4.1.2. The evolution of single stars with $M > 12\ M_o$ 76
4.2. The effect of large convective core overshooting 85
5. Comparison between observations and evolutionary computations of non-rotating single stars 87
5.1. Evolutionary computations with small convective core overshooting 87
5.1.1. The core hydrogen burning phase 87
5.1.2. The position of the RSGs and WR stars in the HR diagram 93
5.1.3. The progenitor of SN 1987A 94
5.1.4. The progenitor of SN 1993J 95
5.1.5. The overall SN morphology of massive single stars 95
5.1.6. The Blue Hertzsprung Gap (BHG) 96
5.1.7. Chemical abundances in early B type supergiants 96
5.1.8. The distribution of RSGs in the SMC 97
5.1.9. Comparison between theoretical prediction of degenerate stars and observations 98
5.2. Large convective core overshooting: is there a need? 100
6. The effect of rotation on the evolution of massive single stars 101

6.1. Rotation-induced mixing in massive stars 101
6.2. Evolutionary computations of rotating stars 103
6.3. Comparison to observations 104
6.4. Summary 105

Chapter III: MASSIVE CLOSE BINARIES 106

7. General 106
8. The Roche model 106
8.1. Circularization and synchronization 107
8.2 Roche equipotentials 108
8.3. The Roche lobe overflow process 110
8.4. Limitations of the Roche model 111
8.5. The mass loss rate of the mass loser during the RLOF 111
8.6. The different types of unevolved massive close binaries 113
9. The mass transfer process during RLOF 116
9.1. Direct hit versus the formation of a Keplerian disc 116
9.2. The spin-up of the mass gainer due to mass transfer 117
9.3. Computation of the evolutionary behaviour of a mass gainer ... 118
9.3.1. The standard accretion model 120
9.3.2. The accretion induced full mixing model 123
10. The evolution of the binary period in MCBs 124
10.1. One or both components are losing mass by stellar wind only. 124
10.2. The variation of the period during RLOF 125
10.2.1. The RLOF mass loss dominates over a possible SW mass loss 126
10.2.1.1. Conservative RLOF 126
10.2.1.2. Non-conservative RLOF 127
10.2.1.3. Mergers 131
10.2.2. The RLOF and the SW mass loss rates are comparable 132
11. The effect of the supernova explosion of one of the components 132
12. Evolutionary computations of massive close binaries 137
12.1. The evolution of the primary before and during its RLOF 139
12.1.1. Massive primaries with initial mass ≤ 40-50 M_o: case B_r with q > 0.2 139
12.1.2. Massive primaries with initial mass ≤ 40-50 M_o: case B_r with q ≤ 0.2, case B_c and case C 145

12.1.3. Massive primaries with initial mass ≤ 40-50 M_o: case A.... 147
12.1.4. Very massive primaries 150
12.1.5. The binary after RLOF: a CHeB+OB binary or a merger? 151
12.2. The evolution of the primary after RLOF: CHeB+[OB or low mass companion] binaries 153
12.2.1. The mass of the CHeB component ≥ 5 M_o 154
12.2.2. The mass of the CHeB component < 5 M_o 156
12.2.3. The final fate of primaries of MCBs 158
12.3. The evolution of the secondary in a MCB 159
12.4. The formation and evolution of contact binaries 163
12.5. The evolution of a MCB after the collapse of the core of the primary 165
12.6. The effect of rotation on the evolution of a MCB 169
12.6.1. The primary 169
12.6.2. The secondary 170
13. Massive binaries: observations 170
13.1. Massive OBA+OBA binaries 171
13.1.1. Pre-RLOF OBA+OBA MBs 171
13.1.2. Massive OBA+OBA binaries during or after RLOF 179
13.2. WR+OB binaries 181
13.3. X-ray binaries 184
13.3.1. High mass X-ray binaries (HMXB). 184
13.3.1.1. HMXBs with an OBe type optical component 184
13.3.1.2. Standard HMXBs 185
13.3.1.3. HMXB candidates with a WR type component 187
13.3.2. Low mass X-ray binaries (LMXB) with a MCB history 187
13.3.3. The formation of X-rays with energies ≥ 1 keV 188
13.4. Radio pulsars 195
13.4.1. General 195
13.4.2. Binary radio pulsars 195
13.5. Massive star runaways 198
13.5.1. OB type runaways 198
13.5.2. Runaways other than OB type runaways 199
14. Comparison between the predicted evolution of MCBs and observations 200

14.1. The binaries HD 163181, HD 12323, HD 14633, HD 193516, HD25638, HD 209481, Φ Per and υ Sgr 200
14.2. WR+OB binaries 207
14.2.1. WR+OB systems where RLOF/mass transfer did not occur 209
14.2.2. WR+OB systems where RLOF/mass transfer did occur: V 444 Cyg 211
14.2.3. WR stars with a less luminous OB type companion 218
14.2.4. WR stars with a normal low mass companion 219
14.2.5. Late O/early B type stars with a (too) faint WR type component 221
14.3. The formation of carbon enhanced OB type stars 221
14.4. The descendants of CHeB+OB binaries 222
14.4.1. HMXBs with an OB-type component: Vela X-1, Wray 977, Cyg X-1 223
14.4.2. The OB type runaways ζ Oph and ζ Pup 230
14.4.3. Critical remarks 235
14.5. The descendants of OB+cc binaries: the CHeB (WR)+cc candidates HD 50896, HD 197406 and Cyg X-3 236
14.5.1. The formation of X-rays in WR+cc binaries 237
14.5.2. The X-ray opacity in the wind in HMXBs 238
14.5.3. The expected X-rays in WR+cc binaries 245
14.5.4. The formation of WR stars with a cc in their center 251
14.6. The evolution of LMXBs: evidence for MCBs with very low mass ratio 251
14.7. SN-type of a MCB component: SN 1987A 255
14.8. Summary: the overall MCB evolutionary model 257

Chapter IV: MASSIVE STAR POPULATION NUMBER SYNTHESIS 262

15. General 262
16. Observations used as input in a population model 262
16.1. The initial mass function (IMF) 263
16.2. The IMF of starburst regions 264
16.3. The IMF of massive single stars and of primaries of MCBs 265
16.4. The overall MCB frequency in the solar neighbourhood 265

16.5. The observed mass ratio and period distribution of pre-RLOF MCBs in the solar neighbourhood 269
16.5.1. The observed mass ratio distribution 269
16.5.1.1. Method of determination 269
16.5.1.2. O-type primaries 271
16.5.1.3. B-type primaries 272
16.5.2. The observed period distribution 273
16.5.3. The implications of selection effects 274
16.6. The runaway velocities of single pulsars 275
17. The observations which need to be reproduced by a massive star population model 279
17.1. The B/R, WR/O and WC/WN number ratios 279
17.1.1. Continuous star formation regions 279
17.1.2. Starburst regions 280
17.2. WR+OB binaries 284
17.2.1. The WR+OB frequency 284
17.2.2. The q and P distribution of WR+OB 284
17.3. O-type runaways 286
17.4. The number of WR stars with a compact companion 286
17.5. The radio pulsar population 286
17.5.1. Solitary radio pulsars with a binary history 286
17.5.2. The formation rate of double neutron star systems in the Galaxy 287
17.6. The SN rate 288
18. The population number synthesis (PNS) model 289
18.1 The overall evolutionary model of single stars and MCBs 290
18.1.1. Single stars 290
18.1.2. MCBs 294
18.2. Method of computation 297
18.3. The parameters of the PNS model 298
19. PNS model calculations 301
19.1. The expected evolution of the observed galactic WR+OB sample 301
19.2. The predicted distribution of stars in regions of continuous star formation with a realistic binary formation frequency 303

19.2.1. The large number of stars observed in the Blue Hertzsprung Gap (BHG) 303
19.2.2. The distribution of O-type stars predicted by the PNS model 304
19.2.3. The WR population 307
19.2.4. The early B-type star population predicted by the PNS model 314
19.2.5. The theoretically expected number of binary pulsars 316
19.2.6. The theoretically expected number of single pulsars with a MCB history 319
19.2.7. The runaway velocity of single pulsars 319
19.3. The predicted distribution of stars in starburst regions with a realistic binary formation frequency 320
19.3.1. The population of B3-B0 type rapid rotators in starbursts... 322
19.3.2. The population of O and WR type stars in starbursts 322
19.3.3. The population of RSGs and WR type stars in starbursts... 326
19.3.4. Cooking recipe 327
19.4. The predicted SN type $I_{b,c}$/type II ratio in regions of continuous star formation with a realistic binary formation frequency 328

Concluding remarks 330
References 331
Index 343

Introduction

The evolution of galaxies is governed mainly by the evolution of massive stars whereas the evolution of a massive star depends primarily on its mass, chemical composition and on whether or not the star is a single object or a binary component. To study the evolution of galaxies, it is therefore essential to know how stellar masses are distributed at birth, how many stars are formed in binaries, and what the mass ratio and orbital period distribution of binaries look like.

Massive stars are intrinsically the brightest stars, so that it may be possible to discover their properties in distant groups provided that large telescopes can be used for basic stellar observations. However, until now the observations of massive stars have been reasonably complete only for a small region of our own Galaxy ($\leq$ 3 kpc from the Sun). One hopes that the conclusions resulting from these observations hold for the whole Galaxy, for the whole cosmos.

With 'The Brightest Stars' of De Jager (1980) in mind, the present monograph is an addendum and an update in which we discuss the observations of 'The Brightest Binaries' in the framework of stellar evolution.

A small or intermediate mass star close to the Sun may look brighter than a massive one far away. However, within volume limited star samples, the massive stars are on average also the brightest ones. In the present monograph (similarly as in the work of De Jager), bright means massive.

The book consists of four main chapters.

In chapter I, we first discuss general observations of massive stars that are important for the evolution of single stars and binary components.

It is unthinkable to study evolutionary properties of massive binaries without studying the evolution of single stars and vice versa. Chapter II therefore deals with the stellar structure processes and the evolution of massive single stars. By comparing the evolutionary computations with the general observations of chapter I, one can possibly restrict uncertain parameters that enter the mathematical formalism of the physical processes describing the structure and evolution.

Binaries are considered in chapter III. We discuss the processes that, in addition to the structure equations introduced in chapter II, govern binary evolution. The latter is then studied in detail with special emphasis on the uncertainties and their effect on the computational results.

We present a fairly complete summary of the available massive binary observations that are important to understand binary evolution. By comparing computations with the observations of binaries that are representative for the class they belong to, general conclusions are proposed and uncertainties are restricted.

In chapter IV we combine the results of massive single stars and massive binaries in order to make predictions concerning the content of stellar populations and the rates of different types of supernovae. We investigate to what extent uncertainties in binary statistics (frequency, mass ratio and period distribution) affect the results.

Two overall conclusions:

- By combining as many observations as are available of massive single stars and of massive close binaries, it is possible to present a consistent evolutionary model for both that is able to explain the stellar content and the overall stellar properties within the solar neighbourhood ($\leq$ 3kpc from the Sun).

- Knowledge of the distribution and the evolution of 'The Brightest Binaries' throughout the Galaxy and the cosmos, is essential in order to understand the population of 'The Brightest Stars' and the rates of the different types of supernovae in different galaxies.

List of abbreviations

AGB	Asymptotic Giant Branch
BC	Bolometric Correction
BH	Black Hole
BHG	Blue Hertzsprung Gap
BSC	Bright Star Catalogue
cc	compact companion
CCB	Core Carbon Burning
CHB	Core Hydrogen Burning
CHeB	Core Helium Burning
DAO8	Eighth Catalogue of the Orbital Elements of Spectroscopic Binary Systems
EUV	Extreme Ultra Violet
HST	Hubble Space Telescope
HMXB	High Mass X-ray Binary
HRD	Hertzsprung-Russell Diagram
IMF	Initial Mass Function
IRAS	Interactive Radar Analysis Software
IUE	International Ultraviolet Explorer
L	Ledoux (criterion)
LAOS	Opacity tables of Cox and Tabor, 1976
LBV	Luminous Blue Variable
LMC	Large Magellanic Cloud
LMXB	Low Mass X-ray Binary
M-L	Mass-Luminosity (relation)
MB	Massive Binary
MC	Magellanic Cloud
MCB	Massive Close Binary
MKK	Morgan, Keenan and Kellman - Atlas (1943).
(N)LTE	(Non) Local Thermodynamic Equilibrium
NS	Neutron Star
OBC	OB-type star with strong Carbon lines
OBN	OB-type star with strong Nitrogen lines
OPAL	Opacity Project tables of Igglesias et al., 1992
PNN	Planetary Nebula Nucleus
PNS	Population Number Synthesis
RLOF	Roche Lobe Overflow
ROSAT	ROntgen SATellite
RSG	Red Supergiant

SB	Spectroscopic Binary
SB1	Single line SB
SB2	Double line SB
SH	Schwarzschild and Härm (criterion)
SFR	Star Formation Rate
SMC	Small Magellanic Cloud
SN	Supernova
SW	Stellar Wind
TAMS	Terminated Age Main Sequence
T_{eff}	Effective Temperature
TZO	Thorne-Zytkow Object
VMCB	Very Massive Close Binary (primary mass > 40-50 $M_\odot$)
WC	WR of the Carbon sequence
WCD	Wind Compressed Disc
WD	White Dwarf
WN	WR of the Nitrogen sequence
WNE	Early type WN
WNL	Late type WN
WO	WR of the Oxygen sequence
WR	Wolf-Rayet (star)
YSG	Yellow Supergiant
ZAMS	Zero Age Main Sequence

Constants

The following values for physical and astronomical constants are used throughout this book.

Physical constants

Gravitation constant	$G = 6.6732\text{x}10^{-8}\text{dyne.cm}^2.\text{g}^{-2}$
Planck constant	$h = 6.626196\text{x}10^{-27}\text{erg.s}$
Radiation density constant (Stefan)	$a_R = \dfrac{8\pi^5 k^4}{15c^3h^3} = 7.56464\text{x}10^{-15}\text{erg.cm}^{-3}.\text{K}^{-4}$
Stefan- Boltzmann constant	$\sigma_R = \dfrac{a_R c}{4} = 5.66956\text{x}10^{-5}\text{erg.cm}^{-2}.\text{K}^{-4}.\text{s}^{-1}$
Universal Gas constant	$\Re = 8.31434\text{x}10^7\text{erg.g}^{-1}.\text{K}^{-1}$
Proton mass	$m_H = 1.672661\text{x}10^{-24}\text{g}$
Electron mass	$m_e = 9.1046\text{x}10^{-28}\text{g}$

Astronomical constants

Solar mass	$M_o = 1.989\text{x}10^{33}\text{g}$
Solar radius	$R_o = 6.9598\text{x}10^{10}\text{cm}$
1 astronomical unit	$1\text{a.u.} = 1.49597892\text{x}10^{13}\text{cm}$
1 parsec	$1\text{pc} = 3.085678\text{x}10^{18}\text{cm}$
Solar effective temperature	$T_{eff,o} = 5770\text{K}$
Solar absolute bolometric magnitude	$M_{bol,o} = 4.75$
Solar absolute luminosity	$L_o = 3.826\text{x}10^{33}\text{erg.s}^{-1}$

CHAPTER I

MASSIVE STARS

1. Definition

A star (single or binary component) is a massive star when it performs all nuclear burning phases non-degenerately up to the formation of an iron-nickel core.

The relation to the initial stellar mass depends on a number of stellar structure and evolutionary parameters, and on the fact that the star is a single star (section 4) or a component of a binary (section 12). We will show that one can distinguish evolutionary phases where the star has properties corresponding to observed properties of certain stellar subclasses. In this way it is possible to identify the following subclasses as massive stars:

- stars with luminosity class V or IV and spectral type earlier than B3
- stars with luminosity class III and spectral type earlier than B4
- stars with luminosity class II and spectral type earlier than B5
- the OBA stars with luminosity class Ib, Iab, Ia and IaO
- the Luminous Blue Variables (LBVs)
- the Yellow Supergiants (YSGs), Red Supergiants (RSGs) and the Hypergiants
- the Wolf-Rayet (WR) stars.

2. General observations

We first summarize general observations that are important for (or must be explained by) massive star evolution. The specific binary observations will be considered in section 13.

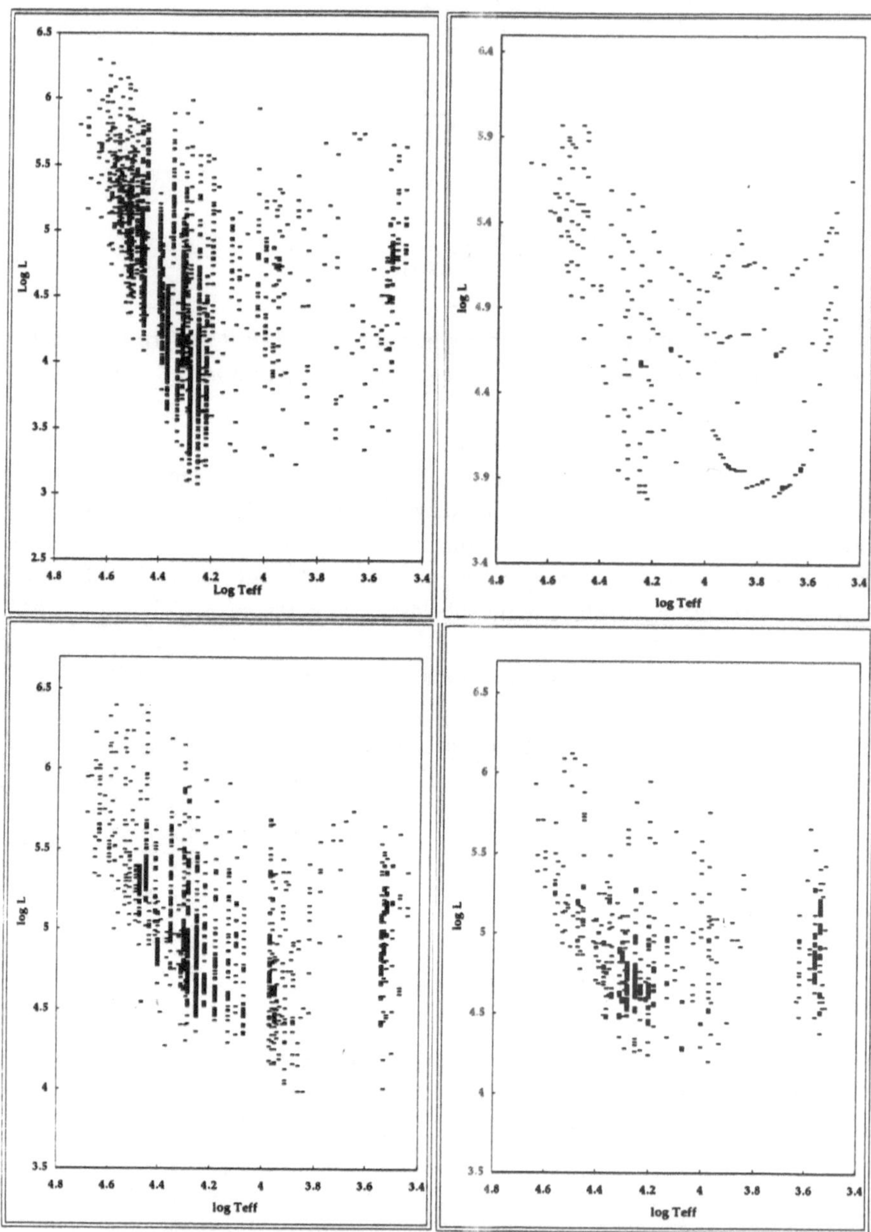

Figure 2.1: The HR diagram of 2400 stars belonging to stellar aggregates in the solar neighbourhood (upper-left), of ~2700 massive galactic field stars in the solar neighbourhood (upper-right; since a [M_{bol}-T_{eff}-spectral type-luminosity class] calibration is used, one point in the diagram may represent many stars), of ~1300 LMC stars (lower-left), of ~500 SMC stars (lower-right).

CHAPTER I

2.1. The HR diagram of massive stars with spectral type O3 - M5

The Hertzsprung-Russell diagram (HR diagram or HRD) is a diagram in which the star's luminosity L is plotted as a function of its effective temperature T_{eff}. Figure 2.1a shows the galactic HRD of all known massive O3-M5 stars that are members of a cluster or association (some 2400 stars, Humphreys and McElroy, 1984). In the following 3 subsections we discuss in detail how the T_{eff} and L of the stars in the HRD are obtained leading to an easy [spectral type - luminosity class - luminosity - T_{eff} - photometric (B-V) index] calibration of massive stars (table 2.1).

Humphreys and McElroy also collected data for ~2700 known galactic field stars (i.e., stars not belonging to aggregates), ~1300 stars in the Large Magellanic Cloud (LMC) and ~500 stars in the Small Magellanic Cloud (SMC) with MKK spectral type and UBV photometry. Figures 2.1b,c,d illustrate these HRDs as well. One assumes that the same calibration applies as for galactic aggregate stars. For the Magellanic Clouds the distances are known and thus the real M_V is used.

2.1.1. THE EFFECTIVE TEMPERATURE AND BOLOMETRIC CORRECTION OF MASSIVE STARS

The most straightforward method to derive L and T_{eff} is to construct a series of models that most readily represent observations such as continuum radiation and line spectrum. To get a real coherent picture a NLTE-code is required including all important chemical species where, at least for stars showing the effects of mass loss, the atmosphere is treated hydrodynamically. This has been performed only for those cases where other methods fail completely, i.e., for LBVs, WR stars, hypergiants where the effects of stellar wind in the spectrum are dominant. They will be considered separately.

Instead, observers (as well as theoreticians) prefer to convert their easy data (magnitudes, colours, spectral types) into L and T_{eff} values. For the determination of T_{eff}, the "angular diameter method" is widely used.

The angular diameter method

In this section we use the symbols and definitions as given by Mihalas (1978). $\mathcal{F}_\nu(r)$ for the monochromatic flux of radiation, $\mathcal{F}(r)$ for the flux of radiation integrated over all frequencies, f_ν for the monochromatic flux received by the observer on earth, f for the integrated flux received by the observer on earth. The astrophysical flux F is related to the stellar flux $\mathcal{F}$ by: $\mathcal{F} = \pi F$.

If the radius R of the layer from which the continuum radiation emerges is independent of wavelength, and if the thickness of this layer is very small compared to the radius (non-extended atmosphere), the observational effective temperature T_{eff} is defined by

$$L = 4\pi R^2 \sigma_R T_{eff}^4 \tag{2.1}$$

Consider a star with radius R, seen at a distance D that exceeds R by many orders of magnitude. One defines the observed angular diameter of the star Θ as:

$$\Theta \approx tg\Theta = \frac{2R}{D} \tag{2.2}$$

From the spherical symmetric definition of $\mathcal{F}_\nu$ at any distance r from the stellar center, simple geometry gives the following relation

$$f_\nu = \frac{\Theta^2}{4} \mathcal{F}_\nu(r = R) \tag{2.3}$$

The total integrated flux at earth distance is thus

$$f = \int_0^\infty f_\nu d\nu = \frac{\Theta^2}{4} \int_0^\infty \mathcal{F}_\nu(r = R) d\nu = \frac{\Theta^2}{4} \mathcal{F}(r = R) = \frac{\Theta^2}{4} \sigma_R T_{eff}^4 = \frac{\Theta^2}{4} \frac{L}{4\pi R^2} \tag{2.4}$$

Suppose that f_ν (corrected for interstellar reddening) has been measured in the interval $[\nu_1 - \nu_2]$. It follows then that

$$\sigma_R T_{eff}^4 = 4\Theta^{-2} F_{model} \int_{\nu_1}^{\nu_2} f_\nu d\nu \tag{2.5}$$

with

$$F_{model} = 1 + \frac{\int_0^{\nu_1} \mathcal{F}_\nu(r = R) d\nu + \int_{\nu_2}^\infty \mathcal{F}_\nu(r = R) d\nu}{\int_{\nu_1}^{\nu_2} \mathcal{F}_\nu(r = R) d\nu} \tag{2.6}$$

A reasonable value for the parameter F_{model} can be estimated by using model atmospheres, provided that the observed $[\nu_1-\nu_2]$-interval covers a wide range of frequencies.

Since F_{model} is an implicit function of T_{eff}, equation (2.5) becomes an implicit equation of T_{eff} that can be solved by iteration if the angular diameter Θ is known.

Hanbury Brown et al. (1974) measured the angular diameter by means of the intensity interferometer for 3 O-type stars and 16 early B-type stars whereas Code et al. (1976) used these data to derive the T_{eff} of these stars applying the method outlined above. The results can be considered as mainly empirical; however they are too scarce to allow any kind of calibration.

For those cases where an independent measurement of the angular diameter does not exist, an alternative method was developed by Blackwell and Shallis (1977). Model atmospheres predict fluxes in the Rayleigh-Jeans tail of the spectrum that only slightly depend on the details of how to construct the model. They primarily depend on the adopted T_{eff}. This means that it may be expected that in the tail of the spectrum (the infrared or visual if massive stars are concerned) equation (2.3) is still valid if $\mathcal{F}_\nu$ is replaced by the monochromatic fluxes of a model atmosphere. If equation (2.3), applied in the tail of the spectrum, is used as definition of the angular diameter (always assuming that the angular diameter is independent of frequency), Θ becomes an implicit function of T_{eff} and thus again equation (2.5) becomes an implicit equation of T_{eff} which can be solved by iteration. Blackwell and Shallis tested this suggestion by using the observed infrared fluxes of the stars observed by Hanbury Brown et al. and concluded that the computed values of Θ are very similar to the observed ones by means of interferometry. This idea has then been applied by Underhill et al. (1979) and Underhill (1982) in order to derive T_{eff} values for about 42 O-type stars and 142 B-type stars.

Note, however, that since the method suggested by Blackwell and Shallis uses the Rayleigh-Jeans tail of the spectrum, the T_{eff} values of the hottest O-type stars (O5 and earlier) determined by the angular diameter method are highly uncertain.

A thorough discussion of the temperatures of massive A-M supergiants was presented by Flower (1977). By combining empirically determined T_{eff}-values (although few in number) with the theoretical scale of Böhm-Vitense (1972), a relation was derived between T_{eff} and the B-V colour that can be used as a standard.

The bolometric correction (BC)

The Blackwell-Shallis method uses the observed flux f_ν in a range $[\nu_1-\nu_2]$. The total flux f is then obtained by adding an estimate of the unobserved flux outside the range $[\nu_1-\nu_2]$ using model atmospheres, in other words by making a correction similar to the BC and which would equal it if $[\nu_1-\nu_2]$ was the range of the V-filter. Effective temperatures and bolometric corrections are thus closely related. The choice of a T_{eff} for a model atmosphere predicts a radiative flux in the V-filter as well as outside the V- filter. This defines the bolometric correction, i.e. $BC=M_{Bol}-M_V$.

2.1.2. THE ABSOLUTE VISUAL MAGNITUDE M_V OF MASSIVE STARS

Individual parallax determinations obviously give the most reliable visual magnitude M_V of a galactic star. However the number of massive stars to which this method can be applied is very limited and restricted to the region within ~500 pc from the Sun. For stars at larger distances, a different procedure has to be followed:

- from a number of stars with well defined spectral type, luminosity class and distance (parallax), it is possible to establish a fairly unique relation between these parameters and the UBV photometric properties
- this 'calibration' together with the standard extinction law allows the definition of a photometric distance for each star with MKK spectral type and UBV photometry
- the determination of M_V then follows from the distance.

This works quite well for main sequence stars or giants with spectral type B or later, but is less secure for the O-type stars or supergiants (too few O-type stars within 500 pc from the Sun). For these types one may proceed as follows:

- one tries to establish the possible membership to a cluster or association with B-star members
- the distance of the aggregate is determined by the B-stars
- from the distance, M_V of the O-type star can be derived

In this way Humphreys and McElroy (1984) determined M_V values for ±2500 galactic massive stars in the solar neighbourhood with known spectral type and which are member of an aggregate.

For spectral types later than B3, Petrie (1952, 1965) established a relationship between the equivalent width of Hγ and M_V that was updated by Balona and Crampton (1974) using data for 1932 OB-type stars.

The determination of M_V values of stars in the LMC and SMC is obviously much easier.

2.1.3. THE SPECTRAL TYPE - LUMINOSITY CLASS - M_{bol} - T_{eff} - (B-V) CALIBRATION OF MASSIVE STARS

Table 2.1 summarizes the present state of the MKK calibration spectral type-luminosity class-M_V-T_{eff}-BC-$(B\text{-}V)_0$. The relation is visualized in figure 2.2. where the absolute luminosity has been calculated with:

$$\log(\frac{L}{L_o}) = 0.4[4.75 - M_{Bol}]. \tag{2.7}$$

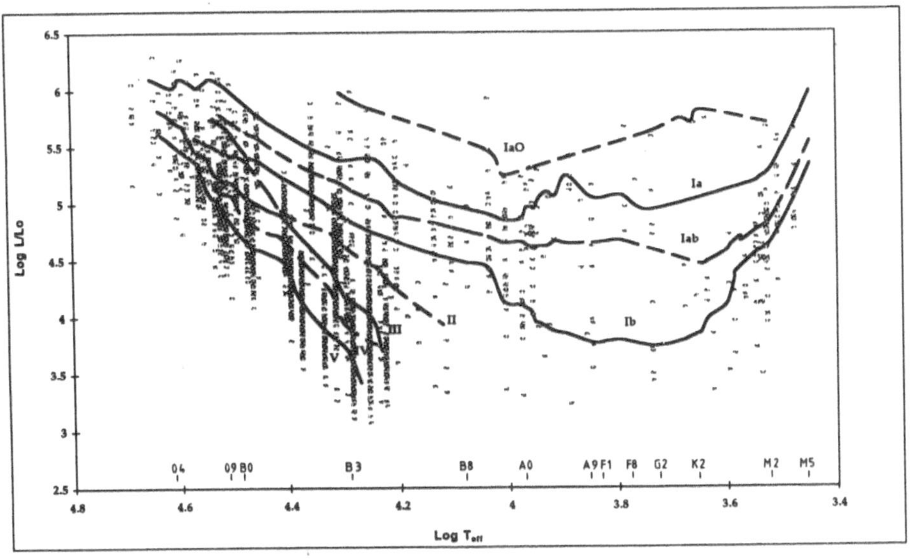

Figure 2.2: The [L-T_{eff}-luminosity class] calibration superposed on the sample of aggregate stars within the solar neighbourhood.

Remarks

- In certain spectral ranges small changes of the photometric $(B-V)_o$ parameter imply large T_{eff} differences, i.e. a small but systematic error in $(B-V)_o$ may significantly affect the HRD hot star distribution.

- For some spectral subclasses, the M_V standard deviation may be quite large, i.e. the average value holding for the subsample may differ significantly from the real value of an individual star. This has been discussed by Lamers et al. (1997). Using Hipparcos distance determinations, they compared the M_V values of stars with similar spectral types and found differences of more than one magnitude. The key could be rotation, the fast rotators being the brighter ones.

- Interpreting relation (2.5) around the V-filter and scaling to the Sun, one obtains a formula which enables a photometric determination of T_{eff},

$$\log T_{eff} + 0.1BC = 4.2207 - m_{v_0} - 0.5\log\Phi \tag{2.8}$$

If one observes the dereddened apparent visual magnitude m_{v_0} and the apparent angular diameter Φ (in arc milliseconds) one obtains thus an expression for $\log T_{eff} + 0.1BC$.

Using angular diameters measured by interferometric techniques (Hanbury Brown et al., 1974) and by Lunar occultation techniques, Barnes and Evans (1976) found that the right hand side of equation (2.8) is a function F_V strongly correlated with $(B-V)_o$, $(R-I)_o$ and $(V-R)_o$ over the entire range of spectral types (O4-M8).

For example, the dependence of the surface brightness parameter F_V on $(B-V)_o$ is given by Barnes et al. (1976) with the following simple relations:

$$F_v = 3.897 - 1.010(B-V)_o \quad \text{if} \quad -0.32 \leq (B-V)_o \leq -0.10$$

$$F_v = 3.964 - 0.333(B-V)_o \quad \text{if} \quad -0.10 \leq (B-V)_o \leq 1.35$$

Combining these relations with the BC- and $(B-V)_o$-calibrations of table 2.1. leads to T_{eff}-values that differ in the mean -for all stellar types mentioned in table 2.1- less than 0.03 dex (never exceed 0.07 dex).

TABLE 2.1. M_V -respectively according to (B-C) Balona & Crampton (1974) and (H-M) Humphreys & McElroy (1984)- are given for different MKK-spectral types. The BC and T_{eff}-scale are the same as used in H-M . Where necessary, some interpolations have been made in the H-M-data. The $(B-V)_o$ values are the mean values from different sources quoted in (Conti and Underhill , 1988), (Underhill and Doazan, 1982), Lang (1978, 1992), Allen (1973); $(B-V)_o$ values followed by a (*) means that the value quoted by Allen has been disregarded.

Luminosity class V: main sequence stars

Type	M_V B-C	M_V H-M	BC	T_{eff}	$(B-V)_o$
O3V		-5.5	-4.3	48700	
O4V		-5.5	-4	45000	
O5V	-5.4	-5.5	-3.9	43500	-0.330
O5.5V		-5.25	-3.8	41800	
O6V	-5.3	-5	-3.7	40500	-0.325
O6.5V		-5	-3.65	39000	
O7V	-5.1	-5	-3.6	37500	-0.320
O7.5V		-5	-3.55	37000	
O8V	-4.8	-4.8	-3.5	36250	-0.315
O8.5V		-4.5	-3.4	35000	
O9V	-4.5	-4.5	-3.3	34000	-0.307
O9.5V		-4.2	-3.25	32750	-0.300
B0V	-3.8	-3.8	-3	29600	-0.299
B0.2V		-3.8	-2.9	28100	
B0.5V		-3.8	-2.6	25400	-0.280
B1V	-3.2	-3.2	-2.5	24150	-0.264
B1.5V		-2.8	-2.2	21800	-0.250
B2V	-2.5	-2.5	-2	19700	-0.243
B3V	-2	-2.2	-1.8	18700	-0.206

Luminosity class IV: subgiants

Type	M_V B-C	M_V H-M	BC	T_{eff}	$(B-V)_o$
O8IV		-4.9	-3.2	34500	-0.31
O8.5IV		-4.9	-3.1	33600	
O9IV	-4.8	-4.9	-3.05	32800	-0.31
O9.5IV		-4.9	-3	31900	-0.30
B0IV	-4.3	-4.5	-2.9	30300	-0.30
B0.2IV		-4.5	-2.6	27300	
B0.5IV		-4.5	-2.5	25900	-0.28
B0.7IV		-4.2	-2.2	23400	
B1IV	-3.7	-3.8	-2	21100	-0.26
B1.5IV		-3.3	-1.8	20000	-0.25
B2IV	-3.1	-3.1	-1.6	18000	-0.24

B2.5IV		-3.1	-1.55	17550	-0.22
B3IV	-2.6		-1.5	17100	-0.20

Luminosity class III: normal giants

Type	M_V B-C	M_V H-M	BC	T_{eff}	$(B-V)_o$
O5III		-6	-3.8	43500	-0.320
O5.5III		-6	-3.7	42000	
O6III		-6	-3.5	40200	-0.320
O6.5III		-6	-3.45	39300	
O7III		-5.5	-3.4	37400	-0.320
O7.5III		-5.5	-3.25	35400	
O8III		-5.2	-3.2	34500	-0.310
O8.5III		-5.2	-3.1	33600	
O9III	-5	-5.2	-3.05	32800	-0.310
O9.5III		-5.2	-3	31900	-0.300
B0III	-4.6	-5	-2.9	30300	-0.298
B0.2III		-5	-2.6	27300	
B0.5III		-5	-2.5	25900	-0.280
B0.7III		-4.7	-2.2	23400	
B1III	-4.1	-4.3	-2	21100	-0.263
B1.5III		-3.9	-1.8	20000	-0.250
B2III	-3.7	-3.7	-1.6	18000	-0.245
B2.5III		-3.5	-1.55	17550	-0.220
B3III	-3.1	-3	-1.5	17100	-0.203
B4III	-2.6		-1.4	16700	-0.180

Luminosity class II: bright giants

Type	M_V B-C	M_V H-M	BC	T_{eff}	$(B-V)_o$
O7II		-5.8	-3.4	37400	-0.320
O7.5II		-5.8	-3.25	35400	
O8II		-5.8	-3.2	34500	-0.310
O8.5II		-5.8	-3.1	33600	
O9II	-5.1	-5.8	-3.05	32800	-0.310
O9.5II		-5.8	-3	31900	
B0II	-5	-5.8	-2.9	30300	-0.295
B0.2II		-5.4	-2.6	27300	
B0.5II		-5.1	-2.5	25900	-0.280
B0.7II		-5.1	-2.2	23400	
B1II	-4.7	-5.1	-2	21100	-0.260
B1.5II		-5	-1.8	20000	-0.250
B2II	-4.5	-4.8	-1.6	18000	-0.240

B2.5II		-4.8	-1.55	17550	-0.220
B3II	-4.2	-4.7	-1.5	17100	-0.200
B4II	-3.9	-4.65	-1.4	16700	-0.180
B5II	-3.6	-4.6	-1.3	16300	-0.160

Luminosity class Ib: less luminous supergiants

Type	M_V B-C	M_V H-M	BC	T_{eff}	$(B-V)_o$
O7Ib		-6.3	-3.3	35000	-0,310
O7.5Ib		-6.3	-3.25	34000	
O8Ib		-6.3	-3.2	33500	-0.290
O8.5Ib		-6.3	-3.15	33000	
O9Ib	-5.7	-6.3	-3.1	32600	-0.285
O9.5Ib		-6	-2.8	29900	-0.270
B0Ib	-5.6	-6	-2.7	28600	-0.245
B0.2Ib		-6	-2.4	26300	
B0.5Ib		-6	-2	23100	
B0.7Ib		-5.9	-1.8	21200	
B1Ib	-5.5	-5.8	-1.7	20260	-0.200
B1.5Ib		-5.8	-1.55	19400	
B2Ib	-5.4	-5.8	-1.35	18000	-0.18
B2.5Ib		-5.8	-1.25	17150	
B3Ib	-5.3	-5.8	-1.15	16300	-0.14
B4Ib	-5.2	-5.8	-1.05	15600	
B5Ib	-5	-5.8	-0.82	13700	-0.100
B6Ib	-4.9	-5.8	-0.73	12850	-0.080
B7Ib	-4.7	-5.8	-0.64	12000	-0.060
B8Ib	-4.5	-5.8	-0.51	10900	-0.040
B9Ib	-4.2	-5.2	-0.38	10250	-0.030
A0Ib		-5.2	-0.3	9500	-0.005
A1Ib		-5.2	-0.235	9300	+0.030
A2Ib		-5.2	-0.17	9100	+0.040
A3Ib		-5	-0.14	8950	+0.060
A4Ib		-5	-0.1	8800	
A5Ib		-5	0	8500	+0.090
A8Ib		-5	0.09	8200	+0.140
F0Ib		-5	0.14	7800	+0.19(*)
F5Ib		-4.8	0.13	7000	+0.33
F6Ib		-4.8	0.1	6600	
F8Ib		-4.8	0.08	6200	+0.56
G0Ib		-4.5	-0.1	5500	+0.76 (*)
G2Ib		-4.5	-0.12	5100	+0.86
G3Ib		-4.5	-0.14	5000	
G5Ib		-4.5	-0.193	4833	+1.00
G6Ib		-4.5	-0.22	4750	

G8Ib		-4.5	-0.307	4583	+1.14
G9Ib		-4.5	-0.35	4500	
K0Ib		-4.5	-0.387	4433	+1.20 (*)
K2Ib		-4.74	-0.46	4300	+1.33
K3Ib		-4.8	-0.67	4000	+1.46
K4Ib		-4.98	-0.84	3900	
K5Ib		-5.1	-1	3750	+1.59 (*)
K7Ib		-5.2	-1	3750	+1.61 (*)
M0Ib		-5.3	-1.3	3550	+1.64 (*)
M1Ib		-5.3	-1.4	3450	+1.68
M2Ib		-5.3	-1.5	3350	+1.68
M3Ib		-5.3	-1.72	3250	+1.69
M4Ib		-5.3	-2.5	3000	+1.75
M5Ib		-5.3	-3.3	2800	

Luminosity class Iab: supergiants

Type	M_V B-C	M_V H-M	BC	T_{eff}	$(B-V)_o$
O8Iab		-6.5	-3.2	33500	-0.29
O8.5Iab		-6.5	-3.15	33000	
O9Iab		-6.5	-3.1	32600	-0.27
O9.5Iab		-6.5	-2.8	29900	
B0Iab	-6	-6.5	-2.7	28600	-0.24
B0.2Iab		-6.5	-2.4	26300	
B0.5Iab		-6.5	-2	23100	
B0.7Iab		-6.5	-1.8	21200	
B1Iab	-6.2	-6.5	-1.7	20260	-0.19
B1.5Iab		-6.5	-1.55	19400	
B2Iab	-6.2	-6.5	-1.35	18000	-0.17
B2.5Iab		-6.5	-1.25	17150	
B3Iab	-6.1	-6.3	-1.15	16300	-0.13
B4Iab	-6	-6.4	-1.05	15600	
B5Iab	-5.9	-6.5	-0.82	13700	-0.10
B6Iab	-5.7	-6.5	-0.73	12850	-0.08
B7Iab	-5.6	-6.5	-0.64	12000	-0.05
B8Iab	-5.4	-6.5	-0.51	10900	-0.03
B9Iab	-5.2	-6.5	-0.38	10250	-0.02
A0Iab		-6.6	-0.3	9500	-0.005
A1Iab		-6.6	-0.235	9300	+0.02
A2Iab		-6.6	-0.17	9100	+0.03
A3Iab		-6.6	-0.14	8950	+0.06
A4Iab		-6.7	-0.1	8800	
A5Iab		-6.8	0	8500	+0.09
A8Iab		-7	0.09	8200	+0.14
F0Iab		-7	0.14	7800	+0.17 (*)

F5Iab	-7	0.13	7000	+0.32
F6Iab	-7	0.1	6600	
F8Iab	-7	0.08	6200	
K0Iab	-6	-0.387	4433	+1.25 (*)
K2Iab	-6	-0.46	4300	+ 1.36
K3Iab	-6	-0.67	4000	+1.46
K4Iab	-6	-0.84	3900	
K5Iab	-6	-1	3750	+1.60 (*)
K7Iab	-5.9	-1	3750	+1.63 (*)
M0Iab	-5.3	-1.3	3550	+1.67 (*)
M1Iab	-5.3	-1.4	3450	+1.69
M2Iab	-5.3	-1.5	3350	+1.71
M3Iab	-5.3	-1.72	3250	+1.69
M4Iab	-5.3	-2.5	3000	+1.76
M5Iab	-5.3	-3.3	2800	+1.80 (*)

Luminosity class Ia: luminous supergiants

Type	M_V B-C	M_V H-M	BC	T_{eff}	$(B-V)_o$
O3Ia		-6.5	-4	45000	
O4Ia		-6.5	-3.8	41000	
O5Ia		-6.8	-3.7	39800	-0.31
O5.5Ia		-6.8	-3.6	38000	
O6Ia		-6.8	-3.5	37000	-0.31
O6.5Ia		-7	-3.4	36000	
O7Ia		-7.2	-3.3	35000	-0.31
O7.5Ia		-7.2	-3.25	34000	
O8Ia		-7.2	-3.2	33500	-0.29
O8.5Ia		-7.2	-3.15	33000	
O9Ia	-6.8	-7.2	-3.1	32600	-0.28
O9.5Ia		-7.1	-2.8	29900	
B0Ia	-7	-7	-2.7	28600	-0.24
B0.2Ia		-7	-2.4	26300	
B0.5Ia		-7	-2	23100	
B0.7Ia		-7	-1.8	21200	
B1Ia	-7	-7	-1.7	20260	-0.19
B1.5Ia		-7.4	-1.55	19400	
B2Ia	-7	-7.4	-1.35	18000	-0.16
B2.5Ia		-7.4	-1.25	17150	
B3Ia	-7	-7.2	-1.15	16300	-0.12
B4Ia	-7	-7.1	-1.05	15600	
B5Ia	-7	-7	-0.82	13700	-0.08
B6Ia	-7	-7	-0.73	12850	-0.06
B7Ia	-7	-7	-0.64	12000	-0.04
B8Ia	-7	-7	-0.51	10900	-0.02

B9Ia	-7	-7	-0.38	10250	0.00
A0Ia		-7.1	-0.3	9500	+0.01
A1Ia		-7.4	-0.235	9300	+0.02
A2Ia		-7.5	-0.17	9100	+0.03
A3Ia		-7.5	-0.14	8950	+0.05
A4Ia		-7.75	-0.1	8800	
A5Ia		-8	0	8500	+0.09
A8Ia		-8	0.09	8200	+0.14
F0Ia		-8.5	0.14	7800	+0.17 (*)
F5Ia		-8	0.13	7000	+0.31
F8Ia		-8	0.08	6200	+0.56
G0Ia		-7.5	-0.1	5500	+0.75 (*)
M0Ia		-6.9	-1.3	3550	+1.67 (*)
M1Ia		-6.9	-1.4	3450	+1.69
M2Ia		-6.9	-1.5	3350	+1.71
M3Ia		-6.9	-1.72	3250	+1.69
M4Ia		-6.9	-2.5	3000	+1.76
M5Ia		-6.9	-3.3	2800	

Luminosity class Ia-O: extreme luminous supergiants

Type	M_V H-M	BC	T_{eff}
B1Ia-O	-8.5	-1.7	20260
B1.5Ia-O	-8.5	-1.55	19400
B2Ia-O	-8.5	-1.35	18000
B8Ia-O	-8.4	-0.51	10900
B9Ia-O	-8	-0.38	10250
A3Ia-O	-8.4	-0.14	8950
G0Ia-O	-9.2	-0.1	5500
G5Ia-O	-9.4	-0.193	4833
G8Ia-O	-9.2	-0.307	4583
K0Ia-O	-9.4	-0.387	4433
M2Ia-O	-8	-1.5	3350

2.2. The stellar wind of massive OB type stars

Since the early discovery that massive stars lose mass by a more or less spherical symmetric stellar wind, and the original suggestion that these winds are radiatively line-driven (Lucy and Solomon, 1970), despite the enormous efforts to improve the observational data or to update the theory, we still are faced with a factor two uncertainty in the mass loss rates of OB type stars. Unfortunately for the massive stars with $M > 20\ M_o$, this factor 2 sometimes makes the difference between '*mass loss by stellar wind of OB type stars is unimportant for stellar evolution'* and *'mass loss by stellar wind of OB type stars significantly alters stellar evolution'*.

Let us summarize the present state of the art.

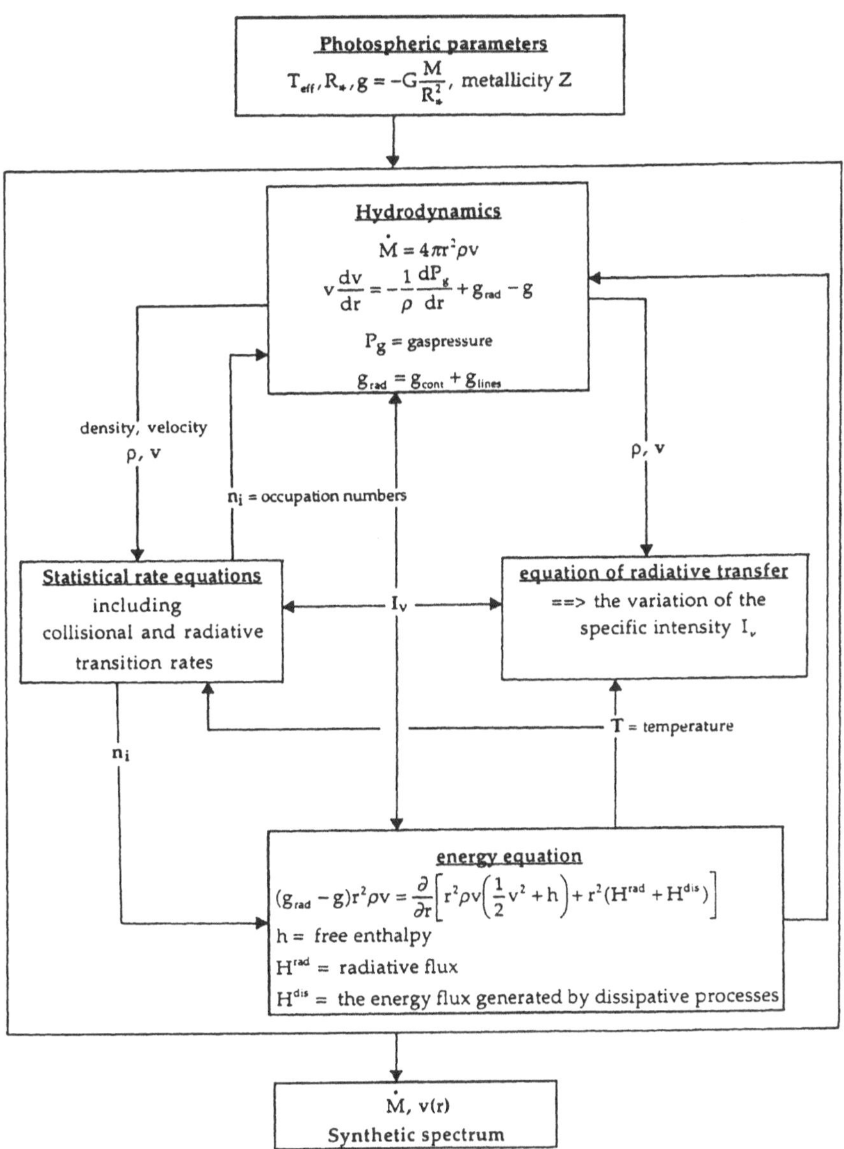

Figure 2.3: The concept to solve the full set of integro-differential equations, describing a stationary outflowing atmosphere.

The full set of integro-differential equations, describing a stationary (outflowing) atmosphere and the concept to solve them is sketched in figure 2.3. It is assumed that the only important forces in the equation of motion are the gas pressure force, the radiation force and gravity, i.e. possible magnetic forces and rotation are ignored. The radiative term g_{rad} must account for all lines of all chemical species present. The abundance of these species must be determined by the statistical rate equation, which depends on the hydrodynamics of the plasma. Together with the equation of transfer and the energy equation determining the temperature variation in the atmosphere it is in principle possible, given T_{eff}, the gravity g, the radius R at some specified optical depth and the metallicity Z, to produce a full spectrum of the star, its hydrodynamic structure and its mass loss rate $\dot{M}$ This means that pasting such a detailed atmosphere on top of a stellar interior model would in theory be sufficient to allow the computation of stellar evolution with stellar wind mass loss.

In practice outflowing atmosphere models are used in order to determine the hydrodynamic structure and the mass loss rate of a star by detailed comparison of the observed spectrum with the predicted one. One then tries to establish a relation between the mass loss rate and stellar parameters such as luminosity, T_{eff} and mass. This relation is then used in evolutionary computations.

Needless to say that the numerical solutions of the hydrodynamic problem and the inclusion of as many line transitions as possible is an enormous task (Pauldrach et al. 1994). Until now this sophisticated model has been applied to a handful of OB-stars only (Puls et al. 1996) by neglecting the rotation.

In order to make the mathematical problem tractable, some simplifications are justified. One assumes the stellar atmosphere to be composed of two parts:

a) A photosphere where radiative equilibrium and hydrostatic equilibrium hold. In stars where mass loss is not too large the photosphere includes the region where the optical continuum is formed. In most computations the photospheres are treated assuming a plan-parallel stratification and assuming that its structure is entirely determined by hydrogen and helium. For the other ions, the rate and transfer equations are solved independently of H and He. This may be correct considering their low abundance, relative to the abundance of H and He,

b) An outflowing (inhomogeneous?) outer mantle on top of the photosphere. The idea of line-driven winds has been worked out in a self-

consistent way by Castor et al. (1975) (see also Abbott, 1982). Despite many simplifications these studies were very promising and inspired the Münich group to develop the theory further and to combine hydrodynamics, extended rate equations, equation of radiative transfer and the energy equation. The latter equation that makes the problem extremely complex, is usually substituted by a simplified (ad hoc) temperature structure such as T=constant throughout the wind = 0.3-0.6 T_{eff}.
The crucial term in the problem is then the radiative acceleration g_{rad} with contributions from continuous absorption, scattering and line absorption. Castor et al. (1975) showed that the line absorption term can be expressed by the so-called line-force multiplier, characterized by three parameters, k, α, δ. In principle k, α and δ values can be computed for each star, given the luminosity, the effective temperature, the mass, the chemical composition (or equivalently T_{eff}, log g, R, Z). However this can only be done with a reasonable degree of certainty, provided that complete lists of all necessary atomic data for all transitions of the chemical elements of interest are available.

When one is interested primarily in the stellar mass loss rate of a star for which high S/N observations are available, one of the two following methods can be used:

method 1: $\dot{M}$ *from emission lines and the infrared excess.*

One assumes that the stellar parameters are known from the photosphere analysis so that they can be kept fixed in the analysis. The line absorption term is replaced by the line-force multiplier. The parameters k, α (and δ) are varied in order to obtain the correct value of the stellar wind velocity at infinity v_∞, the correct profile of the Hα-emission and of the He II 4686 emission line (and possibly the infrared excess).

The most sophisticated model up to now (the model of the Münich group, Gabler et al. 1989, Puls et al. 1996) combines hydrodynamics with a detailed statistical equilibrium treatment of hydrogen and helium and the correct spherical radiative transfer equation. The temperature throughout the wind is kept constant and rotation is neglected. The latter may result into a significant overestimation (typically 20-30% and possibly up to a factor 2) of the mass loss rate of a particular star (Petrenz and Puls, 1996).

method 2: $\dot{M}$ *from radio data*

If part of the mantle, emitting the radio spectrum, can be considered as a uniformly expanding, homogeneous sphere of hydrogen gas at constant temperature, $\dot{M}$ can be expressed as (see also Wright and Barlow, 1975)

$$\dot{M} = -0.095 \frac{\mu v_{\infty} S_{\nu}^{3/4} D^{3/2}}{z\sqrt{\gamma g_{\nu} \nu}} \tag{2.9}$$

v_{∞} = velocity of the gas in km/s
μ = the mean atomic weight of the gas
S_{ν} = the flux at radio frequency ν in Jansky
D = distance of the star in kpc
z = average charge on each ion
γ = average number of electrons per ion
g_{ν} = Gaunt factor at frequency ν

This $\dot{M}$ formalism gives probably the most reliable results; however owing to uncertainties in D, μ ,z, γ and g_{ν} one still has to admit a factor two of uncertainty in $\dot{M}$

The results:

B-main sequence stars show very little evidence for significant (significant for evolution) outflow, unlike what is found for the BA-type supergiants. Their mass loss rates range from a few times 10^{-7} M_o/yr for late B-type and A-type supergiants (Ib) to a few times 10^{-6} M_o/yr for early B-type supergiants (Ia).

Howarth and Prinja (1989) analyzed IUE-data of about 200 galactic OB-stars and derived a relation (although with poor correlation) between $\dot{M}$ (in M_o/yr) and L (in L_o),

$$\log(-\dot{M}) = 1.69 \log L - 15.4 \tag{2.10}$$

Using method 1 discussed in this section, Puls et al. (1996) obtained $\dot{M}$ values for 24 galactic stars with spectral type ranging between O3 and O9.5 with high-quality data. De Jager et al. (1988) presented a compilation of the available $\dot{M}$ observations of stars all over the HR diagram. We selected the O- and the B-type stars and updated the $\dot{M}$ values of the 24 stars also listed by Puls et al.. The following relation between $\dot{M}$, L and T_{eff} (L and T_{eff} resulting from calibration in most of the cases) can be deduced (with a coefficient of correlation of 0.86)

$$\log(-\dot{M}) = 1.67 \log L - 1.55 \log Teff - 8.29 \qquad (2.11)$$

It may be concluded that the two relations discussed above give very similar evolutionary results for massive stars.

Let us finally note that the radiation driven wind theory predicts the following dependence of the $\dot{M}$ on the metallicity Z

$$\dot{M} \propto (Z/Z_o)^{\zeta} \qquad (2.12)$$

Z_o is the solar metallicity; ζ ranges between 0.5 and 1.

2.3. The chemical abundances in massive OBA type stars

The chemistry of massive OBA-type stars is still poorly known. Basically the spectra of galactic O-type stars appear to indicate normal solar abundances except the ON- and OC-stars. However only detailed comparison between high quality spectra and the complete stellar atmosphere models can give reliable answers. The situation prior to 1990 was reviewed by Kudritzki and Hummer (1990). Herrero et al. (1992) made a detailed analysis of 25 galactic OB-stars with NLTE-atmospheric models. Abundances of a sample of 39 stars with spectral types between O9 and B3 were determined by Gies and Lambert (1992) using both LTE and non-LTE model atoms. a NLTE-study of 46 galactic B-supergiants was presented by Lennon et al. (1992,1993), whereas Fitzpatrick (1991) and Fitzpatrick and Bohannan (1993) surveyed about 100 B-supergiants in the LMC (60 have spectral type between B0.7 and B3) and measured CNO-line strengths for a number of strong absorption lines. Walborn (1971, 1976) identified the OBN/OBC class as a small group of OB-stars, showing anomalously strong nitrogen lines (N-class) or strong carbon lines (C-class). A NLTE study of a number of these stars was performed by Schönberner et al. (1988) and by Lennon et al. (1992, 1993). Venn (1994) investigated the chemistry of some 22 A-type supergiants.

Important to notice is that in most of the studies cited above, the effect of microturbulence has been neglected. The latter effect has been investigated by McErlean et al. (1998) and Smith et al. (1998).

The picture that emerges from all these studies can be summarized as follows.

- Except for a few ON stars the majority of the main sequence OB-type stars and the slowly rotating O-type giants and supergiants have normal (solar) composition.

- Atmosphere models where the effect of microturbulence is not included predict that the most luminous O-type stars (Of, Ia, Iab stars) and the fast rotators show significant helium enhancements [ε = N(He)/(N(H)+N(He)) > 0.15]; the carbon, nitrogen and oxygen-abundances correspond to CN or CNO equilibrium values (section 3.3.1). However, the inclusion of microturbulence significantly reduces the ε-values required to explain the He lines; a H-He mixture close to solar cannot be excluded.

- OBN stars show inordinately strong N lines and weak C lines; OBC stars show just the opposite. OBN stars are found among the dwarfs (class V) and among the supergiants (class I). OBC stars however are all very luminous objects and are found only among the giants/supergiants. Also oxygen seems to be enhanced in the latter objects. Walborn (1976) postulates that OBC stars could be supergiants with normal abundances, OB supergiants without the N enhancement characteristics could then be slightly CN enhanced whereas the OBN stars are the extremes.

- Except for the few (~7) BC stars, most B supergiants (also the early types studied by Fitzpatrick and Bohannan, 1993) exhibit evidence for moderate (Ib) to large (Ia) N enhancement and CO depletion, corresponding to respectively CN and CNO equilibrium abundances.

- A-type supergiants seem to have moderately altered CNO surface abundances.

2.4. Rotational properties of OB type stars

Using the list of Howarth and Prinja (1989) and the current SIMBAD Bright Star Catalogue, figure 2.4 shows the $v_e$sini (v_e = equatorial rotation velocity, i = inclination angle of the equatorial plane) distribution $\phi(v_e \sin i)$ of the O-type stars and of the early B type stars (earlier than B3). We separately consider the 'normal' OB type stars and the peculiar OBe types.

The v_e distribution $\varphi(v_e)$ can be determined as follows:

a. correct the observed distribution $\phi(v_e \sin i)$ for the effect of observational errors and obtain the distribution $\psi(v_e \sin i)$. If the errors on the measured

$v_e$sini follow a Gaussian distribution with variance σ^2, it follows that (see also Eddington, 1913)

$$\phi(v_e \sin i) = \int \psi(\xi) P(v_e \sin i|\xi) d\xi$$

with (2.13)

$$P(v_e \sin i|\xi) = \frac{1}{\sqrt{2\pi}\sigma} e^{-\frac{(v_e \sin i - \xi)^2}{2\sigma^2}}$$

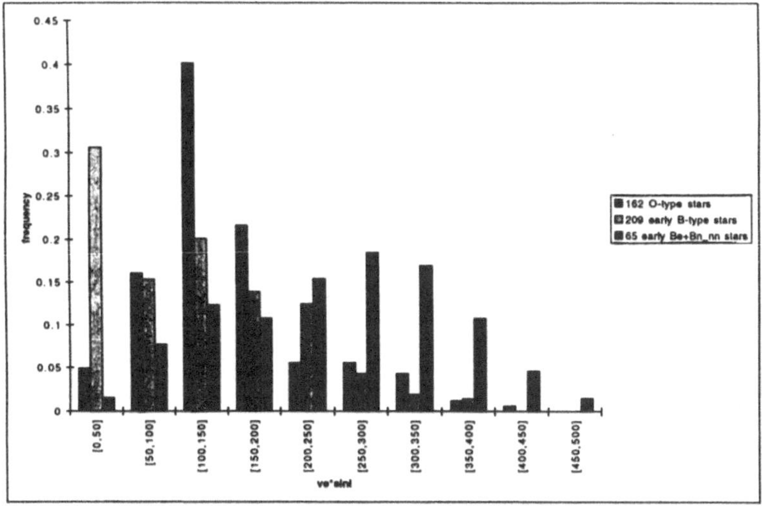

Figure 2.4: The distribution of $v_e$sini of O-type stars, normal early B-type stars and early Be+Bn_nn stars.

b. from the corrected distribution $\psi(v_e \sin i)$ we determine the distribution $\varphi(v_e)$ by solving the integral equation (see also Chandrasekhar and Münch, 1950)

$$\psi(v_e \sin i) = \int \varphi(\xi) P(v_e \sin i|\xi) d\xi$$

with (2.14)

$$P(v_e \sin i|\xi) = \frac{v_e \sin i}{\xi} \frac{1}{\sqrt{\xi^2 - v_e^2 \sin^2 i}} H(\xi - v_e \sin i)$$

(the function H is the classical Heavyside function)

In order to solve the integral equations (2.13) and (2.14) an iterative technique has been developed by Lucy (1974). Compared to other techniques (a.o. Deutsch, 1970; Balona, 1975) it has the big advantage that no assumptions about ψ or φ have to be made other than the obvious constraints

$$\begin{aligned} &\int \psi(\xi)d\xi = 1 \text{ and } \psi(\xi) \geq 0 \\ &\int \varphi(\xi)d\xi = 1 \text{ and } \varphi(\xi) \geq 0 \end{aligned} \tag{2.15}$$

2.4.1. 'NORMAL' OB TYPE STARS

The O-type stars

Figure 2.5 gives the v_e distribution (normalized) for O type stars. We assumed a standard error of 25 km/s. It can be concluded that:

- the rotational velocity distribution of O-type stars is bimodal; a first group (about 80 % of the O-type stars) with an average $<v_e> \approx 160$ km/s and a standard deviation $\sigma \approx 50$ km/s, and a second group (about 20% of the O-type stars) with $<v_e> \approx 330$ km/s and $\sigma \approx 60$ km/s; we can compare this with the critical break-up velocity

$$v_{e_{crit}} \approx 437\sqrt{\frac{M}{R}} \tag{2.16}$$

(M and R in solar units, the velocity in km/s); using the evolutionary computations discussed in section 4, it follows that the break-up velocity for most O-type stars is ≥ 500 km/s.

Remark:

Penny (1996a) reanalysed 120 O-type stars from the data set of Conti and Ebbets (1977) and, in addition, studied newly obtained spectra of 57 objects from the list of Howarth and Prinja (1989). In figure 2.5 we compare the v_e distribution resulting from the data set of Penny and of Howarth and Prinja. As can be noticed, the former predicts a bimodal distribution as well. However the average v_e-value of the subset of slower rotators (more than 80% of the total sample) ≈ 100 km/s which is ~60 km/s smaller than the value resulting from the data set of Howart and Prinja.

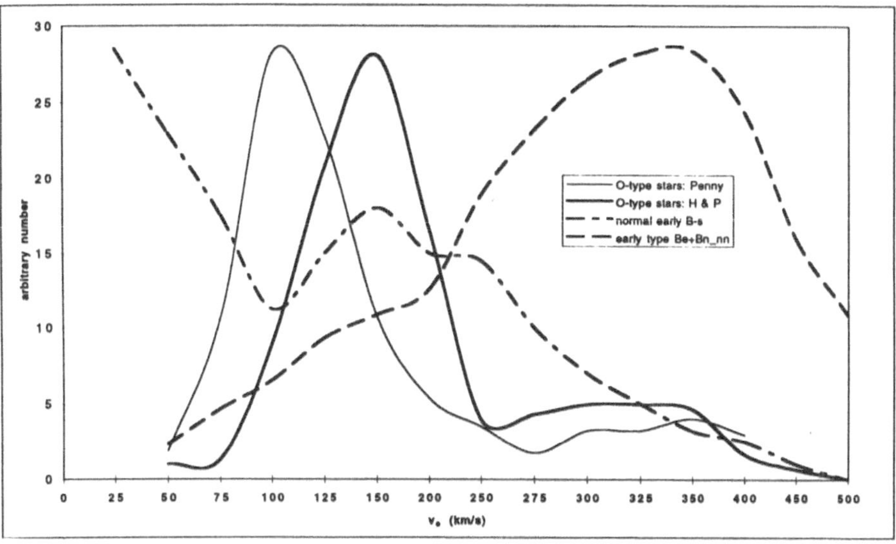

Figure 2.5: The statistically expected v_e distribution of O-type stars (using the data set of Howarth and Prinja and of Penny), of the normal early B-type stars and of the early Be and Bn_nn stars. All curves are normalized in a way that the maxima are at the same height.

The early B-type stars

Figure 2.5 also displays the probable v_e distribution of the B-type stars with spectral type earlier than B3 (Be and Bn_nn stars excluded). It looks as if the distribution is bimodal although the effect is not so pronounced as among the O-type stars. However also here is a tail of 'rapidly' rotating normal B-type stars.

2.4.2. OBe TYPE STARS[1]

About 15-20 % of all the early B type stars and few late O type stars show emission features (Hα) which are typical for the presence of a (rotating) disc. OB type stars showing these features are classified as OBe types. In figure 2.4 we show the $v_e$sini distribution of the Be stars with spectral type earlier than

---

1 Literature: I.A.U. Symposium 162, '*Pulsation, Rotation and Mass loss in Early-Type stars*', 1994, eds. L.A. Balona, H.F. Henrichs, J.M. Le Conte.

B3 (we included the Bn nn stars, since they may be physically related as far as the rotation is concerned) whereas figure 2.5 gives the v_e distribution. When compared to the rotation of the normal early B type stars, Be stars rotate on the average twice as fast.

Two remarks:

- the emission feature in some OBe stars may disappear temporarily,
- when we consider stars with $v_e \geq 300$ km/s, there are as many normal B-type stars as Be-types; among the stars with $v_e \geq 250$ km/s, the normal B-types outnumber the Be-types by ~50%.

As a consequence, it can be concluded that

'Rotation is a necessary but probably not sufficient condition to show the e-feature'.

Among the proposed OBe mechanisms, the wind-compressed disc (WCD) model of Bjorkman and Cassinelli (1993) is interesting. Early B type and late O type stars have small stellar winds. Since they are less luminous than O type stars, the radiation force driving the wind is smaller. In this case rotation may cause the fluid streamlines to fall toward the equator and form a disc.

In its present form the model is not yet entirely satisfactory. The disc densities predicted by the model are two orders of magnitude smaller than the observed values (Bjorkman, 1994) hence effects other than rotation have to be considered.

2.5. LBVs[2]

LBVs are very hot, unstable and very luminous OB-type supergiants. Most of them have enhanced nitrogen, depleted carbon/oxygen atmospheres and they are losing mass by a more or less steady stellar wind ($\dot{M} \approx 10^{-7}$-10^{-4} M_o/yr) and by eruptions where the star may lose mass at a rate of 10^{-3}-10^{-2} M_o/yr. Typically, outflow velocities are higher than a few 100 km/s.

Two famous historic LBVs are η Carinae and P Cyg (a HST picture of η Carinae is shown in figure 2.6; the circumstellar dust cloud is the brightest infrared source in the sky outside the solar system).

2 Literature: '*Luminous Blue Variables: Massive Stars in Transition*', 1997, eds. A. Nota, H.J.G. Lamers, ASP Conf. Ser., which collects the papers presented at the meeting where LBVs were the main topic.

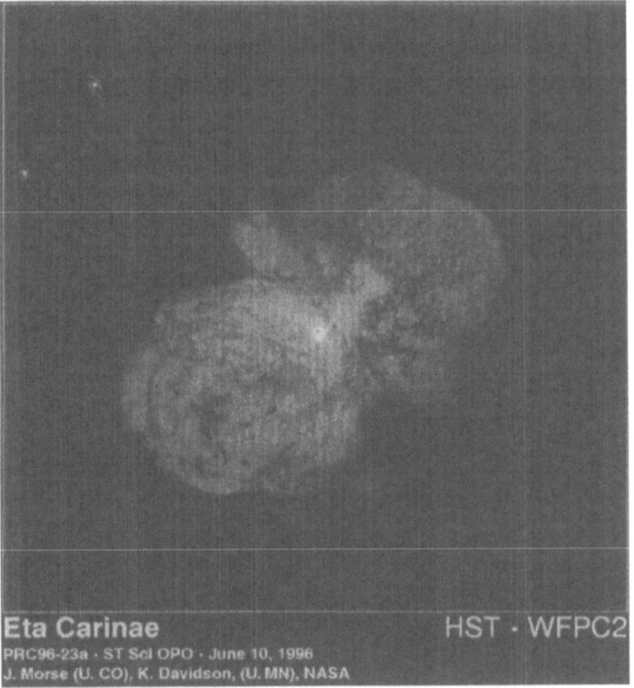

Figure 2.6: η Carinae seen by the HST.

The luminosity of LBVs can be determined by using a similar procedure as for the normal OB type stars and at least for some of them T_{eff} has been determined by detailed modeling with NLTE computer codes.

There are two types of LBVs (Humphreys and Davidson, 1994): those brighter than the observed upper luminosity limit for RSGs ($M_{bol} \approx -9.5$, section 2.6) (a.o. η Car, P Cyg) and the fainter ones (a.o. R71, R101). The latter have smaller amplitudes in variability, show lower mass loss rates and are presumably in a post-RSG phase of stellar evolution.

HD 5980 is a particularly interesting case. It is a close binary with an orbital period of 19.3 days and consists of a WR type star (section 2.7) and an O-type companion (Breysacher and Perrier, 1991). In 1994, one of the two components (it is as yet unclear which one) experienced a LBV outburst (Barba et al., 1995) accompanied by a visual brightening of 2.3 mag which makes it comparable to (or even more extreme than) η Car. It is interesting to remark here that since the binary is a very close binary, it is doubtfull that one of the two components has ever been a RSG, i.e.

the binary nature of HD 5980 may provide indirect evidence that the LBV phase of a massive star is not always preceded by a RSG phase.

2.6. YSGs, RSGs and hypergiants[3]

YSGs and RSGs are massive stars with log $T_{eff} \leq 3.7$ and luminosities extending to log L ≈ 5.4. Not much information is available concerning their surface abundance but they suffer mass loss from a stellar wind with relatively small outflow velocity (of the order of 10 km/s). The observed rates are quite uncertain. Probably the best relation between $\dot{M}$ and stellar parameters can be derived as follows:

Jura (1987) proposed a mass loss rate formula for RSGs using infrared data, i.e.,

$$-\dot{M} = 1.7 \cdot 10^{-5} v_{15} d^2 L^{-1/2} S_\nu(60) \lambda_{10}^{1/2} \tag{2.17}$$

where v_{15} = outflow velocity in 15 km/s units
λ_{10} = average wavelength of flux distribution in 10 μm units
d = distance in kpc
L = the luminosity in L_o
$S_\nu(60)$ = IRAS 60 μm flux

Using IRAS data, Reid et al. (1990) determined the stellar wind mass loss rate using Jura's formalism for 16 RSGs in the LMC with $M_{bol} \leq -7$. Independent of the fact that they made the simplifying assumptions $v_{15} = \lambda_{10} = 1$ and $S_\nu(60) = 0.5 S_\nu(25)$, their rates are uncertain by at least a factor two.

Using an average distance modulus to the LMC of 18.55 mag, the following equation gives a surprisingly well defined relation between the mass loss rate and the luminosity of these RSGs, i.e.,

$$\log(-\dot{M}) = 0.8 \log L - 8.7 \tag{2.18}$$

$\dot{M}$ in M_o/yr, L in L_o.

---

3 Literature: '*Instabilities in Evolved Super- and Hypergiants*', eds. C. de Jager and H. Nieuwenhuijzen, 1992.

Equation (2.18) is valid for the LMC. If the metallicity dependence of the stellar wind mass loss expressed by equation (2.12) also applies for RSGs, using Z = 0.008 for the LMC and Z = 0.02 for the Galaxy, the mass loss rates for Galactic RSGs may be 1.6 times larger than predicted by equation (2.18).

Yellow and red hypergiants are the stars located in the HRD on top of the RSGs. They all show evidence of instability including light and spectral variability, they are losing mass at very high rates ($\dot{M} \approx 10^{-4}$ - 10^{-3} M_o/yr) and many (or all?) of them exhibit shell ejections. There is an upper luminosity limit ($M_{bol} \approx -9.5$) above which no RSGs or hypergiants are observed. Such an upper limit is present in the Galaxy as well as in the Magellanic Clouds. Anticipating, $M_{bol} \approx -9.5$ corresponds to stars with initial mass 40-50 M_o.

The T_{eff} and the luminosity of the yellow and red hypergiants indicate that they may be closely related to the fainter LBVs discussed in the previous section.

2.7. The WR stars[4]

WR stars have a spectrum dominated by emission lines of helium and nitrogen (WN stars), helium and carbon (WC stars), or helium and oxygen (WO stars) and it is now widely accepted that the majority of them are hot hydrogen deficient core helium burning stars. Within the limitations of NLTE atmosphere models, WN stars show CNO equilibrium abundances, whereas the WC and WO types expose the products of the 3α-process (section 3.3.1). There are a few WR stars showing WC and WN characteristics together: they are classified as WNC stars.

Depending on the relative strengths of NIII, NIV, NV in WN stars and CIII, CIV in WC stars, WR stars are further subdivided ranging from WN2 to WN9, WC4 to WC9 (see Smith et al., 1996, and references therein for more details). Most of the late type WN stars show hydrogen in their spectrum whereas the majority of the early types are hydrogen deficient. For evolutionary purposes, Vanbeveren and Conti (1980) introduced the WNE and WNL notation meaning respectively WN stars without hydrogen (mostly early types) and WN stars with hydrogen (mostly late types). WC and WO stars do not have hydrogen in their atmosphere.

---

4 Literature: I.A.U. Symposium 163, '*Wolf-Rayet Stars: Binaries, Colliding Winds, Evolution*', eds. K.A. Van der Hucht and P. Williams, IAU Symp. 163, 1995.

The WR area in the HR diagram (figure 2.7) corresponds to the NLTE analyses of Hamann et al. (1995), Koesterke and Hamann (1995), Koesterke et al. (1992), Crowther et al. (1995) (for a more extended discussion, see Hamann and Koesterke, 1996, Hillier, 1996; Schmutz, 1996). The determination and definition of T_{eff} for WR stars, however, are not straightforward. Due to the large stellar wind mass loss rates, the outflowing mantle is very dense and all photospheric information is obscured. The WR spectrum is the spectrum of the mantle, thus the derived temperature is the temperature of this mantle. The real T_{eff} may be much larger than the one resulting from the NLTE analyses, may be up to 100000K (see also section 3.7). In summary, WR stars occupy the upper-left corner of the HR diagram, i.e., Log $T_{eff} \geq 4.4$ and Log $L/L_o \geq 4.5$.

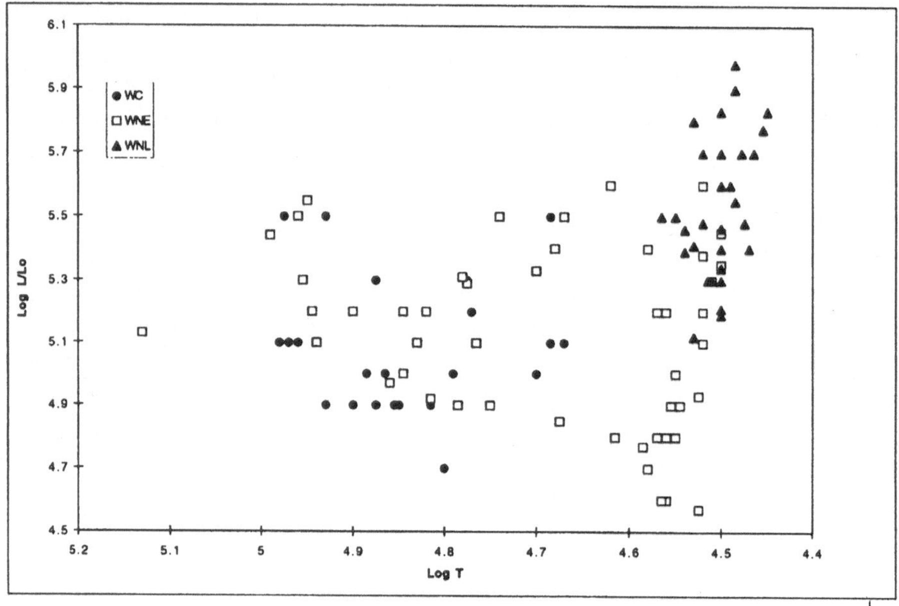

Figure 2.7: The HRD area occupied by WR stars.

In order to discuss their relation to other massive stars, one needs to consider volume limited samples and therefore, also for WR stars, distance determinations are necessary. When the WR star is a confirmed member of a (well populated) stellar aggregate, the distance determination is straightforward. When the WR star does not belong to a cluster or association, one has to rely on WR subtype-absolute visual magnitude calibrations. Van der Hucht (1996) used the galactic WR stars that are members of a stellar aggregate, to derive such a calibration. Vacca and Torres-Dodgen (1990) averaged data of galactic WR stars with LMC data. Both calibrations are given in table 2.2. The relatively large differences between both calibrations of certain subtypes imply large uncertainties (~50%) in the individual distance determinations of WR field stars and may thus affect the estimated numbers in volume limited samples.

TABLE 2.2: WR subtype - M_V calibrations. Columns headed "no." give the number of stars determining M_V. When this number is more than 1, the standard deviation is given.

Type	M_V	s.d.	no.	M_V	s.d.	no.
	Van der Hucht (1996)			Vacca and Torres (1990)		
WN2	-2.4		1			
WN2.5				-4.5	0.1	2
WN3	-2.2		1	-3.4	0.4	4
WN4	-3.3	0.3	2	-3.5	0.4	6
WN4.5	-4.1	0.7	3	-4.2		1
WN5	-3.7	0.4	2	-4.7		1
WN5.5				-4.5		1
WN6	-4.8	0.4	6	-5.6	1.1	6
WN7	-6.5	0.6	8	-6.3	0.8	6
WN8	-6.7		1	-5.7		1
WN9	-6.7	extrapolated		-6.5		1
WC4	-3.0	extrapolated		-4.1	0.7	6
WC5	-3.6	0.2	2	-3.7		1
WC6	-3.7	0.4	3	-3.9		1
WC7	-4.1	0.4	5			
WC8	-4.4	0.3	3	-4.6		1
WC9	-4.6	0.3	2			
WC10	-4.8	extrapolated				
WO2	-2.7		1	-1.8		1

Stellar wind mass loss rates for a number of galactic WR stars have been determined semi-empirically, using similar techniques as for the OBA-type stars. The following formula is a linear best fit although, due to the large observational scatter, with very poor statistical significance (Hamann, 1994):

$$\log(-\dot{M}) = 1.5\log L - 12 \qquad (2.19)$$

$\dot{M}$ in M_o/yr, L in L_o.

The dependence of the WR mass loss rates on metallicity is unclear at present.

The mass loss rates where the relation above is based on were determined by interpreting WR spectra with atmosphere models that assume homogeneity of the stellar wind. However, there is increasing evidence that WR winds are inhomogeneous (Moffat, 1996; Hillier, 1996), i.e. WR winds consist of clumps. First order estimates of the effect of a clumpy wind on the $\dot{M}$ determination have been reviewed by Hillier (1996). In general, it can be concluded that homogeneous models overestimate the $\dot{M}$ at least by a factor 2-3 and thus the same applies for equation (2.19).

2.8. Circumstellar shells around massive stars

Most of the massive RSGs, the yellow and red hypergiants and the LBVs are associated with ring nebula. They are called 'rings' because of their appearance, however dynamical studies show that they are in fact expanding shells.

LBVs are usually surrounded by small ejecta nebula (e.g., η Car, P Cyg, AG Car, HR Car) with diameters in the range 0.2-2 pc, masses between 0.1 M_o and 4 M_o (Nota et al., 1995).

IRAS studies of a large sample of RSGs, yellow and red hypergiants, show that ~25 % possesses circumstellar shells (Stencel et al. 1989).

As will be discussed in detail in sections 4 and 12, the overall evolution of a massive O-type star can be summarized as follows:

single star:	OB phase --->	LBV and/or RSG phase	-------->	WR phase
binaries:	OB phase --->	RLOF and/or LBV and/or RSG phase	-------------->	WR phase

(RLOF stands for Roche lobe overflow, sections 8, 9 and 10).

The formation of circumstellar shells may therefore occur in three stages. The fast flowing SW during the OB phase sweeps up the surrounding interstellar gas to form a wind-blown bubble. During a non-conservative RLOF, a LBV or RSG mass loss phase, a star will lose a large amount of mass on a short time scale. The observed outflow velocities in LBVs and in RSGs are quite small (between 10 km/s and a few 100 km/s). As a consequence, one expects the formation of shell structures with a restricted diameter around the central object (a.o. Garcia-Segura et al. 1996). Finally, the fast WR wind will interact with the structure created during the previous phase.

The observations of such structures around WR stars, their dynamics, masses and chemistry may thus provide very important information to understand the overall evolution of massive stars and their transformation into WR-types.

Large cool IRAS shells (diameter or the order of 10 pc) are found in more than 30% of the WR stars. They are most likely related to the SW phase during the previous OB phase (Marston, 1995). Optically observable shell structures with smaller diameter (< a few pc) are found in 31% (resp. 16%) of the galactic (single) WN stars (resp. WC stars) (Marston et al. 1994). It can be expected that at least some of them are due to LBV and/or RSG SW mass loss. It is interesting to note that when only the WR stars are considered above a galactic latitude of 2°, the latter percentage rises to 63%.

Large cool IRAS shells are also detected around some WR+OB systems. This is understandable since also the OB+OB progenitors went through an OB SW mass loss phase. However, small-scale shell structures seem to be very rare, at least among the short period WR binaries. This could indicate that the OB companion acted in the past as an efficient vacuum cleaner and that the RLOF process in WR+OB progenitors is (quasi)-conservative.

We finally want to mention the particularly interesting WNE star HD 50896 (EZ CMa). The star is associated to the nebula S 308, which has an expansion velocity ~60 km/s (Chu et al., 1982). The nebula is enriched in N by a factor 5, depleted in O by a factor 5 and the helium abundance is about 2 times solar (Esteban et al., 1992). Furthermore, the star seems to be one of the fastest runaways of our Galaxy (section 13.5.2) and it exhibits periodic variations with a period ~3.7 days (Firmani et al., 1980) which are attributed to binarity or to rotation (St.-Louis et al., 1993a).

2.9. The SN explosion of a massive star

At the end of the life of a massive star, the FeNi-core collapses and becomes degenerate (section 3.2). Two possibilities:

- the collapse results in the formation of a neutron star (further abbreviated as NS) and is accompanied by a supernova (SN) explosion with ejection of the mass layers outside the FeNi-core;

- the collapse results in the formation of a black hole (further abbreviated as BH); it is expected that the rest of the star is accreted by the BH (Wijers, 1996) and no SN occurs.

After the SN explosion, layers become visible that have been nuclearly processed in an earlier evolutionary stage. It is therefore obvious that the study of SN events is very important for massive star evolution.

Supernovae are classified as Type I when no hydrogen is present in the spectra, and Type II when hydrogen is visible. Both types are further subdivided in what could be called standard subtypes: type Ia, Ib, Ic, II-P, II-L. Figure 2.8 illustrates typical light curves of the different types of SN events. By comparing the observations with theoretical models of SN explosions, it is possible to propose some general properties concerning the mass of the exploding star and its chemical composition (see e.g. the reviews of Nomoto et al., 1995; Wheeler and Benetti, 1997).

Summarizing:

Type Ia: are observed in all types of galaxies, also in older stellar populations; they are characterized by elements of intermediate mass, O, Mg, S, Si, Ca and iron-peak elements (showing up typically 20 d after optical maximum). It is widely accepted that a type Ia SN is the result of a thermonuclear explosion of a CO white dwarf that has accreted mass from a Roche lobe overflowing companion (Nomoto and Yamaoka, 1992; Hachisu et al., 1996; Wheeler, 1996); no remnant is left after the explosion,

Type Ib: associated with young stellar populations; strong He lines are present in their spectra. They are most probably related to massive stars which collapse into NSs. The stars have lost all their hydrogen rich layers and have a helium rich envelope,

Type Ic: as type Ib they are associated with young populations. However contrary to type Ib they seem to be very He deficient. Also type Ic is related to massive stars however in this case the star should also have lost most of its helium rich envelope, so that carbon and oxygen are the main elements left,

Type II-P: are associated with population I stellar environments; after maximum light, the luminosity remains nearly constant (a plateau) for weeks to months. The light curve can most easily be explained by the explosion of a massive star with a large radius where the mass of the layers containing hydrogen is large ($> 5\ M_o$),

Type II-L: are also associated with population I stellar environments; after maximum light, the luminosity declines linearly. Also here the light curve points towards a star with large radius, however the mass of the layers still containing hydrogen must be small (~1-2 M_o, certainly < 5 M_o).

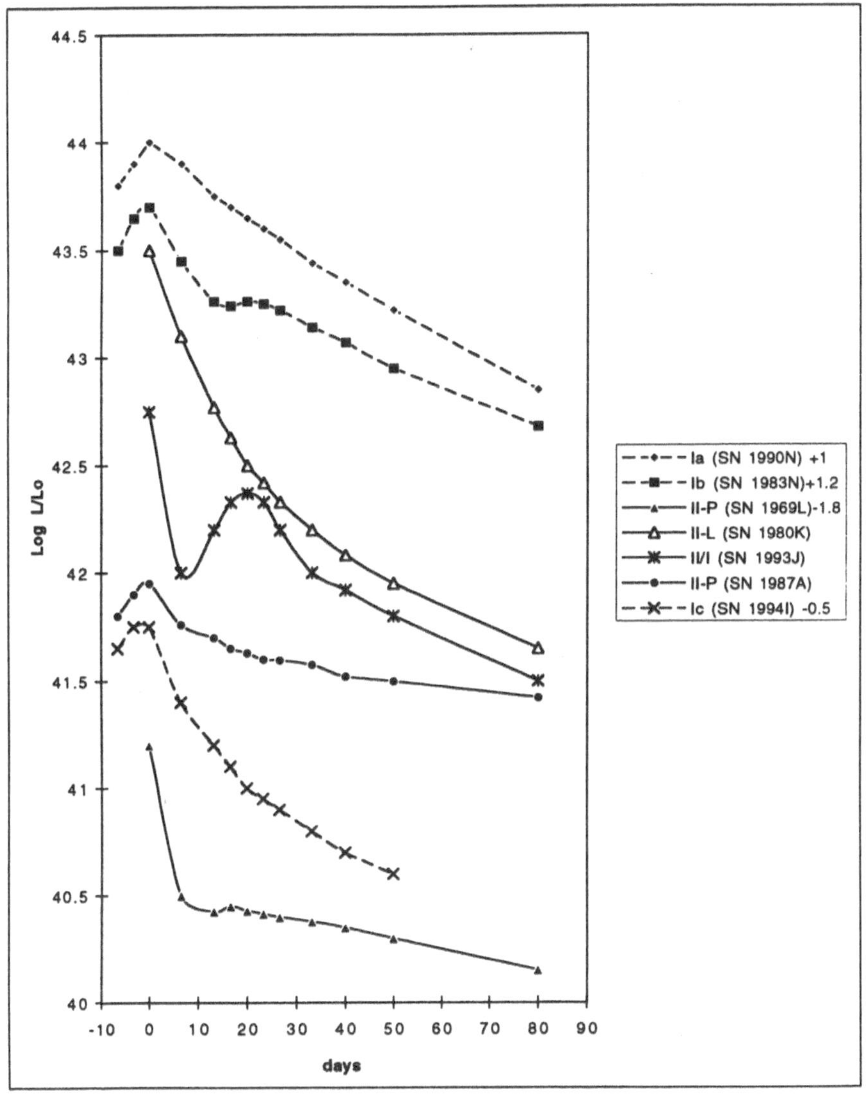

Figure 2.8: Typical light curves of the different types of SN events.

Three special cases:

<u>SN 1987A</u>

SN 1987A in the LMC is a peculiar Type II-P, providing important constraints for the evolution of massive stars, i.e.

- The progenitor of SN 1987A was a blue supergiant (a B3 I star, Rousseau et al., 1978),
- the circumstellar nebula around SN 1987A (figure 2.9) has the morphology of a ring rather than that of a shell (Jakobsen et al., 1991); nitrogen is overabundant whereas carbon and oxygen are underabundant compared to solar (Fransson et al., 1989).

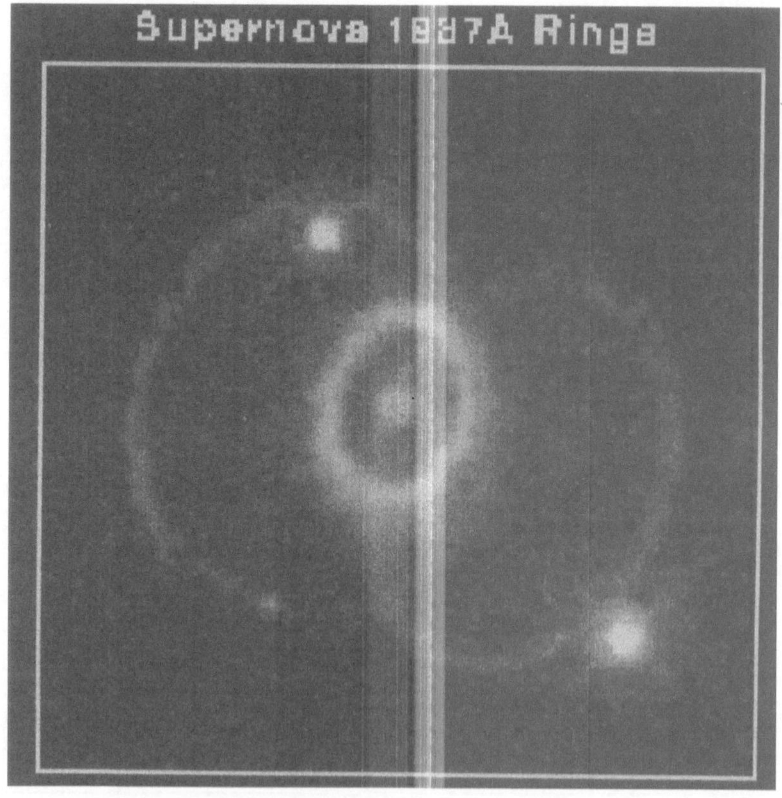

Figure 2.9: SN 1987A seen by the HST.

- A few weeks after the SN explosion, the spectra showed an overabundance of barium and other s-processed material (Mazzali et al., 1992).
- in order to explain the light curve, theoretical SN models need a star with a radius of ~40 R_o (which corresponds to a radius of a B3 supergiant), a He core mass of ~6 M_o (corresponding to the He core mass of a 20 M_o star, section 4) and a hydrogen rich envelope with mass larger than 5 M_o.

SN 1993 J

SN 1993 J in M81 was originally identified as a type II-L however after a few months the SN began to show features that are typical for a type Ib. It is therefore classified as a type II/I.

In order to explain its light curve, the progenitor star must have been a RSG with a He core mass of ~3-4 M_o (corresponding to the He-core mass of a star with initial mass of ~15 M_o, section 4) and an envelope with hydrogen but with a mass smaller than 1 M_o (Nomoto et al., 1993). This means that the 15 M_o star should have lost most of its hydrogen rich layers, but not all.

From X-ray observations and translating them into properties of the circumstellar matter around the exploding star, Suzuki et al. (1993) estimated that the progenitor RSG must have had a SW with $\dot{M} \approx 4\text{-}5.10^{-5}$ M_o/yr. We note that this high value is not inconsistent with the $\dot{M}$ of RSGs predicted by equation (2.18).

SN 1994 I

SN 1994 I in M51 has been classified as a type Ic. In order to explain the light curve, Nomoto et al. (1995) propose a helium deficient C-O star with a mass < 3 M_o where the ejected mass is less than ~0.9 M_o

2.10. Summary

Accounting for the discussion in the previous subsections, the overall HR diagram of the massive stars in the Galaxy is shown in figure 2.10. Although the observations of the LMC and SMC are scarcer, it looks as if the HRD position of the different types of stars is similar as for the Galaxy. Therefore showing the most probable position of the progenitor of SN 1987A in the same diagram seems meaningful. Notice that it has a luminosity that is larger than or equal to the minimum luminosity of WR stars whereas it is located between the RSG and WR area. As will be discussed in section 5.1.3, this observation is difficult to explain if the progenitor were a normal single star.

Let us finally note that similarly as in the Galaxy and the LMC, also the SMC contains a large number of RSGs. It will be demonstrated (section 5.1.8) that this restricts the value of a number of uncertain parameters that enter the mathematical description of the physical processes that govern stellar evolution.

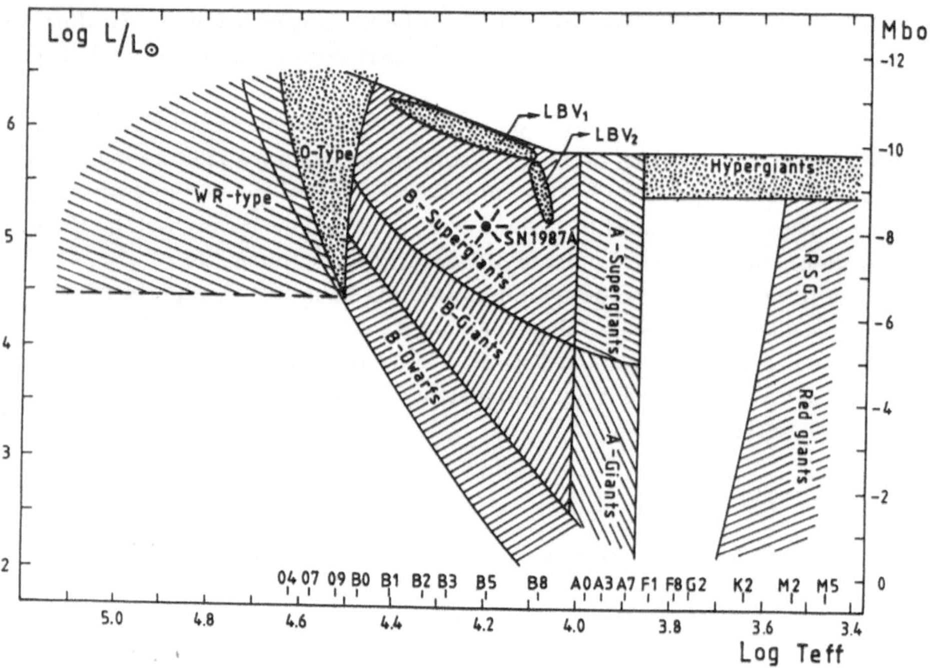

Figure 2.10: The overall HRD of the massive stars in the Galaxy together with the probable position of the progenitor of SN 1987A; LBV_1 (resp. LBV_2) is the area occupied by the more violent (resp. less violent) LBVs (section 2.5).

CHAPTER II

MASSIVE SINGLE STARS

3. The equations of stellar structure of a non-magnetic, non-rotating single star

The equations of stellar structure have been discussed in detail by Chandrasekhar (1957), Schwarzschild (1958), Cox and Giuli (1968), Clayton (1968), Kippenhahn and Weigert (1989), De Loore and Doom (1992). In the present work we summarize only those processes which are important for massive stars.

The structure and evolution of a non-magnetic, non-rotating sphere of gas can be described by the following physical principles and processes:

the continuity of mass
the physical state of the plasma
the energy equation
the energy transport mechanism
the time variation of the nuclear species X_i
the time variation of the total stellar mass

These principles and processes yield six differential equations, describing the stellar structure. Before discussing the equations we will list the parameters used to describe the physical conditions inside stars:

r: since we consider spherical models, all parameters depend only on the distance r from the star's center. The total radius of the star is denoted by R. We will use the radius of the sun, R_o, as basic unit;

ρ: the density of the stellar matter in g/cm^3;

μ: the molecular weight;

M_r: the total mass, contained in a sphere with radius r. We will use the mass of the sun, M_o, as basic unit. The total stellar mass M_R, is denoted by M;

T: the temperature of the stellar material;

P: the total pressure in the stellar interior i.e. the sum of the gas pressure and the radiation pressure; in general P is a function of density, temperature and chemical composition,

L_r: the parameter L_r denotes the amount of energy passing across a sphere with radius r. The energy at the stellar surface, L_R, is denoted by L. We will use the total luminosity of the sun, L_o, as basic unit.

$\dot{M}$: the mass loss/mass gain rate of a star, expressed in M_o/yr.

Obviously, all parameters depend on r and on time t.

3.1. The continuity of mass

From the definition of M_r, one immediately derives the first structure equation:

$$\frac{\partial M_r}{\partial r} = 4\pi\rho r^2 \tag{3.1}$$

3.2. The physical state of the plasma

3.2.1. HYDROSTATIC EQUILIBRIUM

In the stationary case, the equation of motion of a stellar plasma can be written as

$$v\frac{\partial v}{\partial r} = -\frac{1}{\rho}\frac{\partial P}{\partial r} - G\frac{M_r}{r^2} \tag{3.2}$$

In most circumstances of normal stellar evolution, the velocities of the mass layers in stellar interiors are very small, and the acceleration may be neglected: the star is in hydrostatic equilibrium. It is only in advanced burning stages, or when stellar pulsations are present, where the variation of the stellar parameters is very rapid, that the 'hydrodynamic' term becomes important.

Assuming hydrostatic equilibrium, the pressure equation (3.2) takes the form:

$$\frac{\partial P}{\partial r} = -G\frac{M_r\rho}{r^2} \tag{3.3}$$

The pressure P is the sum of the total gas pressure and the radiation pressure $P_{rad} = \frac{1}{3} a_R T^4$. The total gas pressure is the sum of the electron pressure and the ion pressure and has to be determined from the equation of state which depends on whether the plasma is degenerate or not.

3.2.2. NON-DEGENERACY AND DEGENERACY: DEFINITION

Free electrons follow the Fermi-Dirac statistics. They are mutually indistinguishable and at most two electrons -differing from each other in their spin- can be simultaneously present in the same cell of the phase-space.

The second condition is a consequence of the Heisenberg uncertainty relation which states, in one dimension, that

$$\Delta x . \Delta p_x \geq \hbar \tag{3.4}$$

When the density is small enough, the velocity distribution of the free electrons is given by Maxwell's law: the plasma is non-degenerate. However if due to contraction, the density becomes so high that one forces too many electrons into the same space they will leave their Maxwellian velocity distribution in order not to violate rule (3.4): the initially ideal gas turns into a degenerate gas.

3.2.3. THE EQUATION OF STATE OF A NON-DEGENERATE PLASMA

The equation of state is a relation between the total gas pressure of a plasma, its density, temperature and chemical composition.

When the gas is non-degenerate, the total gas pressure is given by

$$P = NkT = \frac{\Re \rho T}{\mu} \tag{3.5}$$

with N the number of free particles per cubic centimeter. When the plasma is completely ionized (corresponding to the largest part of the star), the following relation holds with very high accuracy:

$$\frac{1}{\mu} = 2X + \frac{3}{4} Y + 0.56Z \tag{3.6}$$

where (X, Y, Z) = weight fractions of (hydrogen, helium, heavier elements), and thus X + Y + Z = 1.

Note that equation (3.6) does not hold for the outermost layers of a star where the temperature is smaller than ~ 500000 K so that the ionization is not complete. In these layers the Saha equations can be used to compute the degree of ionization.

3.2.4. THE EQUATION OF STATE OF A DEGENERATE PLASMA

We first consider the electron pressure P_e. If the electron degeneracy is complete (i.e. precisely two electrons per cell in the phase-space) one has n(p) = $2.h^{-3}$ per unit volume in the momentum (p) space. The electron number density n_e (cm^{-3}) is then

$$n_e = \int_0^{2\pi}\int_0^{\pi}\int_0^{p_{max}} \frac{2}{h^3} p^2 dp \cos\theta d\theta d\phi = \frac{8\pi}{3h^3} p_{max}^3 \tag{3.7}$$

Assuming that the plasma is completely ionized (a situation which is valid in most cases of degeneracy in astrophysics)

$$n_e = \frac{1}{2}(1+X)\frac{\rho}{m_H} = \frac{\rho}{\mu_e m_H} \tag{3.8}$$

and thus

$$\frac{\rho}{\mu_e} = \frac{8\pi}{3}\frac{m_H}{h^3} p_{max}^3 \tag{3.9}$$

a. non-relativistic degeneracy

Let us assume that the electron velocities are small when compared to the velocity of light c. Since in that case $p_x = m_e v_x$, whereas $p_x^2 = \frac{1}{3}p^2$, it follows that

$$P_e = \frac{2}{h^3}\int\int\int p_x v_x dp_x dp_y dp_z = \frac{8\pi}{15 m_e h^3} p_{max}^5 \tag{3.10}$$

Using (3.9) and replacing the constants by their numerical values (in cgs-units) one obtains in the completely degenerated non relativistic case the polytropic relation

$$P_e = C\rho^k = 9.915\text{x}10^{12}\left(\frac{\rho}{m_H}\right)^{\frac{5}{3}} \tag{3.11}$$

b. relativistic degeneracy

Here one assumes that most of the electrons attain the velocity of light c. Similar to (3.10) but valid for the relativistic case:

$$P_e = \frac{2\pi c}{3h^3} p_{max}^4 \tag{3.12}$$

and from (3.9)

$$P_e = C\rho^k = 1.231\text{x}10^{15}\left(\frac{\rho}{m_H}\right)^{\frac{4}{3}} \tag{3.13}$$

The transition between a non relativistic degenerate medium and a relativistic degenerate medium occurs when (3.11) equals (3.13), i.e. when $\rho = 1.916 \cdot 10^6 \mu_e\,(\text{g.cm}^{-3})$. These very high densities are independent of temperature and are only attained in the core during the final stages of stellar evolution.

To obtain the total gas pressure we still have to add the ion (or nuclei) pressure to the electron pressure. Because of their higher mass, nuclei have available $(m_{nucl}/m_e)^{3/2}$ times more cells in momentum space than electrons. Thus nuclei-degeneracy will occur only when the density is hundred thousand times larger than for degeneracy of electrons, a situation which only occurs in NSs. When the nuclei are not degenerate their pressure can be computed with an equation similar to equation (3.5). The latter is very small compared to the pressure of the degenerate electrons so that to good accuracy the degenerate electron pressure P_e covers the total gas pressure almost entirely .

3.2.5. THE CHANDRASEKHAR MASS LIMIT OF DEGENERATE STARS

Since P_e (and thus P) does not depend on the temperature in the case that the plasma becomes degenerate [equations (3.11) or (3.13)], the solution of the hydrostatic equation (3.3) [when combined with the continuity equation (3.1)] is independent from the temperature variation. Results are shown in table 3.1 for $\mu_e = 1$ (from Sobolev, 1990). As can be noticed there exists a situation in which the plasma is relativistically degenerate and where the pressure can no longer sustain gravity: the star collapses. This occurs when $M/M_o = 5.75\,\mu_e^{-2}$ = the Chandrasekhar mass M_{Ch}. During the final phases of stellar evolution X = 0 in the degenerate core, thus $\mu_e = 2$ and thus $M_{Ch} = 1.44\ M_o$.

TABLE 3.1. Relation between mass, radius and mean density of a White Dwarf. If μ_e differs from one, values of M have to be divided by μ_e^2; values of R have to be divided by μ_e and values of the mean density have to be multiplied by μ_e.

M/M_o	R [in cm]	$<\rho>$ [in g/cm^3]	M/M_o	R [in cm]	$<\rho>$ [in g/cm^3]
5.75	0.00	∞	2.95	1.51×10^9	4.04×10^5
5.51	4.13×10^8	3.70×10^7	2.45	1.72×10^9	2.29×10^5
5.32	5.44×10^8	1.57×10^7	2.02	1.93×10^9	1.34×10^5
4.87	7.69×10^8	5.08×10^6	1.62	2.15×10^9	7.70×10^4
4.33	9.92×10^8	2.10×10^6	0.88	2.79×10^9	1.92×10^4
3.54	1.29×10^9	7.90×10^5	0.00	∞	0

3.2.6. NEUTRON STAR MASSES

As discussed in the previous subsection, when the degenerate star has a mass larger than the Chandrasekhar limit $M_{Ch} = 1.44\ M_o$, it cannot resist the collapse and acquires densities at which the neutrons become degenerate. When the mass of the core is smaller than some maximum $(M_{NS})_{max}$ the degeneracy pressure of the neutrons is able to stop the collapse: the star is called a neutron star.

Using the appropriate equation of state of degenerate nuclei, the maximum NS gravitational mass has been found by Bethe and Brown (1995) to be $(M_{NS})_{max} \approx 1.56 M_O$. This requires the collapse of a degenerate plasma with baryonic mass[5] $(M_b)_{max} \approx 1.76 M_O$.

[5] *To convert the gravitational mass M_{NS} -which would have been measured by binary motion if the neutron star is a binary member- into the baryonic mass of the core M_b, one has to add the gravitational binding energy E of the neutron star e.g. by using the formula of Lattimer and Yahil (1989):* $E \approx 0.084 M_O (\frac{M_{NS}}{M_O})^2$ *where* $M_{NS} = M_b - E$; *the difference*

Note, however, that the equation of state is the result of a number of assumptions and therefore the maximum values quoted above are still uncertain.

At present, there is no reason why a degenerate plasma with mass larger than $(M_b)_{max}$ would not exist. Since in this case also the degeneracy pressure of neutrons is not able to stop the contraction, the plasma will collapse into a BH.

3.3. The energy equation

Denoting by ε the net energy production by nuclear reactions per unit of mass and per unit of time, by ε_v the energy lost by neutrinos (very important at late evolutionary stages. They will not be the scope of the present work; a detailed discussion was given by Kippenhahn and Weigert, 1989), and by Q the heat content of the stellar matter per gram, so dQ=TdS, with S the entropy, we can write the energy equation:

$$\left.\frac{\partial L}{\partial M_r}\right|_t = -T\left.\frac{\partial S}{\partial t}\right|_{M_r} + \varepsilon - \varepsilon_v \qquad (3.14)$$

3.3.1. THE ENERGY GENERATION RATE BY NUCLEAR REACTIONS ε

The enormous energy emitted by stars of all masses during several millions of years can be explained only by nuclear processes. We distinguish processes where hydrogen is converted into helium (H-->He), helium into carbon and oxygen (He-->C-O), carbon into neon and magnesium (C-->Ne-Mg), neon into magnesium (Ne-->Mg), oxygen into silicon (O-->Si), silicon into iron (Si-->Fe). It is common to use the terms hydrogen burning, helium burning, ...They have been studied in detail by Fowler et al. (1975) and Caughlan and Fowler (1988).

$M_b - M_{NS}$ *is caused by the difference in gravitational energy before and after the contraction into a neutron star. This difference is radiated away.*

Hydrogen burning

Stars which are burning hydrogen in their cores are called 'Main Sequence Stars'. Stars at the beginning of this core hydrogen burning phase (further abbreviated as CHB) are called ZAMS (zero age main sequence stars).

Hydrogen can be converted into helium either by the proton-proton-reaction chain (p-p), or by the CNO-cycle.

The p-p-process is important as energy source only in solar-type stars (mass $\leq 2M_{o}$), and can be neglected in massive stars.

The complete CNO-reaction chain consists of 3 cycles: the CN-cycle where only carbon and nitrogen are involved, the ONI and ONII-cycles where only nitrogen and oxygen isotopes enter. Omitting the very fast (immediate) β-decay of the unstable nuclei, the chain can be written as:

<u>CN-cycle</u>

1. $^{12}C+^{1}H \dashrightarrow ^{13}C + e^{+} + \nu$
2. $^{13}C+^{1}H \dashrightarrow ^{14}N$
3. $^{14}N+^{1}H \dashrightarrow ^{15}N + e^{+} + \nu$
4. $^{15}N+^{1}H \dashrightarrow ^{12}C + ^{4}He$

<u>ONI-cycle</u>

5. $^{15}N+^{1}H \dashrightarrow ^{16}O$
6. $^{16}O+^{1}H \dashrightarrow ^{17}O + e^{+} + \nu$
7. $^{17}O+^{1}H \dashrightarrow ^{14}N + ^{4}He$

<u>ONII-cycle</u>

8. $^{17}O+^{1}H \dashrightarrow ^{18}O + e^{+} + \nu$
9. $^{18}O+^{1}H \dashrightarrow ^{15}N + ^{4}He$

Calling P_{ij} the probability per unit time that a nucleus of type i will be transformed into one of type j, calling X_i (i=1,4,12,13,14,15,16,17,18) the abundance (by weight) of the nucleus of type i, it follows that the time variation of X_i due to nuclear reaction is given by

$$\frac{\partial X_i}{\partial t} = \sum_j X_j P_{ji} - X_i \sum_k P_{ik} \qquad (3.15)$$

The first term describes the creation of elements; the second one the destruction. Obviously an element reaches its equilibrium abundance when its time derivative is zero. Using the reaction rates of Fowler et al. (1975) it is a straightforward matter to determine these equilibrium values and the time needed to reach this equilibrium. Although we will return to this in more detail for conditions appropriate for massive stars, it may be concluded that all mass layers which have ever been in the hydrogen burning core, have CN-equilibrium abundances, i.e.

$N \approx 0.01$ (primarily ^{14}N)
$C \approx 0.00027$ ($^{12}C \approx 0.0002$ and $^{13}C \approx 0.00007$)

Mass layers which have been in the hydrogen burning core for a significant part of the whole CHB lifetime will have full CNO-equilibrium abundances. For temperatures which are typical in the hydrogen burning cores of massive stars, these equilibrium abundances (by weight) are

$N \approx {}^{14}N \approx 0.02$
$C \approx {}^{12}C + {}^{13}C \approx 0.0003$
$O \approx {}^{16}O \approx 0.0006$

The energy rate ε has to be computed by considering all reactions of the CNO-cycle in detail. In practice however it is sufficient to consider the rates of the CN-cycle only. The reason is that in terms of probability, the ON cycle occurs 10000 times less frequently than the CN-cycle.

Helium burning

When all hydrogen fuel is consumed, nuclear energy production in the core will come to an end. As a consequence the radiation pressure, which was able to keep the mass of the core in equilibrium, drops, and the core starts contracting. The central temperature increases and when it exceeds 10^8K, helium burning starts.

The most important reactions of helium burning are:

<u>3α-process</u>

$3\ {}^4He \rightarrow {}^{12}C + \gamma$

compact notation : ${}^4He(2\alpha,\gamma){}^{12}C$

<u>the carbon chain</u>

$^{12}C + ^{4}He \rightarrow ^{16}O + \gamma$

compact notation : $^{12}C(\alpha,\gamma)^{16}O$

$^{16}O + ^{4}He \rightarrow ^{20}Ne + \gamma$

compact notation : $^{16}O(\alpha,\gamma)^{20}Ne$

and subsequent reactions $^{20}Ne(\alpha,\gamma)^{24}Mg(\alpha,\gamma)^{28}Si$
which are rare in typical stellar-interior environments.

<u>the nitrogen chain</u>

$^{14}N + ^{4}He \rightarrow ^{18}O + \gamma$

compact notation : $^{14}N(\alpha,\gamma)^{18}O$

and subsequent reactions $^{18}O(\alpha,\gamma)^{22}Ne(\alpha,\gamma)^{26}Mg(\alpha,\gamma)^{30}Si$

Obviously, a relation similar to (3.15) determines the variation as a function of time of the abundance X_i of the nucleus i.

The bulk of energy in a massive core helium burning (further abbreviated as CHeB) model is produced by the well known 3α-process, and to a lesser extent by the first two reactions of the carbon chain. The nitrogen chain is unimportant as a nuclear energy source.

The main importance of the carbon chain is the trade-off between carbon- and oxygen which is important for further nuclear burning stages. The $^{12}C(\alpha,\gamma)^{16}O$ reaction is still subject to some uncertainty; e.g. compared to the rate given by Caughlan and Fowler (1988), Azuma et al. (1994) propose a rate which is ~2 times larger. It should be repeated that the uncertainty mainly concerns the final constitution of the core at the end of helium burning, and to a much lesser extent, the evolution of a massive star through the helium burning phase.

The transformation of ^{14}N into ^{18}O under conditions appropriate for helium burning is very rapid; i.e. in layers where He-burning occurs, all ^{14}N produced previously by the CNO-cycle is rapidly converted entirely into ^{18}O and heavier elements. Therefore, stars showing the products of helium burning at their surface (Wolf-Rayet stars of type WC, section 2.7) are NOT expected to have ^{14}N.

Subsequent burning phases

The nuclear burning phases after core helium burning are very fast processes (i.e. they constitute less than 1% of the total stellar lifetime). The study of these phases and a correct treatment of the nuclear processes are important in order to study the morphology of supernova events and the chemical enrichment of galaxies. A detailed discussion has been presented by a.o. Arnett (1996). They will not be the scope of the present work.

N.B. we further use X, Y, Z for the weight abundances of hydrogen, helium and the ensemble of all other elements.

3.3.2. THE GRAVITATIONAL ENERGY GENERATION

Let $\dot{M}(t)$ be the possible mass loss/mass gain rate of the star. It follows then that

$$\left.\frac{\partial S}{\partial t}\right|_{M_r} = \left.\frac{\partial S}{\partial t}\right|_m - \dot{M}\frac{M_r}{M}\left.\frac{\partial S}{\partial M_r}\right|_t \qquad (3.16)$$

$$\text{with } m = \frac{M_r}{M}$$

The first term describes the work exerted by an expanding or contracting mass layer. During steady nuclear burning phases the structure of each mass layer varies very slowly and the $\left.\frac{\partial S}{\partial t}\right|_m$ - term can be neglected. In between two successive burning phases the stellar core contracts and the structure of the core mass layers varies much more rapidly. In this case the $\left.\frac{\partial S}{\partial t}\right|_m$ - term proves to be a non-negligible source of energy in equation (3.14).

The second term in equation (3.16) is called the homologous term and it represents the energy loss/gain when the star loses/gains matter at a rate $\dot{M}$. As will be shown later, this term is of fundamental importance for the evolution of a star through phases where its mass changes drastically (e.g. the Roche lobe overflow in close binaries, the LBV phase, or the hypergiant phase in the most massive stars).

3.4. Energy transport

The main processes governing the energy transport in stellar interiors are radiation and/or convection, i.e. the total energy flux is the sum of the radiation flux and of the convective flux,

$$\frac{L_r}{4\pi r^2} = F_{rad} + F_{conv} \tag{3.17}$$

3.4.1. RADIATIVE ENERGY TRANSPORT

The radiative flux is given by the well known relation

$$F_{rad} = -\frac{4a_R c}{3}\frac{T^3}{k\rho}\frac{\partial T}{\partial r} \tag{3.18}$$

The opacity k (cm^2/g) in stellar interiors is a complex function of temperature, density and chemical composition. Several decades ago, when calculations of stellar structure and stellar evolution started, a Kramers-type function for the stellar opacity was used. As computer power increased, opacity tables were implemented, starting with tables produced by Cox and Stewart (1969), later on Los Alamos (Cox and Tabor, 1976) LAOS-tables, and recently the OPAL-tables (Iglesias et al. 1992). The main difference between the latter two occurs around T = 300000 K where the OPAL opacities are up to a factor 4 larger than the Los Alamos ones.

The effect of the opacity on stellar evolution will be discussed in more detail in section 4.1. It is important to realize that iron is the main contributor to the opacity in massive stars, while the other heavy elements play only a minor role. Hence it is essential to know exactly the iron-abundance of a particular stellar environment in order to compare stellar evolution predictions with observations. The chemical composition of some particular mass layer may change in the course of the evolution (primarily the CNO-abundances), however iron remains the same. This means that although in stellar evolution all kinds of mixtures occur, in practice it is sufficient to specify the hydrogen abundance X and the iron abundance Fe. For the solar neighbourhood (resp. LMC and SMC) one uses generally X=0.7 (resp. 0.74 and 0.76), a total metallicity weight abundance Z=0.02 (resp. 0.008 and 0.002). The fractional iron abundance Fe is taken as 0.02164 (Anders and Grevesse, 1989).

3.4.2 CONVECTIVE ENERGY TRANSPORT AND MIXING

The time variation of a nuclear species X_i is given by the diffusion equation

$$\frac{dX_i}{dt} = \frac{\partial X_i}{\partial t} + \frac{\partial}{\partial M_R}\left[(4\pi r^2 \rho)^2 D \frac{\partial X_i}{\partial M_r}\right] \tag{3.19}$$

The $\partial X_i/\partial t$ term is the variation of X_i due to nuclear processes (section 3.3.1). The second term in equation (3.19) describes the diffusion of X_i in a plasma where X_i may vary as function of the mass coordinate M_r. The typical time scale t of the process is determined by the diffusion coefficient D, i.e. $t \approx 1/D$.

The value of D and of F_{conv} [equation (3.17)] depends on the temperature, the density and on whether the plasma is stable against convection. Criteria which allow the description of this stability are based on the following three gradients (Kato, 1966):

$$\nabla_{rad} = \frac{3kPL_r}{16\pi a_R c G M_r T^4} = \text{radiative temperature gradient} \tag{3.20}$$

$$\nabla_{ad} = \left.\frac{\partial \ln T}{\partial \ln P}\right|_S = \frac{\Gamma - 1}{\Gamma} = \text{adiabatic temperature gradient} \tag{3.21}$$

Γ = the adiabatic exponent (Chandrasekhar, 1957)

$$\nabla_{Led} = \nabla_{ad} + \frac{\beta}{4 - 3\beta}\frac{\partial \ln \mu}{\partial \ln P} = \text{Ledoux gradient (Ledoux, 1947)} \tag{3.22}$$

β = gas pressure/total pressure)

Mass layers where $\nabla_{rad} < \nabla_{ad}$ are stable against convection; this means that the total energy is transported by radiation only ($F_{conv} = 0$ and $D = 0$) .

A plasma is convectively unstable when $\nabla_{rad} > \nabla_{Led}$ (usually referred to as the Ledoux-criterion for convection). Expressions for D and for F_{conv} can be derived by describing convection by means of the mixing length theory of Böhm-Vitense (1958). According to this theory F_{conv} and D depend explicitly on λ (= mixing length = the average vertical distance which a convective element travels before it dissolves into its surrounding). λ is of order of the pressure scale height, but its value cannot be determined exactly from first principles. The uncertainty in λ introduces significant uncertainty in the treatment of convective layers situated just below the stellar photosphere (in massive stars this occurs during the red supergiant phase, see section 4). For convection in the deep interior of stars the uncertainty is less critical. This can be understood as follows: in convective cores the diffusion coefficient is very large, independently of the uncertainty in λ. This means that the mixing time scale is always much smaller than the corresponding evolutionary time scale (the nuclear time scale, section 4). Approximating the diffusion process in these layers, by means of instantaneous and full mixing, is therefore justified and this drastically simplifies the computations. Furthermore, using temperatures and density values, typical for these regions, even with the assumption that the whole energy flux is carried by convection, the difference between the actual temperature gradient and the adiabatic temperature gradient is extremely small, so that with a large degree of accuracy it is possible to use the adiabatic expression [instead of equation (3.17)]. Consequently the uncertainty in λ is entirely removed.

The central part of a star where nuclear burning occurs, is always convective (= convective core). It is a general property that as nuclear burning proceeds, the mass extent of the convective core decreases. In this way a composition gradient is left behind in the stellar interior (this is illustrated in figure 3.1).

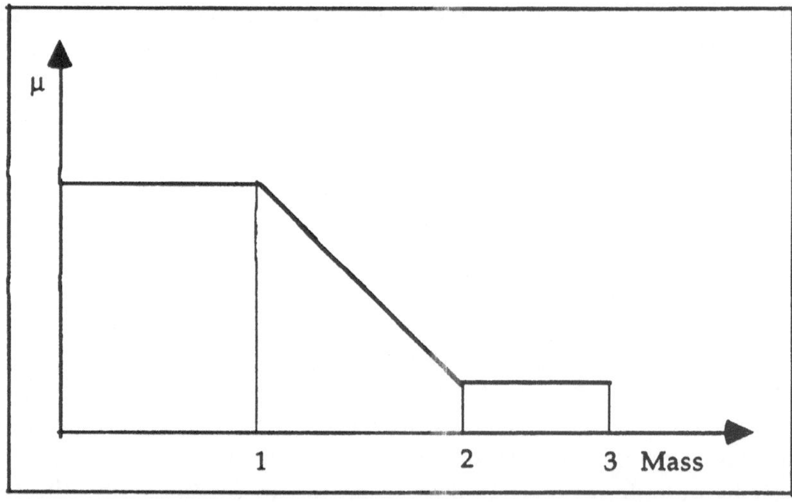

Figure 3.1: The variation of the μ-gradient in the interior of a massive star in the second half of its CHB. On the mass axis, in arbitrary units, (1) = the mass of the present convective core, (2) = the mass of the convective core when the star was a ZAMS star and (3) = the present total mass of the star.

Especially in massive stars a situation occurs where in this zone with varying μ-gradient $\nabla_{ad} < \nabla_{rad} < \nabla_{Led}$. As shown by Kato (1966) such layers are dynamically stable, but vibrationally unstable; this may initiate 'slow' mixing, a process commonly known as 'semi-convection'. The appropriate diffusion coefficient for this type of mixing satisfies

$$D = \alpha \frac{k}{6c_p\rho} \frac{\nabla_{rad} - \nabla_{ad}}{\nabla_{Led} - \nabla_{rad}} \tag{3.23}$$

The proportionality constant α is a parameter whose value cannot be determined from first principles. The actual temperature gradient in these semi-convective layers can be computed consistently with the energy flux similarly as in the case of super-adiabatic convection (see e.g. Langer et al. 1985) but here the parameter α enters explicitly as well. Two limiting cases are of some interest.

If α=0, i.e. no mixing in the semi-convective region, the energy is transported entirely by radiation (this case is generally referred to as the Ledoux-criterion for semi-convection, the L criterion, Ledoux, 1947).

For values of α large enough ($\alpha \gg 1$) so that the mixing time scale in the semi-convective region becomes much smaller than the evolutionary time scale, the mixing process can be treated as an instantaneous process just like in convective regions. Since any mixing changes the instability condition, it can readily be understood that in this case the diffusion process will continue until a situation is reached in which $\nabla_{rad} = \nabla_{ad}$. In this way the classical Schwarzschild and Härm (1958) criterion for semi-convection is recovered (the SH criterion).

It is clear that the computational procedure is largely simplified in both limiting cases since the diffusion equation no longer needs to be solved explicitly.

3.4.3. THE BOUNDARY BETWEEN CONVECTIVE AND RADIATIVE LAYERS.

If we were able to determine the diffusion coefficient D from first principles for temperature, density and composition appropriate for stellar interiors, the problem of where to place the boundary between convective, semi-convective and radiative layers would not exist; one only needs to solve explicitly the diffusion equation. Unfortunately D contains uncertain parameters, thus different approaches are needed.

As discussed in the previous subsection, a star can be divided into three regions:

$\nabla_{rad} < \nabla_{ad}$	$\nabla_{ad} < \nabla_{rad} < \nabla_{Led}$	$\nabla_{Led} < \nabla_{rad}$
region I	region II	region III
radiative equilibrium	semi-convection	convection

These inequalities are strict ones, and it is tempting to determine the boundaries between the regions using equalities. Since in fully convective regions, $\frac{\partial \mu}{\partial r} = 0$, this would mean that the boundary between a convective region and an adjacent non-convective region is determined by the condition $\nabla_{rad} = \nabla_{Led} = \nabla_{ad}$. However convective motions do not necessarily stop where the condition for convection is no longer fulfilled. The problem is commonly known as 'convective penetration' or 'convective overshooting'. Penetration exists in laboratory experiments, is observed in the earth's atmosphere and in oceans. The problem is not whether convective penetration exists or not, but what is its extent in stellar interiors. It is obvious that any uncertainty in the extent of convective overshooting is not a problem for envelope convection (red giant or red supergiants); overshooting from convective cores, however, may imply a serious uncertainty in the evolution of stars in general.

Convective core overshooting.

A discussion of the problem has been given e.g. by Roxburgh (1978, 1989, 1992), who proposed an integral condition for the radius r_c of the convective core. However, the latter seems to be incorrect (Eggleton, 1983).

For evolutionary computations it is advisable to parameterize overshooting and to study its effect. The best way to do so is to use the radius r_s of the Schwarzschild core (i.e. the radius of the layer where $\nabla_{ad} = \nabla_{rad}$) and to use a fraction αr_s as the overshooting distance. Compared to other parameterizations which appeared in literature this has the advantage that overshooting only occurs there where a convective region is present, which is obviously a condition sine qua non to have overshooting. Detailed comparison with observations will hopefully restrict the possible values of the parameter α.

3.5. The time-variation of mass in evolutionary computations

To determine the variation of the mass of a massive star during its evolution one can use the following semi-empirical mass loss rate relations:

- Although the OB type stars where $\dot{M}$ values have been determined cover only a part of the CHB phase of massive stars, equation (2.10) or (2.11) is used for the whole phase. Both give very similar results.

- When a star becomes a RSG, equation (2.18) can be used.

- When the star evolves into a star resembling a WR star, WR-like mass loss rates have to be applied. One could be inclined to use equation (2.19) however as discussed in section 2.7, due to inhomogeneities of the WR stellar wind the latter overestimates the $\dot{M}$ possibly by a factor 2-3. Therefore, in order to study the effect of the WR wind on stellar evolution, one may use a formalism that satisfies the following observational facts and/or criteria:

  - the mass loss rate of the WNE component of the binary V444 Cyg resulting from the observed orbital period variation ~ 1.1±0.5 10^{-5} M_o/yr (Khaliullin et al., 1984; Underhill et al., 1990). The orbital mass of the WR star is very well known and equals ~9 M_o (section 13.2) implying a luminosity log $L/L_o \approx 5$ (section 12.2.1),

  - a detailed wind model of the WN5 star HD 50896 including clumping has been performed by Schmutz (1997). The model yields a luminosity of log L/L_o = 5.6-5.7 and a SW mass loss rate ≈ (4±1).10^{-5} M_o/yr,

  - the discovery of very massive BHs in X-ray binaries (section 13.3.2) indicates that stars with initial mass > 40 M_o may end their life with mass ≥ 10 M_o (= the mass of the star at the end of CHeB),

  - the observed star number ratio of WN/WC ≈ 1 in the Solar neighbourhood ; the theoretically predicted one is largely dependent on the adopted $\dot{M}$ formalism; a $\dot{M}$ that is too large (resp. too small) predicts a too small (resp. too large) WN/WC ratio.

  Anticipating, a WR $\dot{M}$ relation that meets these constraints (within the errors) is given by

$$\log(-\dot{M}) = \log L - 10 \tag{3.24}$$

$\dot{M}$ in M_o/yr, L in L_o,

predicting a WR $\dot{M}$ that is indeed 2-3 times smaller (for Log $L/L_o \geq 4.5$) than the value obtained when equation (2.19) is used.

- When the star becomes a LBV with $M_{bol} \leq -9.5$, the SW mass loss becomes very large although the exact rate is very uncertain. Due to the fact that no RSGs are observed brighter than $M_{bol} = -9.5$, a working hypothesis could be

  'the $\dot{M}$ during the LBV phase of a star with $M_{bol} \leq -9.5$ must be sufficiently large to suppress a large expansion, hence to prohibit the redward evolution in the HRD'

  or relaxing somewhat the foregoing criterion

  'the $\dot{M}$ during the LBV and RSG phase of a star with $M_{bol} \leq -9.5$ must be sufficiently large to assure a RSG phase which is short enough to explain the lack of observed RSGs with $M_{bol} < -9.5$'.

  The meaning of both formalisms will be illustrated in section 4.1.2.

- To study the metallicity dependence of the mass loss rate during CHB, one can use relation (2.12). This dependence is less clear during the RSG, the LBV or the WR phase.

3.6. Summary

In the computation of the evolution of a non-rotating, non-magnetic single star one has to adopt a value for the convective core overshooting, a value for the mixing length parameter λ for the treatment of surface convection, a value for the proportionality constant in the diffusion coefficient in semi-convective regions, and then to solve a set of six non-linear differential equations, i.e.

$$\frac{\partial P}{\partial r} = -G\frac{M_r \rho}{r^2}$$

$$\frac{\partial M_r}{\partial r} = 4\pi\rho r^2$$

$$\left.\frac{\partial L}{\partial M_r}\right|_t = -T\left.\frac{\partial S}{\partial t}\right|_{M_r} + \varepsilon - \varepsilon_\nu$$

$$\frac{L_r}{4\pi r^2} = F_{rad} + F_{conv} \qquad (3.25)$$

$$\frac{dX_i}{dt} = \frac{\partial X_i}{\partial t} + \frac{\partial}{\partial M_R}\left[(4\pi r^2\rho)^2 D\frac{\partial X_i}{\partial M_r}\right]$$

$$\dot{M} = \dot{M}(t)$$

together with the equation of state and with expressions for the absorption coefficient κ which is related to the opacity k through $\kappa=\rho k(T,X,Z)$, for the nuclear energy $\varepsilon=\varepsilon(\rho,T,X,Z)$ and for the energy lost by neutrinos ε_ν.

3.7. Surface layers - Boundary conditions

The solution of differential equations requires boundary conditions. In the center they are obvious.

The surface layers in present day stellar evolution codes are treated assuming a gray atmosphere in hydrostatic equilibrium and LTE. Using the Eddington approximation this would mean that when $T_{2/3}$ and $R_{2/3}$ are respectively the temperature and the radius at optical depth $\tau = 2/3$, the following relations should hold

$$L = 4\pi R_{2/3}^2 \sigma_R T_{2/3}^4$$

$$P = \frac{2}{3} G \frac{M}{R_{2/3}^2} \qquad (3.26)$$

These relations together with the conditions in the center allow the equation set (3.25) to be solved.

However, to what extend do the approximations on which condition (3.26) is based affect stellar evolutionary computations? The evolution of a star is determined by the physical processes in the stellar interior which depend mainly on the total mass of the star and the ensuing gravity. The total mass of the surface layers where in some circumstances the assumptions mentioned above are questionable is negligible compared to the total stellar mass. Therefore these surface layers and their modelation will not be very important for the total stellar structure. This is confirmed by the numerical solution of the coupled set of stellar structure equations which indeed depends hardly or not at all on the treatment of the surface layers. As an example the use of the so-called 'zero boundary conditions' (i.e. at the stellar surface T=0 and P=0, Schwarzschild, 1958) in the integration from center to surface, gives a very similar stellar structure and evolution as the one calculated by using the Eddington approximation (i.e. $T = 0.841T_{2/3}$ at the stellar surface).

It is of course necessary to compare evolutionary computations with observations. Observations allow the positioning of the star in a HR diagram which requires the knowledge of the luminosity of the star and the effective temperature. In evolutionary computations the effective temperature T_{eff} is defined as $T_{eff} = T_{2/3}$. This definition of T_{eff} is compatible with the observations (read spectroscopic definition) provided that the surface layers (read continuum forming region) are not extended (i.e. the variation of the radius in the surface layers is very small).

This restriction is not valid for the most massive stars which are subject to stellar wind mass loss at very large rates (Luminous Blue Variables, Wolf-Rayet stars, hypergiants, ...). The problem is sketched in figure 3.2.

When the SW is large, the layer with continuum optical dept τ around unity is located inside the SW and possibly at a radius $R_1 = R(\tau=1)$ which may be considerably larger than the photospheric radius computed with the hydrostatic stellar structure equations. In between this photosphere and the $\tau = 1$ layer, there is a (unobservable) region which is probably not in hydrostatic equilibrium but where theory is yet unable to predict the variation of T and r. This region is marked with a question mark in figure 3.2.

The value of R_1 can be estimated as follows:

the velocity (v) as a function of radius R in a SW can in most cases be described by a relation

$$v = v_1 + (v_\infty - v_1)\left(1 - \frac{R_1}{R}\right)^\beta \qquad (3.27)$$

where v_∞ = terminal velocity
v_1 = velocity at $\tau = 1$ ($R=R_1$).

The theory of SW and the observations predict β values in the range 0.7 - 1.5 (a.o. Hamann and Koesterke, 1996).

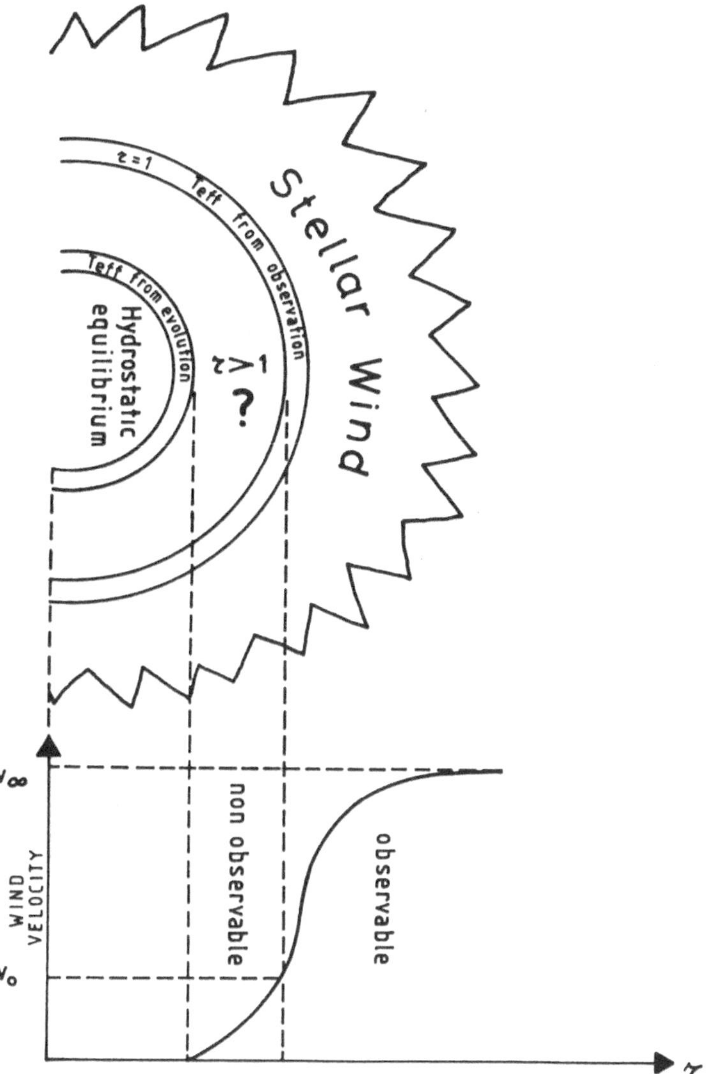

Figure 3.2: The difference between the radius of the layer with optical dept τ = 1 resulting from the hydrostatic stellar structure equations and the realistic one when a mantle caused by a large stellar wind is present around the star.

In massive stars, electron scattering is the main contributor to the continuum opacity ($k \approx k_e$) in the visual and the UV, and therefore the radius R_1 of the layer where τ(visual) ≈ 1 can be approximated by

$$k_e \int_{R_1}^{\infty} \rho dR = 1$$

$$\text{with} \tag{3.28}$$

$$\rho = \frac{\dot{M}}{4\pi R^2 v}$$

This means that the T_{eff} resulting from spectral analysis will roughly satisfy

$$L = 4\pi R_1^2 \sigma_R T_{eff}^4 \tag{3.29}$$

Since the interior structure of a star and its evolution do not or hardly depend on the details of the SW or of the region marked with a question mark in figure 3.2, the most straightforward method to compare evolutionary prediction with observations is to use the luminosity resulting from the integration of the stellar structure equations, to use R_1 from equation (3.28) and compute T_{eff} from equation (3.29).

4. Evolutionary computations of non-rotating massive single stars

Semi-convection and its treatment determines critically the evolution of stars with initial mass larger than ~12 M_o. We therefore separately discuss the evolution of stars with $M \leq 12\ M_o$ and the evolution of stars with larger initial masses.

When the extent of the overshooting region is large, the effect on stellar evolution can be significant. We therefore distinguish the case where convective core overshooting is small (and can be neglected), from the case with large convective core overshooting.

CHAPTER II

4.1. Small convective core overshooting

4.1.1. THE EVOLUTION OF SINGLE STARS WITH 5 $M_o \le M \le 12$ M_o

By numerical integration of the basic structure equations (3.25) one predicts the evolution of a star. Figure 4.1 shows evolutionary tracks of stars with masses of 5, 7, 9, 12 M_o (Schaller et al. 1992) for the Galaxy and for the SMC. Lifetimes are given in table 4.1.

TABLE 4.1: The CHB and CHeB timescale for stars with initial mass 5 M_o, 7 M_o, 9 M_o and 12 M_o (in units of 10^6 yrs), for Z = 0.02 and Z = 0.001.

M_i	τ_{CHB}		τ_{CHeB}	
	Z = 0.02	Z = 0.001	Z = 0.02	Z = 0.001
5	94.5	88.3	12.4	10.9
7	43.2	45.1	4.7	4.7
9	26.4	28.8	2.6	2.7
12	16.0	18.1	1.6	1.6

The assumptions behind these computations are

- semi-convection is treated according to the SH criterion (see section 3.4.2); the overall results and conclusions discussed here are very similar if instead the L criterion is used,
- OPAL-opacities are used for temperatures above 6000K, radiative opacities of Kurucz (1991) including the main molecule lines for low temperatures,
- during helium burning one uses the $^{12}C^{16}O$ nuclear reaction rate from Caughlan et al. (1985),
- the mixing length parameter λ (important for the RSG phase during CHeB) is taken to be 1.6 times the pressure scale height,
- the SW during the RSG phase is assumed to be very small, i.e. too small to affect the evolution during CHeB. Some criticism will be given at the end of this section.

Core hydrogen burning (CHB).

The evolutionary tracks between points 0 (the beginning of core hydrogen burning or ZAMS = zero age main sequence) and 1 correspond to the core hydrogen burning phase which constitutes about 90% of the total stellar lifetime. The star is in thermal equilibrium and all time variations are very small. The entropy term $T\partial S/\partial t$ in the energy equation (3.14) can therefore safely be neglected. The evolutionary time scale during this phase is called the 'nuclear time scale' τ_{nuc} which is the time needed for the star to exhaust its nuclear fuel (which is proportional to the mass M of the star) at its present fuel consumption rate (which is proportional to the luminosity L), i.e.

$$\tau_{nuc} = 10^{10} \frac{M/M_o}{L/L_o} \quad \text{(expressed in year)} \tag{4.1}$$

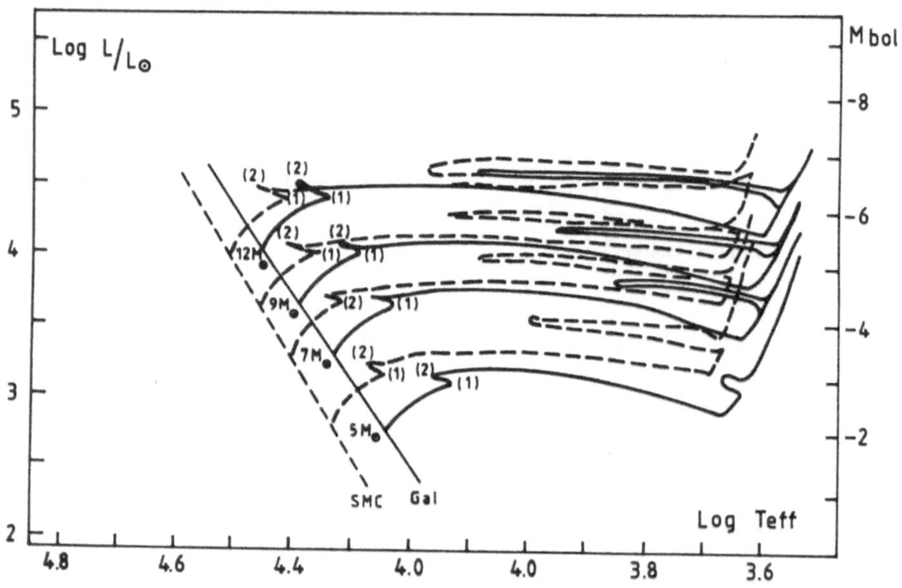

Figure 4.1: Evolutionary tracks of stars with masses of 5, 7, 9, 12 M_o for the Galaxy and for the SMC (from Schaller et al. 1992).

Hydrogen shell burning.

When the central hydrogen abundance X_c drops below ~0.05 the energy production in the convective core is no longer capable of supporting the stellar layers and a short overall contraction phase sets in (points 1 and 2 along the tracks in Figure 4.1). One generally refers to point 1 as 'red point' or TAMS (terminated age main sequence). The energy emitted by the star during this phase is precisely the gravitational energy liberated as a consequence of the overall contraction. An overall contraction implies an overall temperature increase and a situation is reached where just outside the newly formed helium core the temperature increases beyond the minimum value necessary for hydrogen burning, i.e. from point 2, hydrogen is burning in a shell just outside the core (the hydrogen shell burning phase). The energy produced in the shell stops the contraction of the outer layers (one therefore denotes point 2 as 'blue point'); even more, together with the energy liberated by the still contracting core of the star, it is sufficiently high to power a rapid expansion of these outer layers. The stars very rapidly cross the HR diagram towards the red part ($T_{eff} \leq 10000$K). The time scale of this process roughly equals the Kelvin-Helmholtz time scale or thermal time scale τ_{th} (typically a few percent of the corresponding core hydrogen burning lifetime), which is the time it takes for a star with mass M and radius R to emit its thermal energy content (GM^2/R) at its present luminosity L , i.e.

$$\tau_{th} = 3.1 \cdot 10^7 \frac{M^2}{RL} \quad \text{(M, R and L in solar units, } \tau_{th} \text{ in year)} \qquad (4.2)$$

Since this phase is so short one expects to observe only few stars compared to the number of stars in their CHB phase: the blue Hertzsprung gap.

Core helium burning (CHeB).

As the stars move towards lower temperatures, the opacity in the outer layers increases and the radiative gradient [equation (3.20)] becomes larger than the adiabatic one [(equation (3.21)]: envelope convection sets in. Almost at the same time helium burning starts in the core. The helium burning lifetime of a star typically amounts to ~10 % of the core hydrogen burning lifetime: the values are given in table 4.1 as well. The envelope convection extends down to layers which have at some time been in the hydrogen burning core. This leads to the dredging up of CNO processed matter and the stellar atmosphere becomes enhanced in N, depleted in C and O.

Of particular importance are the blue loops after the onset of core helium burning. They are due essentially to changes in the potential of the core and the resulting rate of gravitational energy production in the central region (Lauterborn et al., 1971). As a consequence part of the core helium burning occurs in the red supergiant region, and another part in the blue. The exact ratio depends on the value of the mixing length, and the amount of overshooting (sections 3.4.2 and 3.4.3).

Table 4.2: The mass of the CO core at the end of CHeB as a function of initial mass M_i for two values of the metallicity

M_i	M_{CO}	
	Z=0.02	Z=0.001
7	1.0	1.1
9	1.4	1.6
12	2.1	2.3

Table 4.2 gives the mass of the CO-core at the end of core helium burning as a function of the initial mass M_i.

When the He burning phase comes to an end, the CO-core starts contracting, resulting in an increase of the central temperature and density. The subsequent core evolution (which is largely independent of the mass layers outside the CO-core) may follow two paths:

a. *the CO-core becomes degenerate before the temperature and density are large enough to initiate quiescent carbon burning.*

When the convective core overshooting during CHB and CHeB is small, stars with initial mass smaller than ~8 M_o (and larger than ~3 M_o) are red giants on the Asymptotic Giant Branch (AGB) and develope a degenerate CO-core before the central temperature is large enough for carbon to ignite (for a review, Iben and Renzini, 1983). We distinguish two possibilities:

- ***a1***. The observations reveal that AGB stars lose mass by SW at 'super' rates and a nebula is formed around the star. If these rates are large enough, almost the entire hydrogen rich envelope of the star is lost prior to carbon ignition. The star moves rapidly to the blue and becomes a CO White Dwarf (WD). This type of evolution is illustrated in figure 4.2. The hydrogen stripped star leaves the AGB and evolves towards the WD region in the HRD. When the temperature of the star exceeds ~30000K, the surrounding nebula will become ionized and the star will be observed as a 'Planetary Nebula Nucleus' (PNN). It is interesting to remark that when SW continues and is large enough, the star at this stage may show WR like features, although it has a mass smaller than 1.4 M_o.

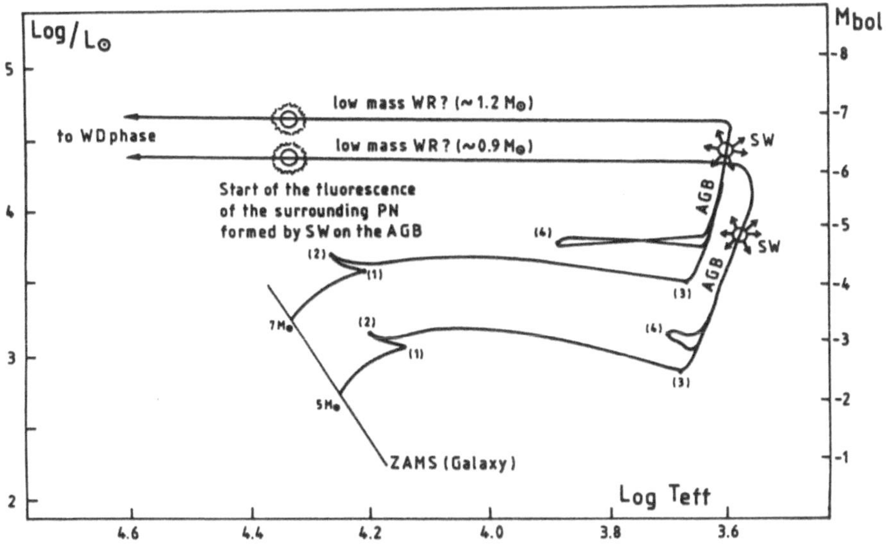

Figure 4.2: The effect of stellar wind mass loss at 'super rates' during the AGB phase on the evolution of a galactic star with initial mass = 5 M_o and 7 M_o.

- *a2*. The scenario sketched above critically depends on the adopted SW mass loss rate of AGB stars, and similarly as with massive stars, these rates are uncertain. This uncertainty has profound consequences for the final evolution of AGB stars. When $\dot{M}$ is not quite large enough, carbon can be ignited. For stars with initial mass ≤ 8 M_o, this process is so violent (carbon flash) that the liberated energy probably disrupts the entire star in a so-called carbon-deflagration supernova and obviously the WD phase is avoided.

The spectroscopic mass determinations of WDs are discussed in section 5.1.9. The largest observed mass is ~1.2 M_o and we therefore conclude that stars which are able to form CO cores with mass up to 1.2 M_o lose all there hydrogen layers before carbon ignition. Since a 1.2 M_o CO core is formed in a star with initial mass ~8 M_o, we expect that scenario a1 is most probable and that a carbon-deflagration supernova is very rare in single stars.

b. the quiescent carbon burning starts before the CO-core becomes degenerate.

Detailed computations of CHeB stars up to the formation of a FeNi-core (Habets, 1986) reveal that all CHeB stars which are able to form a CO-core with mass ≥ 1.4 M_o (i.e. with initial mass larger than 9 M_o), will end their life as a NS or as a BH. When the core collapse results in the formation of a NS, it is accompanied by a supernova explosion which removes the layers outside the collapsed core.

Also in stars with initial mass between 8 M_o and ~9 M_o, carbon ignites quiescently and a O-Ne-Mg core is formed but it becomes degenerate before the subsequent burning phases start. How these stars end their life depends on the details of the further burning processes. Depending on what kind of fine tuning is used, these stars either end their life as O-Ne-Mg WDs, or as a NS.

Remark

The calculations discussed above hold when it is assumed that the SW during the RSG phase is small and does not effect the evolution. However, when equations (2.18) and (2.12) also apply for RSGs with initial mass ≤ 12 M_o, it cannot be excluded that also the 12 M_o star loses a few solar masses. This may not alter significantly the overall core evolution discussed above, nor does it affect the surface chemical abundance. However it may affect the blue loop and especially the lifetimes spend in the red and in the blue.

4.1.2. THE EVOLUTION OF SINGLE STARS WITH M > 12 M_o

for the Galaxy:

Figure 4.3 shows evolutionary tracks of stars with initial mass larger than 12 M_o. Quantitative results are given in table 4.3 which also includes the results for the SMC.

The overall evolutionary behaviour of a star with initial mass larger than 12 M_o is similar to that discussed in section 4.1.1, however stellar wind mass loss and the effects of semi-convection imply a few important differences.

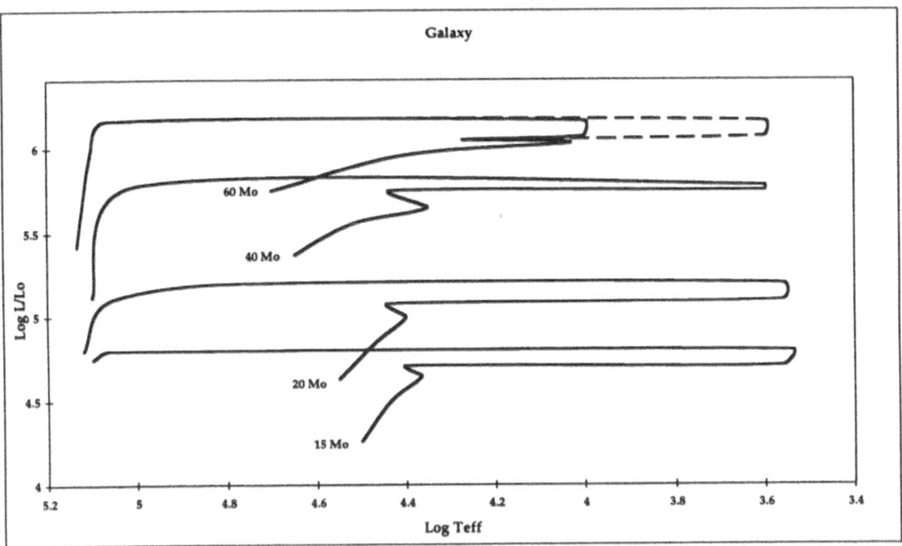

Figure 4.3: Evolutionary tracks of galactic massive single stars with initial mass larger than 12 M_0. The evolution of the 60 M_0 star is first computed by assuming that the LBV stellar wind is large enough to prohibit redward evolution (full track); the dashed evolutionary track corresponds to the case where the LBV mass loss is smaller, the star becomes a RSG and loses its remaining hydrogen rich layers there.

Summarizing:

- Stellar wind mass loss during CHB has only a small effect on the general evolution of a star with initial mass up to ~40 M_0. After CHB, hydrogen burning continues in a shell just outside the He core. The star expands on the thermal time scale [equation (4.2)], and crosses the HR diagram very rapidly. At the beginning of CHeB the star becomes a RSG.

- Stellar wind mass loss during the RSG phase is of primary importance; when equation (2.18) is applied during the whole RSG phase,

  all stars with initial mass between ~15 M_0 and 40 M_0 lose most of their hydrogen rich layers during the RSG phase.

Even if we account for the uncertainty of RSG mass loss rates and we use a 2 times lower rate, it follows that all stars with initial mass between 20 M_0 and 40 M_0 lose all their hydrogen.

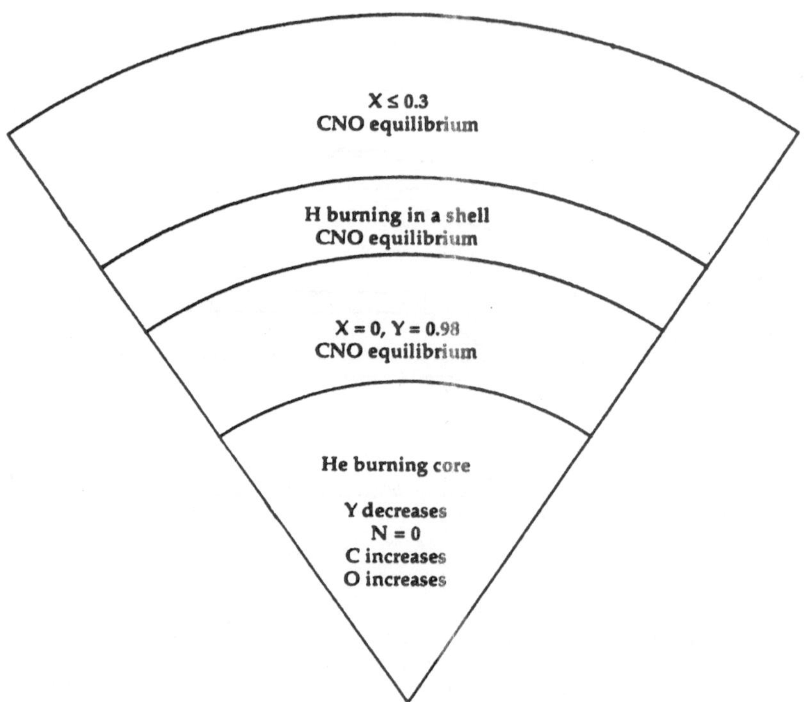

Figure 4.4: The structure of the CHeB remnant after the removal of most of its hydrogen rich outer layers.

- When the atmospheric hydrogen abundance by weight X_{atm} becomes smaller than 0.3-0.4, the star starts contracting, regains thermal equilibrium and moves to the blue part of the HR diagram, i.e. all stars with mass larger than ~15 M_o (20 M_o if the RSG $\dot{M}$ is two times smaller) become hydrogen stripped helium cores. The overall structure of a star at that moment is shown in figure 4.4.

- Table 4.3 gives the lifetime of the CHeB star spent as a RSG (the remaining lifetime then obviously equals the fraction spent as a blue hydrogen stripped CHeB star). The results of the table are computed using equation (3.24) to determine the stellar wind mass loss rate once the star becomes a WR star. A hydrogen deficient helium burning star is considered as WR when its mass is larger than 5 M_o (section 12.2).

TABLE 4.3: The time (in millions of years) spent by a star as a CHB star, a RSG, a WR, the total CHeB lifetime, the mass M_{eCHB} at the end of CHB, the mass M_{bWR} at the beginning of the WR phase, the mass M_{bWC} at the beginning of the WC phase and the final mass M_{final} at the end of CHeB, for two formalisms of the stellar wind mass loss rate during the RSG phase.

Galaxy

$\log(-\dot{M}) = 0.8\log L - 8.7$

M_{ZAMS}	T_{CHB}	T_{RSG}	T_{WR}	T_{He}	M_{eCHB}	M_{bWR}	M_{bWC}	M_{final}
15	14.8	0.75	0	1.12	15	-	-	4.9
20	10.1	0.59	0.26	0.85	19.5	7.1	-	5
25	8.2	0.46	0.27	0.73	24	9.9	5.2	5
30	6.9	0.32	0.28	0.6	28	12.9	7.8	6
40	5.5	0.18	0.29	0.47	36	19.2	12.3	8
60	3.4	-	0.35	0.35	48	32	23	10.6

$\log(-\dot{M}) = 0.8\log L - 9.0$

M_{ZAMS}	T_{CHB}	T_{RSG}	T_{WR}	T_{He}	M_{eCHB}	M_{bWR}	M_{bWC}	M_{final}
15	14.8	1.12	0	1.12	15	-	-	8
20	10.1	0.75	0	0.75	19.5	7.3	-	7.3
25	8.2	0.63	0.02	0.65	24	10	-	8.9
30	6.9	0.52	0.04	0.56	28	13.2	-	11
40	5.5	0.36	0.11	0.47	36	19.8	12.6	12
60	3.4		0.35	0.35	48	32.5	23	10.6

SMC (Z = 0.002)

M_{ZAMS}	T_{CHB}	T_{RSG}	T_{WR}	T_{He}	M_{eCHB}	M_{bWR}	M_{bWC}	M_{final}
15	16.1	0.8	0	1.0	15	-	-	5.4
20	11.5	0.64	0.18	0.82	20	7.9	-	5.7
25	9.2	0.49	0.2	0.69	25	11.5	-	6
30	7.9	0.37	0.2	0.57	29.5	16.5	8.1	6.4
40	6.2	0.15	0.29	0.44	39	26.1	13.7	8.7
60	3.9	-	0.32	0.32	53	36	25.1	11.2

- We started with the assumption that the LBV mass loss is large enough to prohibit redward evolution for stars with initial mass larger than 40-50 M_o (the tracks in full)[6]. Whether or not the large value of $\dot{M}$ that is needed corresponds with observations of LBVs is a matter of faith. If in reality the average rates are lower, also stars with initial mass larger than 40 M_o will evolve to the red part of the HR diagram, similarly as those with lower initial mass. This is illustrated in figure 4.3. As an example, let us consider the evolution of a 60 M_o star. The star loses ~12 M_o during CHB as a consequence of a normal O-type SW. At the end of CHB, it still has a ~16 M_o hydrogen rich envelope. To avoid the redward evolution (i.e. to avoid a RSG phase), this whole envelope has to be removed during the LBV phase. However, if we suppose that the real (observed) SW is two times lower than is needed, only 8 M_o will be lost during the LBV phase and the star becomes a RSG with luminosity $\log L/L_o \approx 6.1$. Applying equation (2.18), the RSG will lose mass by SW at a rate ~2.10^{-4} M_o/yr. It thus takes ~40000 yrs to remove the 8 M_o envelope that was left after the LBV phase. This time scale is sufficiently short to explain the lack of observed RSGs with $M_{bol} < -9.5$.

- Differences in evolutionary behaviour of a massive star due to a different treatment of semi-convection (section 3.4.2) occur mainly for masses M < 15 M_o and have been discussed in detail by Langer and Maeder (1995). Models with the SH criterion start CHeB as RSG whereas during their CHeB burning blue loops occur. The blueward excursion is suppressed when semi-convection is treated with the L-criterion, i.e. the stars remain RSGs during their entire CHeB phase.

---

6 *the suppression of a star's expansion during its hydrogen shell burning phase in terms of mass loss rates means that we have to accept that a star loses most of its hydrogen rich layers on the Kelvin-Helmholtz time scale, i.e. roughly*

$$\dot{M} \approx \frac{M}{\tau_{th}} \approx 3 \cdot 10^{-8} \frac{RL}{M} \qquad (\dot{M} \text{ in } M_o/yr, M, R \text{ and } L \text{ in solar units}).$$

If the LBV phase corresponds to the hydrogen shell burning phase, the RSG phase can be avoided only when the LBV SW mass loss rate > 10^{-4} M_o/yr. Note however, that it can not be excluded that the LBV phase already begins while the star is still burning hydrogen in the core.

- In the larger part of the stellar interior, the opacity is due to electron scattering. Only in the outermost layers, where the plasma temperature becomes smaller than ~ 500000 K and where the heavier elements are only partially ionized, does bound-free opacity become important. For these layers it is required to use the most sophisticated opacity tables. But, since these outer layers constitute only a small amount of the total mass of the star, it is obvious that the overall stellar interior evolution of a massive star is hardly affected by these outermost layers. This explains why evolutionary computations with the older LAOS opacities, are very similar to those calculated with the newer OPAL tables. The main difference between both sets are the T_{eff} values of the models: OPAL opacities are larger, hence the resulting radiation pressure is larger in the outer layers causing the star to be larger. Since the nuclear energy production (and thus the luminosity) depends very little on the opacity in the outer layers, OPAL opacities imply lower T_{eff} values.
  Particularly interesting is the effect of the new OPAL opacities in the most massive stars: in the outer layers where the ionization of the heavy elements is only partial, the radiation force nearly equals gravity, i.e. these outer layers are very unstable. Perhaps this process might be the cause of the LBV phenomenon.

- When due to RSG mass loss, a large fraction of the hydrogen rich layers were removed (i.e. when $X_{atm} \leq 0.4$), the star regains thermal equilibrium, moves to the blue part of the HR diagram and becomes a hydrogen deficient CHeB star. When its mass is large enough ($\geq 5\ M_o$, section 12.2), the star resembles a WR star (a WN star with hydrogen = WNL). Due to WR like SW mass loss, first the layers in which still some hydrogen was left, are removed (the star will be observed as a WN star without hydrogen = WNE). If mass loss by stellar wind continues, layers may appear at the surface which have at some time been in the CHeB core and one expects to see a star with atmospheric chemistry corresponding to the 3α-process. This means that ^{14}N has disappeared completely and an overabundance of C and O is expected: the star has turned into a WC type star.

- During the entire WN phase, the star has atmospheric layers which belonged to the hydrogen burning core during several million years. As outlined in section 3.3.1, these layers should have CNO equilibrium abundances. Evolutionary computations of massive stars therefore predict than all (massive) WN stars should have very similar surface chemistry.

- At the beginning of the WR phase when hydrogen is still present, the luminosity of the star is determined by the CHeB reactions and by the hydrogen burning in the shell just outside the helium burning core. As stellar wind strips the star further down until all hydrogen is removed, the hydrogen shell burning is obviously turned off. This is visible in the tracks of figure 4.3 where at a certain moment the luminosity drops considerably.

- Table 4.4 shows the relation between the initial ZAMS mass, the final mass of the star just before the SN explosion and the final mass of the CO core for galactic stars. For the 20 M_o and 15 M_o star, we also give the results when the RSG SW mass loss rate is a factor 2 smaller than predicted by equation (2.18). Since the evolution of the final CO core should be largely independent from the mass layers outside this core, the final mass of the FeNi core is estimated from the detailed computations of Woosley (1986). We conclude:

  - when the mass of the final FeNi core is compared with the maximum mass of a NS (section 3.2.6), it cannot be excluded that all single stars with mass larger than 20-30 M_o collapse to form a BH and no SN explosion happens,

  - with the SW mass loss rate of WR stars discussed in section 3.5, our evolutionary computations predict the existence of BHs with mass larger than 10 M_o,

- the above discussion and the evolutionary results hardly depend on the treatment of semi-convection.

TABLE 4.4: The pre-SN mass M_{sn}, the CO-core mass M_{CO} and the final FeNi -core mass M_{FeNi} for stars with ZAMS mass between 20 M_o and 60 M_o. For the 15 M_o and the 20 M_o model, we also give the results when the RSG SW mass loss is a factor 2 smaller.

M_{ZAMS}	M_{sn}	M_{CO}	M_{FeNi}
15	5/8	2.5/3	1.5/1.65
20	5/7	3.5/4.3	1.7/2.0
40	8	6.5	2.9
60	10	8	3.1

for the Magellanic Clouds:

The LMC evolutionary tracks are very similar to the galactic ones. Evolutionary tracks for SMC single stars are shown in figure 4.5. Quantitative results are included in table 4.3. The computations depend critically on the treatment of semi-convection. Figure 4.5a and 4.5b (resp. 4.5c) correspond to the case where semi-convection is treated with the L criterion (resp. the SH criterion).

We conclude

- for $Z < 0.004$, when semi-convection is treated with the SH criterion, the stars with initial mass $12 < M/M_o \leq 40\text{-}50$ perform most of their CHeB in the temperature range $10000\ K \leq T_{eff} \leq 20000\ K$ and possibly reach the RSG phase at the end of CHeB. During their short RSG phase, the stars do not lose enough mass to become hydrogen stripped CHeB stars and they remain in the red part of the HRD.
  When the L-criterion is used, all massive stars become RSGs and thus at least some of them could become single WR stars provided that the RSG stellar wind mass loss is large enough. The computations shown in figure 4.5a correspond to the case in which equation (2.18) also applies during the RSG phase of SMC single stars. If the RSG mass loss depends on the metallicity as predicted by the radiation driven wind theory [equation (2.12)] the SMC value could be a factor 3 smaller than in the Galaxy. In this case the stars remain in the red part of the HRD and no WR types are formed (figure 4.5b).

- the evolution of stars with initial mass larger than 40-50 M_o depends on the LBV mass loss and in particular on how Z affects the $\dot{M}$ during this phase. As for the Galaxy, we first assume that the LBV mass loss is large enough to prohibit redward evolution for stars with initial mass larger than 40-50 M_o. We then lower the mass loss rate by a factor 3 and apply the RSG SW formalism. The overall evolutionary behaviour is very similar to that of the Galaxy. When stars are allowed to move to the red, due to the dependence of the SW mass loss on Z, the RSG phase is longer.

- if the $\dot{M}$ during the RSG phase depends on the metallicity, the minimum mass for BH formation is smaller in low metallicity environments (LMC or SMC) compared to the Galaxy.

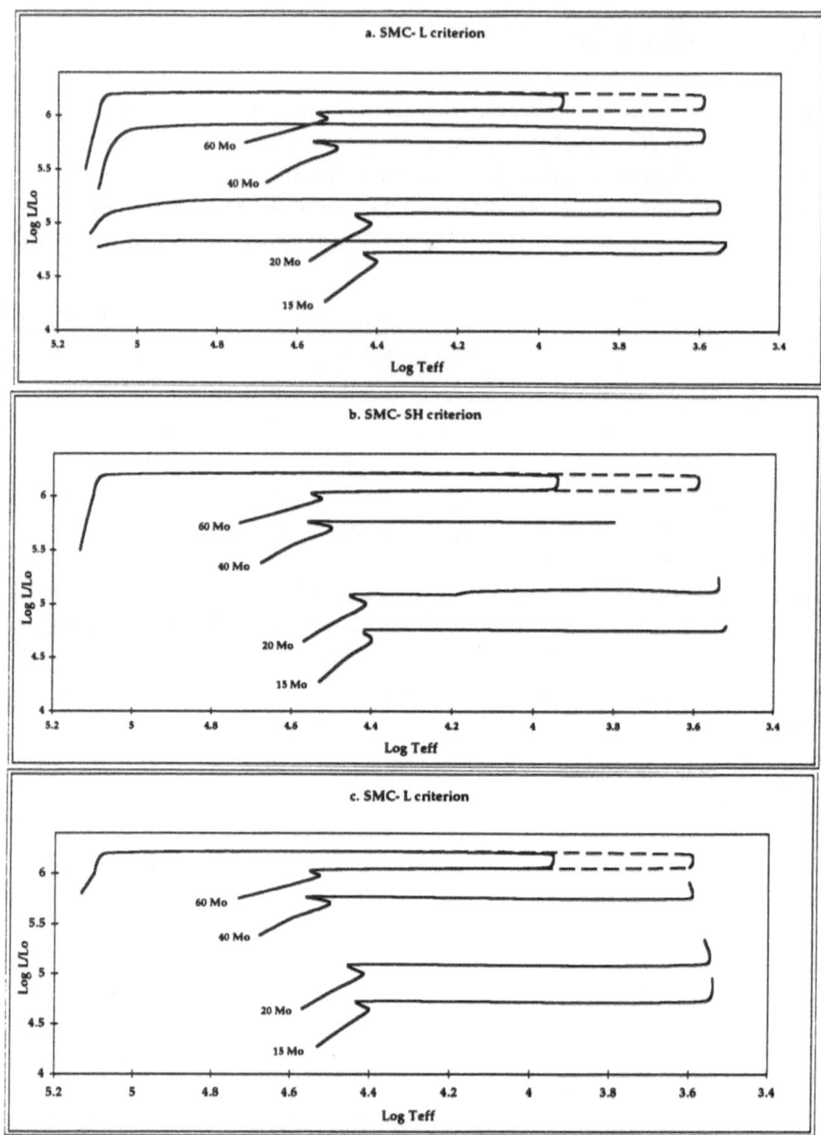

Figure 4.5: Evolutionary tracks of massive single stars with initial mass > 12 M_o, a: with initial abundances holding for the SMC (X=0.76, Z = 0.002), semi-convection is treated as a very inefficient mixing mechanism (Ledoux criterion), the stellar wind mass loss rates during the RSG and LBV phase are assumed to depend on the metallicity as predicted by the radiatively driven wind theory (equation 2.12 with ζ = 0.5); b: is the same as fig a but the RSG and LBV stellar wind mass loss rates are assumed to be independent from the metallicity and the same values apply as for the Galaxy; c: is the same as fig a but semi-convection is assumed to be a fast mixing process (the SH criterion). Since stars with initial mass ≤ 40 M_o do not reach the RSG phase (or reaches the RSG phase at the end of CHeB), it follows that the tracks do not depend on the adopted SW mass loss rate during the RSG phase.

4.2. The effect of large convective core overshooting

Figure 4.6 shows the evolutionary track of a galactic 20 M_o star with increased convective core overshooting.

When compared to the case in which convective core overshooting is small (previous subsections), we notice the following differences:

- owing to the fact that more fuel is available, the CHB lifetime becomes longer for larger convective cores, hence for a larger amount of overshooting,
- the track runs at larger luminosities when the amount of overshooting is larger,
- the CHB phase extends to $T_{eff} \approx 15000$ K when overshooting is large; this is considerably cooler than with small convective core overshooting,
- since the convective core is significantly larger when overshooting is large, the hydrogen rich envelope of the star after CHB is considerably smaller. This means that the amount of mass which needs to be lost from the star to become a hydrogen deficient CHeB star (a WR star?) during the RSG phase, is much smaller compared to the 'small convective core overshooting' case. A 20 M_o star for example only has to lose ~7 M_o during its RSG phase,
- when convective core overshooting during CHB is large, the importance of semi-convection is small, not only for galactic stars but also for stars in the Magellanic Clouds. As a consequence, a massive star becomes a RSG at the beginning of CHeB, whereas blue loops are very much suppressed,

The results and conclusions above are typical for all massive stars, independent of the initial chemical composition, and this leads to a number of more general conclusions:

- since the stars evolve at larger luminosities, a comparison with the luminosities of LBVs leads to a minimum mass of ~30 M_o for the latter (compared to 40-50 M_o when convective core overshooting is small); since the convective core is significantly larger when overshooting is large, the amount of mass that needs to be removed from the star during the LBV phase before the star becomes a WR star is much smaller in the large overshooting case,

TABLE 4.5: Evolutionary properties of a galactic 20 M_o star when convective core overshooting is assumed to be small and with increased overshooting distance; T_{CHB} = CHB lifetime in million years, M_{cc} = the mass of the convective core (when the central hydrogen content = 0.7 and 0.3), TAMS means red point during CHB.

overshooting	T_{CHB}	$M_{cc}(X=0.7)$	$M_{cc}(X=0.3)$	T_{eff}(TAMS)	L/L_o(TAMS)
small	10.1	10.4	7.3	4.38	5
large	13.3	13.5	11.4	4.13	5.3

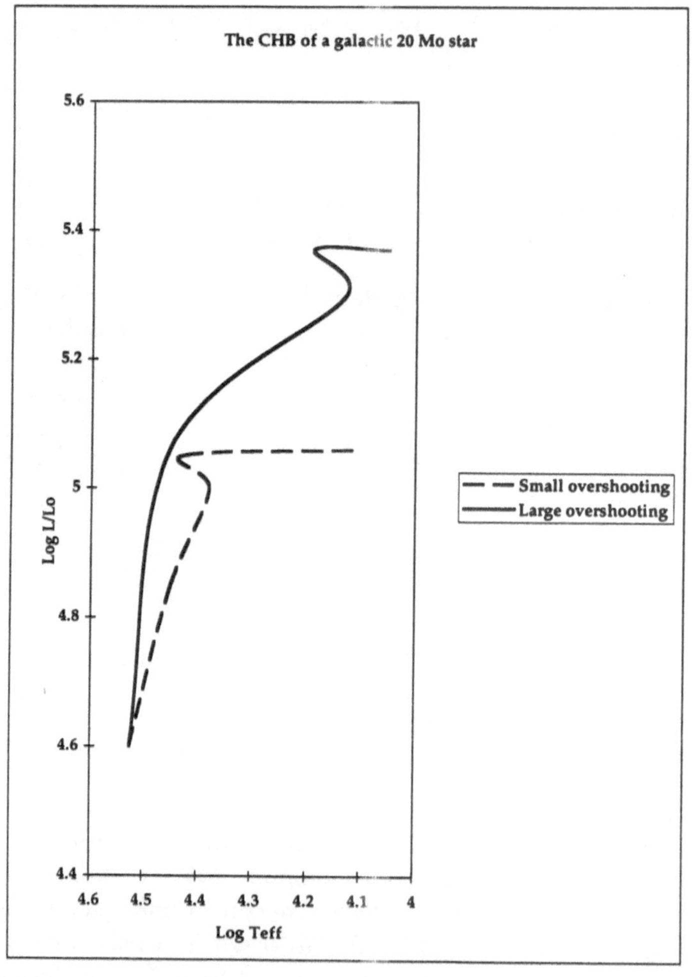

Figure 4.6: The evolution during CHB of a galactic 20 M_o star when convective core overshooting is assumed to be small (dashed track) and when the overshooting distance is increased (full track).

- since the effect of semi-convection is largely reduced when convective core overshooting is large, all stars with initial mass ≤ 30-40 M_o become RSGs independent of the initial metallicty, and blue loops do not occur,
- especially between 10 M_o and 20-25 M_o, the CHB phase extends to $T_{eff} \approx$ 15000 and this is considerably cooler than with small convective core overshooting,
- when convective core overshooting during CHB is large, the minimum mass for a star to end its life as a NS is significantly reduced, i.e. 6 M_o compared to 9 M_o when overshooting is small,

5. Comparison between observations and evolutionary computations of non-rotating single stars.

In this section we will compare the general observations of massive stars summarized in chapter I, with the massive single star evolutionary computations discussed in section 4, i.e. we want to see to what extent single star evolution where rotational effects are neglected can explain some of these observations. We distinguish the case in which evolutionary computations are used with small convective core overshooting from the case in which overshooting is large.

5.1. Evolutionary computations with small convective core overshooting

5.1.1. THE CORE HYDROGEN BURNING PHASE

With the calibration given in table 2.1, table 5.1 gives the relation [present mass - present lifetime - spectral type - luminosity class] of a star in the Solar neighbourhood.

TABLE 5.1. Mass (in M_O) and age (in 10^6 yrs) for galactic stars (Z=0.02) derived from evolutionary tracks of single stars (Schaller et al. 1992) and the calibration of table 2.1. Extremely luminous supergiants (Ia-O) are in the region of the HR- diagram where very massive stars (M > 85 M_O) move hence and forth and are omitted.

Type	Mass (M_O)	Age (in 10^6 yrs)	Type	Mass (M_O)	Age (in 10^6 yrs)
O3 V	69.3	0.1	O4 V	56.1	0.8
O5 V	50.9	1.3	O5.5 V	42.3	1.5

O6 V	37.1	1.8	O6.5 V	35.3	2.4
O7 V	33.7	2.9	O7.5 V	32.7	3.1
O8 V	29.9	3.2	O8.5 V	25.0	3.6
O9 V	23.6	4.1	O9.5 V	21.1	4.6
B0 V	16.7	6.8	B0.2 V	15.5	8.4
B0.5 V	13.5	19.8	B2 V	8.0	27.0
B3 V	7.0	36.1			
O8 IV	26.3	4.0	O8.5 IV	24.6	4.4
O9 IV	23.8	4.8	O9.5 IV	23.1	5.1
B0 IV	19.3	6.2	B0.2 IV	16.9	8.6
B0.5 IV	16.0	9.6	B1 IV	10.8	18.9
B1.5 IV	9.0	24.1	B2 IV	8.0	33.6
B2.5 IV	7.7	36.5	B3 IV	6.8	46.0
O5 III	58.8	1.5	O5.5 III	55.4	1.9
O6 III	49.3	2.3	O6.5 III	46.8	2.5
O7 III	36.4	3.0	O7.5 III	33.5	3.6
O8 III	29.4	4.0	O8.5 III	27.9	4.3
O9 III	26.7	4.6	O9.5 III	25.5	5.0
B0 III	22.4	5.8	B0.2 III	19.5	7.4
B0.5 III	18.4	8.3	B0.7 III	14.7	11.4
B1 III	11.9	15.7	B1.5 III	10.3	20.9
B2 III	9.0	25.9	B2.5 III	8.5	29.7
B3 III	7.5	38.3	B4 III	6.7	46.3
O7 II	41.0	2.9	O7.5 II	36.9	3.3
O8 II	35.9	3.6	O8.5 II	34.5	3.8
O9 II	33.7	4.0	O9.5 II	32.5	4.3
B0 II	30.5	4.7	B0.2 II	21.9	6.6
B0.5 II	18.5	8.4	B0.7 II	15.8	10.7
B1 II	14.7	11.6	B1.5 II	13.5	13.5
B2 II	12.0	15.9	B2.5 II	11.8	16.4
B3 II	11.4	17.9	B4 II	11.0	19.4
B5 II	10.5	21.1			
O7 Ib	47.5	3.0	O7.5 Ib	45.5	3.2
O8 Ib	44.4	3.3	O8.5 Ib	43.0	3.4
O9 Ib	41.3	3.5	O9.5 Ib	32.0	4.6
B0 Ib	30.4	4.8	B0.2 Ib	25.9	5.6
B0.5 Ib	19.9	7.9	B0.7 Ib	17.9	9.2
B1 Ib	16.8	10.0	B1.5 Ib	15.9	10.7

B2 Ib	14.7	11.6	B2.5 Ib	14.4	12.0
B3 Ib	14.1	12.5	B4 Ib	13.9	12.9
B5 Ib	13.0	14.3	B6 Ib	12.7	14.8
B7 Ib	12.4	15.3	B8 Ib	11.9	16.0
B9 Ib	10.0	23.8	A0 Ib	9.5	24.5
A1 Ib	9.4	25.1	A2 Ib	9.2	25.8
A3 Ib	8.7	29.1	A4 Ib	8.6	29.8
A5 Ib	8.4	31.2	A8 Ib	8.3	32.5
F0 Ib	8.2	32.9	F5 Ib	7.9	35.9
O8 Iab	47.8	3.1	O8.5 Iab	46.3	3.2
O9 Iab	45.2	3.3	O9.5 Iab	37.2	3.9
B0 Iab	35.1	4.3	B0.2 Iab	30.0	5.3
B0.5 Iab	23.4	6.5	B0.7 Iab	21.8	7.1
B1 Iab	21.1	7.4	B1.5 Iab	20.0	7.8
B2 Iab	18.5	8.7	B2.5 Iab	18.0	9.1
B3 Iab	16.4	10.3	B4 Iab	16.4	10.3
B5 Iab	15.9	10.7	B6 Iab	15.8	10.8
B7 Iab	15.0	11.3	B8 Iab	14.5	11.9
B9 Iab	14.1	12.6	A0 Iab	14.2	12.4
A1 Iab	14.0	12.8	A2 Iab	13.8	13.0
A3 Iab	13.6	13.3	A4 Iab	13.9	12.9
A5 Iab	13.9	12.8	A8 Iab	14.3	12.2
F0 Iab	14.2	12.4	F5 Iab	14.3	11.7
O3 Ia	88.4	1.4	O4 Ia	67.5	2.2
O5 Ia	72.0	2.2	O5.5 Ia	68.2	2.4
O6 Ia	64.6	2.5	O6.5 Ia	66.9	2.5
O7 Ia	69.0	2.5	O7.5 Ia	67.5	2.6
O8 Ia	65.9	2.6	O8.5 Ia	64.0	2.6
O9 Ia	62.4	2.7	O9.5 Ia	50.8	3.1
B0 Ia	45.7	3.4	B0.2 Ia	36-37	4-4.3
B0.5 Ia	29.5	5.4	B0.7 Ia	26.9	5.9
B1 Ia	25.7	6.1	B1.5 Ia	28.8	5.5
B2 Ia	26.1	6.0	B2.5 Ia	24.9	6.2
B3 Ia	22.0	7.0	B4 Ia	20.4	7.7
B5 Ia	18.2	8.9	B6 Ia	17.8	9.2
B7 Ia	17.3	9.6	B8 Ia	16.7	10.0
B9 Ia	16.1	10.5			

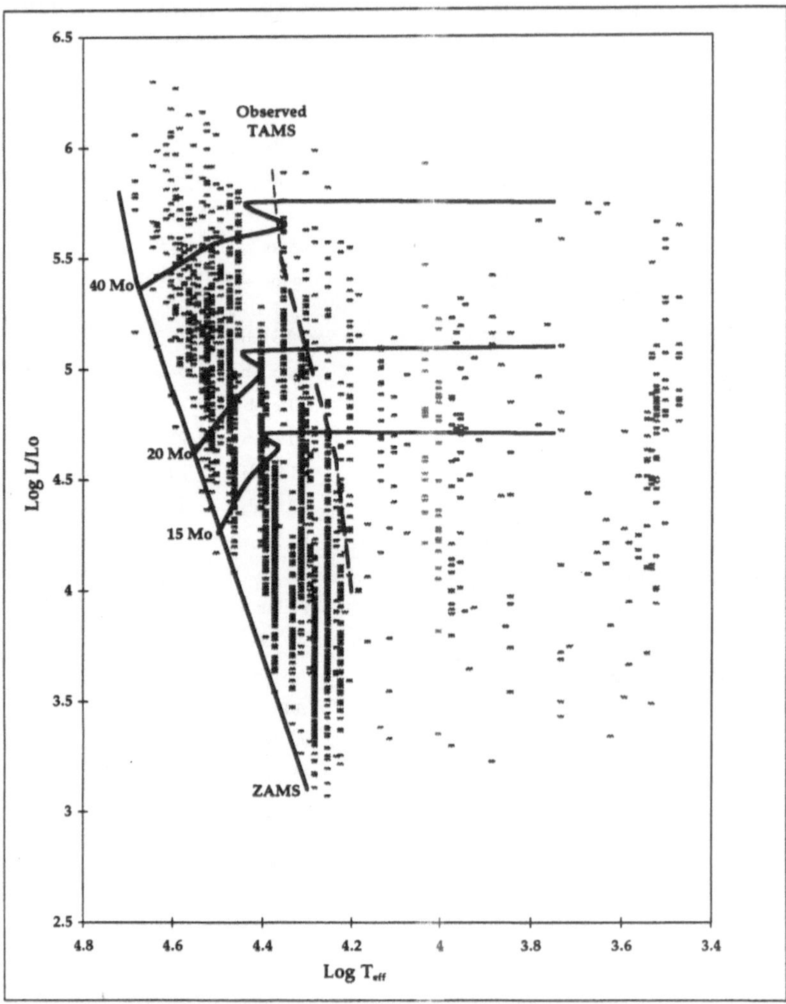

Figure 5.1: Theoretical evolutionary tracks of massive stars compared to the observed HRD of stars in stellar aggregates within the Solar neighbourhood.

In figure 5.1 we compare the observed HR diagram of the stars in associations (figure 2.1a) and the theoretical evolutionary tracks. Two effects are visible:

- when we define the end of the 'observed' CHB sequence in the HRD as the region where the number of stars drops significantly (dashed line in figure 5.1), we have to conclude that at least in the mass range 10 - 25 M_o the observed CHB sequence is broader than the predicted one. Part of the problem may be due to the temperature scale used to construct the HR diagram. The T_{eff} of most of the stars is based on the B-V colour - T_{eff} relation and the difference in B - V is 0.04 only if T_{eff} changes from 30000 K to 20000 K (section 2.1.3). However if we take the HRD position for granted, we are left with a real problem when only single star evolution with small convective core overshooting is considered.

- It is clear that observations of massive stars younger than 1-2 million years are lacking, i.e. we miss ~20-30% of the O-type stars.

The reason for the latter effect could be intrinsic to the formation mechanism of stars, illustrated by the following experiment.

Suppose we have a gas cloud with mass M leading to the formation of a star with mass M. Contraction is initiated at the center, and a core with mass M_c is formed. Now the cloud starts contracting and the mass of the core increases at a rate $\dot{M}$. Typical values for the mass gain rate (mass accretion rate) $\dot{M}$ have been computed by Myers and Fuller (1992) and range between 10^{-5} - 10^{-4} M_o/yr. This means that in order to form a 10 M_o star, accretion has to go on for about one million years (using an accretion rate of 10^{-5} M_o/yr). Since the total CHB lifetime of a 10 M_o star ≈ 30 million years (X = 0.7, Z = 0.02), we can expect that a 10 M_o star will start its evolution almost as a normal ZAMS star. When the accretion rate is ~10^{-4} M_o/yr, also the formation time scale of the more massive stars is considerably smaller than their evolutionary time scale so that the evolutionary tracks are almost entirely similar to those of figure 4.3. The situation changes significantly when the accretion rate is 10^{-5} M_o/yr, e.g. it takes about 4 million years to form a 40 M_o star. Since its CHB lifetime ≈ 4.5 million years, it can be expected that a 40 M_o star does not start its evolution exactly at the ZAMS. To illustrate the effect we first compute a homogeneous ZAMS model of a 10 M_o star with X = 0.7 and Z = 0.02, and we then follow its evolution by adopting an accretion rate of 10^{-5} M_o/yr. When the mass of the star equals 20 M_o, 30 M_o, 40 M_o, 60 M_o, we stop the accretion and the evolution of the star is continued in a normal way, including the effect of stellar wind mass loss. A very simple accretion model is used (similar to the 'standard accretion model' of Neo et al. (1977) also used in binary computations, section 9.3). The model assumes that matter falls gently onto the underlying star with very small velocity and with an entropy which is larger than or equal to the entropy of the envelope of the star. The latter means that accretion of matter does not initiate large scale convection of the envelope, i.e. the evolutionary computations are performed assuming that the outer layers of the massive accreting star are in radiative equilibrium and remain so during the whole accretion phase. Evolutionary tracks are shown in figure 5.2, whereas table 5.2 gives the numbers. It can be seen that the process explains the lack of massive stars close to the theoretical ZAMS. These computations confirm the results of Bernasconi and Maeder (1996).

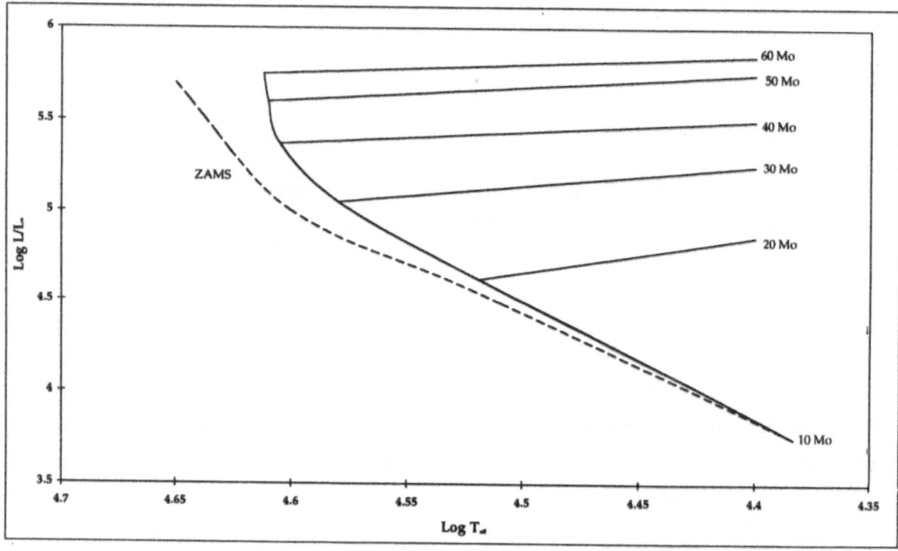

Figure 5.2: Starting from a homogeneous ZAMS model of a 10 M_o star with X = 0.7 and Z = 0.02, we follow its evolution by adopting an accretion rate of 10^{-5} M_o/yr. When the mass of the star equals 20 M_o, 30 M_o, 40 M_o, 60 M_o, we stop the accretion and the evolution of the star is continued in a normal way, including the effect of stellar wind mass loss.

However, the model described above may be too simple. Hydrodynamic computations of protostellar spherical symmetric accretion (Shu and Lubow, 1981) predict that, after passing through a shock front, matter settles down onto the underlying star with significantly reduced entropy. This may initiate convective mixing possibly of the whole star. As an example, we have followed the formation of a 30 M_o star (starting from a 10 M_o star and an accretion rate of 10^{-5} M_o/yr, similar as above) but we have assumed efficient mixing of the whole star. At the end of accretion, the 30 M_o star has a log $T_{eff} \approx$ 4.6 and log $L/L_o \approx 5$ which is very close to the position of a normal ZAMS 30 M_o star. If this model is correct, then the reason why very few stars are observed close to the ZAMS could be that the star, once it is formed, is still embedded into the remains of the protostellar cloud and will therefore only be visible in the infrared.

Particularly interesting is the chemistry of the massive star when the latter formation process applies. The layers of the original core are mixed with the surface layers. The hydrogen/helium abundances are only slightly affected (X = 0.68, Y = 0.3); however the CNO elements are significantly modified (N enriched, CO depleted). The model therefore predicts that a significant fraction of the O-type stars have altered CNO abundances, also the luminosity class V and IV stars. As these stars evolve they will become giants or supergiants with a N enriched/CO depleted atmosphere.

Note

MC tracks are hotter than galactic ones ($\Delta T_{eff} \geq 0.05$ dex). This means that for fixed luminosity, one expects the existence of stars in the MCs that are hotter (thus more compact) than the hottest stars in the Galaxy. One may therefore wonder whether or not the calibration of massive stars in the Galaxy also holds for the MCs.

TABLE 5.2: Starting from a 10 M_o ZAMS star, using an accretion rate of 10^{-5} M_o/yr, a 20 M_o (resp. 30 M_o, 40 M_o) is formed after 1 (resp. 2, 3) million years. The table gives the values of the evolutionary parameters at the moment the stars are formed and at the end of CHB.

Mass (M_o)	t(10^6 yrs)	X_c	Log T_{eff}	Log L/L_o
10	0	0.70	4.39	3.73
20	1	0.68	4.52	4.62
19	10.8	0.00	4.36	4.97
30	2	0.63	4.58	5.07
27.7	7.8	0.00	4.38	5.33
40	3	0.58	4.60	5.38
36.2	6.8	0.00	4.34	5.57

5.1.2. THE POSITION OF THE RSGS AND WR STARS IN THE HR DIAGRAM

In figure 5.3 we compare the overall HRD position of the WR stars and the RSGs with our single star evolutionary tracks. When at least part of the WR stars (especially the WN types) with the smallest luminosities are single stars, it looks as if galactic (and LMC) single stars with initial mass as low as 20 M_o (and even lower) evolve into WR stars. This is the case with our evolutionary models (with both small and large convective core overshooting) where the RSG mass loss is determined by equation (2.18).

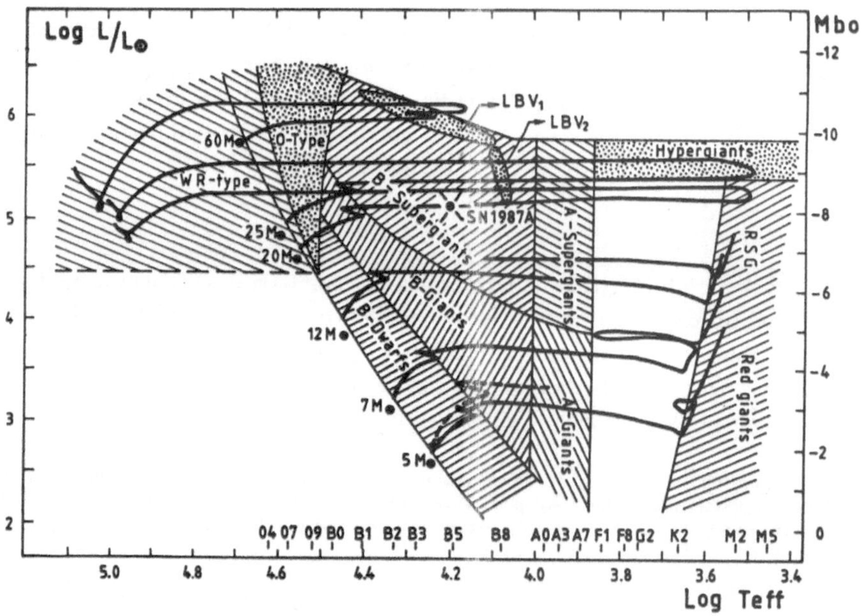

Figure 5.3: The overall observed HRD compared to massive single star evolution for the Solar neighbourhood. Since the HRD and the evolution of massive stars in the LMC and in the Solar neighbourhood are very similar, we also show the position of the blue progenitor of SN 1987A.

One has to notice that after the AGB phase, the evolutionary tracks of intermediate mass stars (figure 4.2) cross the low luminosity part of the observed WR region. Although the crossing time is very small, it cannot be excluded that some of these WR stars are low mass WR stars, descendants of intermediate mass single stars.

5.1.3. THE PROGENITOR OF SN 1987A

The fact that the HRD position of the progenitor of SN 1987A is in between the WR area and the top of the RSG area where the stellar wind mass loss rate is very large (figure 5.3), suggests that if the progenitor had been a 'normal' single star, it would have had a small hydrogen rich envelope. However the SN was a typical type II event and its light curve requires that the progenitor had a large hydrogen rich envelope (section 2.9). Moreover, it is very hard to explain the ring morphology of the circumstellar nebula by single star evolution. Although rotation could produce ring shaped mass loss, a red supergiant is not expected to have a large angular velocity. And last but not least, it is hard to explain the barium and s-process elements in the atmosphere of single stars.

We therefore conclude that within the framework of the present observations, it is unlikely that the progenitor of SN 1987A was a normal massive single star.

5.1.4. THE PROGENITOR OF SN 1993J

As discussed in section 2.9, the progenitor of SN 1993 J must have been a RSG, resulting from a star with initial mass $\geq$ 15 M_o, which has lost a lot of its hydrogen rich layers but retained sufficient hydrogen to stay in the RSG region.

Within the uncertainties of the SW rates of RSGs, it cannot be excluded that a star with mass $\geq$ 15 M_o loses most of its hydrogen rich layers by SW, i.e. binary interaction may have played only a small role during the evolution of the progenitor of SN 1993J.

5.1.5. THE OVERALL SN MORPHOLOGY OF MASSIVE SINGLE STARS

galaxies with large metallicity (Z $\geq$ 0.02):

$M_{BH,min}$ = 20-30 M_o
[15-20 M_o, $M_{BH,min}$] ---> SN I
[8M_o, 15-20 M_o] ---> SN II

galaxies with small metallicity (Z $\leq$ 0.004):

if the RSG SW does not depend on Z, *then* the overall SN morphology of massive single stars is the same as for the Galaxy,

if the RSG SW depends on Z as predicted by the radiatively driven wind theory [equation (2.12) with ζ = 0.5)], *then*

$M_{BH,min} \leq 20$ M_o
no SN I
[8 M_o, $M_{BH,min}$] ---> SN II

The overall SN morphology of massive single stars is summarized above. We define $M_{BH,min}$ as the minimum initial mass of a star that ends its life as a BH escaping a SN event. Bear in mind that due to uncertainties in the equation of state of a degenerate plasma, the value of $M_{BH,min}$ has to be considered with great caution. In chapter IV, we will show that its value affects significantly SN rates predicted by a population synthesis model. When compared to observations of the Galaxy, $M_{BH,min} \approx 20\ M_o$ gives the best correspondence.

5.1.6. THE BLUE HERTZSPRUNG GAP (BHG)

Since the evolution during hydrogen shell burning is very fast, we do not expect to observe many stars redwards of the TAMS. This 'theoretical' region is frequently called the Blue Hertzsprung Gap (BHG). However the observed star distribution in the HRD region occupied by stars with initial mass between ~10 M_o and ~25 M_o appears to extend more or less continuously down to the G-type supergiants (see also figure 2.1a) with a maximum density in the A supergiant region. This effect is visible in the Galaxy as well as in the MCs.

The latter can not be explained by the present single star evolutionary computations.

The existence of the BHG and the difficulty in explaining it, led Alongi et al. (1991) to suggest that convective envelopes in RSGs overshoot at their bottom over a large distance. The main effect is the occurrence of extended blue loops during CHeB for stars with initial mass up to ~20 M_o. However if indeed the SW mass loss rates during the RSG phase of a star are as large as predicted by equation (2.18), blue loops are largely suppressed and the problem of the BHG reappears again. Furthermore, the nitrogen/carbon elemental ratio in the A-type supergiants is roughly ~0.3 dex larger than solar. If these stars would be post-RSG objects, due to the dredge-up process of CNO processed matter during the RSG phase, one would expect a N/C ratio ~0.5-0.6 dex larger than solar.

In chapter III dealing with close binaries, we will discuss an alternative possibility in order to explain the stars in the gap.

5.1.7. CHEMICAL ABUNDANCES IN EARLY B TYPE SUPERGIANTS

All single star evolutionary models discussed in section 4, with small or large overshooting, where semi-convection is treated with the SH-criterion or with the L-criterion, have difficulties in order to explain the CNO enriched B type supergiants (section 2.3), especially the early Bs with spectral type between B0.7 and B3. Even if blue loops occur, they never extend that far (figure 5.3).

A way out may be the suggestion made in section 5.1.1 about the formation of massive stars and the possible efficient mixing of the star during the accretion process. Of course, due to the effect of microturbulence, the interpretation of the observations may be wrong (section 2.3).

5.1.8. THE DISTRIBUTION OF RSGS IN THE SMC

The large number of RSGs with $M_{bol} \geq -9.5$

Massive single star models with $Z \leq 0.004$, with small convective core overshooting and where semi-convection is treated with the SH criterion, spend most of their CHeB phase in the blue and enter the RSG region only at the end. However, when semi-convection is treated as a diffusion process, the RSG part of the CHeB phase becomes significant when the diffusion coefficient D [parameter α in equation (3.23)] is small (or zero, corresponding to the Ledoux criterion of semi-convection).

As illustrated in figure 2.1d, the SMC contains a lot of RSGs brighter than $\log L/L_o = 5.1$ ($M_{bol} \approx -8$). Therefore we are tempted to conclude that

> *the presence of a large number of RSGs in the SMC favors the massive star evolutionary model where semi-convection is treated as a very inefficient (slow) mixing process.*

There is no physical reason why D should be large in the Galaxy when it is small in the SMC. The distribution of stars in the BHG in the Galaxy has been used as an argument in favor of the SH criterion for semi-convection when $Z = 0.02$ (Langer and Maeder, 1995). However a close inspection reveals that neither models with SH nor with the L criterion give a satisfactory picture.

The lack of RSGs brighter than $M_{bol} = -9.5$

As outlined in section 4.1.2, if semi-convection is a slow mixing process and if the LBV SW mass loss depends on the metallicity as predicted by the radiatively driven wind theory, the RSG lifetime of a star in the SMC with initial mass > 40 M_o could be considerably longer than in the Galaxy. However, similarly to the Galaxy and in the LMC, the SMC has no (or very few) RSGs brighter than $M_{bol} = -9.5$. This can only be explained if the formation rate of stars with initial mass > 40 M_o (relative to those with $M \leq 40$ M_o) is considerably smaller in the SMC than in the Galaxy and in the LMC or/and the SW of a LBV with $M_{bol} \leq -9.5$ in the SMC is large enough to assure a RSG time scale which is small enough, similar to that of the Galaxy.

5.1.9. COMPARISON BETWEEN THEORETICAL PREDICTION OF DEGENERATE STARS AND OBSERVATIONS

White Dwarfs

The final remnant at the end of the evolution of a star with initial mass smaller than 9±1 M_o (intermediate mass star) is a degenerate star with mass < 1.44 M_o (section 4) and resembles a White Dwarf.

The mass-radius relation of table 3.1 for degenerate stars with mass < 1.44 M_o predicted by theory can be compared to observations of WDs where the mass can be determined from Kepler's laws because the star is the fainter object of a close binary system. In some of these cases also the radius can be derived from the observed T_{eff} and absolute luminosity.

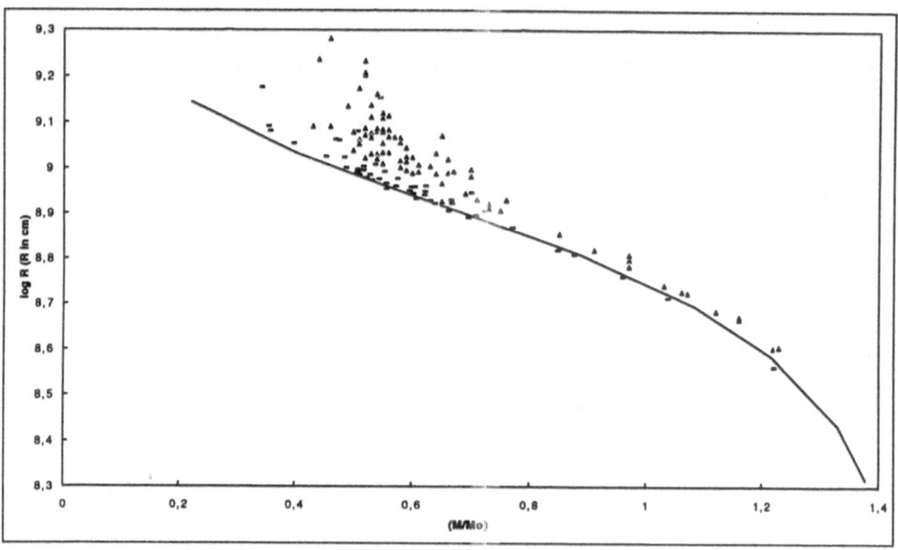

Figure 5.4: The mass-radius relation for 141 white dwarfs. The theoretical Chandrasekhar relation is shown in full line. The 52 white dwarfs from Bragaglia et al. are indicated with dashes, whereas the 89 white dwarfs from the ROSAT Survey are indicated with triangles.

The WD component of Procyon B is observed to be a 0.45 M_o star with a 0.013 R_o radius, whereas for 40 Eridani B one obtains a 0.40 M_o WD with a radius of 0.017 R_o. These values match the relation of table 3.1 very closely.

Atmospheric parameters as T_{eff} and log g for 52 White Dwarfs have been determined by fitting optical spectra to model atmospheric calculations (Bragaglia et al., 1993, 1995) . When the masses of these stars are determined from evolutionary models (Wood, 1992) one can determine the radius from log g and this mass. Moreover, similar data as obtained from the ROSAT EUV All-Sky Survey have been published for 89 White Dwarfs by Marsh et al. (1997).

Figure 5.4 shows the mass- radius relation which is obtained in this way for the 141 White Dwarfs mentioned above.

The Chandrasekhar- relation with μ_e = 2 from table 3.1 fits these data very well.

Remark that WD masses may be as large as 1.2 M_o. In section 4.1.1 this was used as an argument that carbon-deflagration SN should be very rare.

Neutron stars

Obviously, the mass of a NS can only be determined directly if the NS is a component of a binary. There are two classes of binaries where NSs are present: X-ray binaries and binary pulsars. Both types will be discussed in section 13.

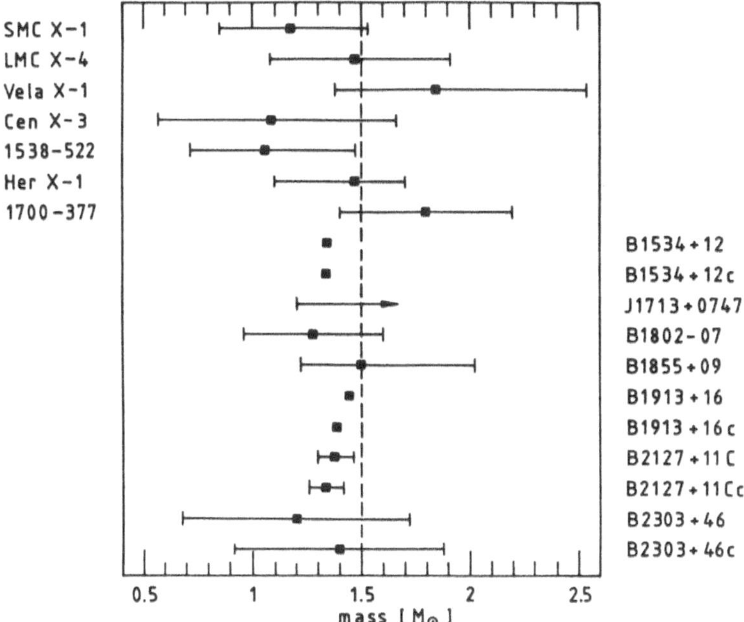

Figure 5.5: Observed mass estimates of 18 neutron stars in binaries (Wijers, 1996).

All measured masses of NSs in binaries (18 in total) are shown in figure 5.5. They are either below 1.5 M_o (15 cases), either their mass is around 1.8 M_o but their error bars extend downwards of 1.5 M_o (2 cases). These mass estimates certainly support the maximum NS mass of section 3.2.6.

5.2. Large convective core overshooting: is there a need?

Pro

Large convective core overshooting models predict a wider CHB band in the mass interval 10-25 M_o, i.e. large convective core overshooting during CHB is capable of explaining the 'observed' width of the CHB sequence (e.g. figure 5.1).

In the next section, we will discuss the effect of rotation on stellar evolution. An important conclusion is that 'large' rotation has the same effect as convective core overshooting, i.e. the difference between the observed width of the CHB sequence and of the one predicted by evolutionary computations of non-rotating stars may be due to the fact that rotation was not included.

Contra

When a region is observed where all stars were formed in a time interval which is short compared to the age of the region (a starburst region), we may expect that the HR diagram of the stars follows more or less a time-isochrone. Of particular interest are starburst regions in which one sees blue and red supergiants, which might be an indication of blue loops during the CHeB phase of a single star. A detailed study of a large number of such regions has been performed by Meynet et al. (1993) and compared to evolutionary predictions. The authors interpret the presence of BSGs and RSGs simultaneously in one starburst as an indication that single stars experience blue loops during their CHeB. Since blue loops are largely suppressed when convective core overshooting is large, they conclude that convective core overshooting during core hydrogen burning is small.

Accounting for the effect of rotation on stellar evolution discussed in the next section, we conclude that

the comparison of observations of single stars and theoretical prediction reveals that at present there seems to be no need for large convective core overshooting during CHB.

6. The effect of rotation on the evolution of massive single stars.

One can distinguish three different kinds of rotational effects on stellar evolution:

a. the centrifugal force reduces the internal pressure in a rotating star. As shown by Strittmatter et al. (1970) and Kippenhahn et al. (1970), this only marginally affects stellar evolution (e.g. the CHB time scale may be enhanced by a maximum of 3%),

b. atmospheres may be more extended by larger rotation; fast rotating stars look cooler (Maeder and Peytremann, 1970, 1971). Since this is an atmospheric effect it hardly has any evolutionary consequences. However, as discussed in section 2.4, a significant number of massive stars rotate quite fast and therefore one has to account for this extention when the observed HRD is compared to the theoretically predicted one,

c. the most important evolutionary effect of rotation is caused by rotational instabilities inducing transport of angular momentum and chemical mixing over some parts of the star (a.o. see Zahn, 1994, for a review).

6.1. Rotation-induced mixing in massive stars

Kippenhahn et al. (1970) investigated the consequences of the non-sphericity of the basic structure equations due to rotation and concluded that this may affect the overall evolution only when the star rotates very fast, i.e. with a velocity near to the break-up value. Therefore, as a first order approximation, the effect of rotation on the evolution of massive stars can be studied by using

a. a spherical symmetric equation for the transport of angular momentum

$$4\pi r^2 \frac{\partial \Omega}{\partial t} = -\frac{1}{r^2}\frac{\partial r^2 F}{\partial r} \tag{6.1}$$

with F the flux of angular momentum within the star and Ω the angular velocity,

b. a model of mixing as a consequence of rotation. During core hydrogen burning, the evolution of a massive star is characterized by a decreasing convective core. As a consequence a molecular weight gradient is built up in the stellar interior. Schwarzschild (1958) discussed the fact that rotation induces meridional circulation which is able to mix matter, provided that the μ-gradient does not inhibit the circulation, i.e. when the Richardson criterion applies

$$\left|\frac{\partial \ln \mu}{\partial \ln r}\right| < \frac{8}{15} R_c \frac{\Omega^2 r}{g} \left|\frac{\partial \ln \Omega}{\partial \ln r}\right|^2 \tag{6.2}$$

with R_c the critical Richardson number (always < 1/4, cfr. Zahn, 1983).

In massive stars, the μ-gradients built up by the retreating CHB core are very large, too large to initiate important mixing, even with very fast rotation (Meynet and Maeder, 1996). However, the regions in massive stars where μ varies may be semi-convective, independent from rotation. This means that the onset of mixing in such zones is not defined by the Richardson criterion, and of course the easiest way to proceed then is to assume that rotation will help the semi-convective mixing, independently of condition (6.2).

If mixing as a consequence of rotation occurs, it can then be treated as a diffusion process with diffusion coefficient

$$D = \frac{4a_R c T^3}{3k\rho^2 c_p} \frac{\Omega^2 r}{g} \left(\nabla_{ad} - \nabla_{rad}\right)^{-1} \tag{6.3}$$

(see also Zahn, 1983).

Since $c_p \propto T^{-1}$, the rotational diffusion coefficient D is proportional to T^4. The more massive the star, the larger is T, so it may be concluded that the rotationally induced mixing efficiency increases with increasing mass.

The function Ω(r,t) has to be computed using equation (6.1). However the present theory of rotation in stellar interiors is unable to predict accurately Ω(r,t) (Chaboyer and Zahn, 1992; Zahn, 1992). Therefore, in order to study the effect of rotation on stellar evolution from a phenomenological viewpoint, assumptions have to be made and parameters have to be introduced. When the theoretically predicted implications of rotation are compared to observations, one may hopefully be able to restrict the possible values of the parameters.

6.2. Evolutionary computations of rotating stars.

The effect of rotation on the evolution of massive single stars has been investigated by Meynet and Maeder (1997) and by Langer and Heger (1997). In the first paper, a parametrized model is used to describe the transport of angular momentum in the stellar interior. The meridional circulation first transports matter from core to the surface along the polar axis. This material then moves across the surface to the equator, increasing its angular momentum. Finally, it transports this momentum inwards along the equatorial plane. In the second paper, the authors assume that rigid rotation is a good approximation to describe the redistribution of angular momentum in the stellar interior during the CHB phase (their assumption is based on the study of Zahn, 1994). In this way angular momentum is transported from the stellar core towards the stellar surface. Despite the differences in physics between both studies, we can put forward the following overall qualitative conclusions:

- a consequence of (fast) rotation is the outward diffusion of helium enriched (and CNO processed) matter, possibly up to the outer stellar layers. Fast rotators could therefore show the products of hydrogen burning in their spectra,

- the extent of the convective core is larger in the fast rotator compared to the non-rotating case. This means that

  - the effect of (rapid) rotation on the evolution of the convective core during CHB is similar to the effect of convective core overshooting.

The latter has a few interesting consequences (see section 4.2):

- rotation may cause a widening of the CHB band,

- the CHB lifetime increases as a consequence of rotation,

- for a given mass, a fast rotating star may be considerably more luminous than a slow rotator, i.e. the calibration of table 5.1 does not apply to fast rotators. As an illustration, when the luminosity and effective temperature of a star are known, when the star is a slow rotator, its mass deduced from evolutionary computations will be larger than when it is a rapid rotator,

- since rotation makes larger cores, the minimum mass of NS progenitors may be smaller for a fast rotating star compared to a slow rotator.

6.3. Comparison to observations

Direct observations of rotation are obviously the $v_e$sini values. However they only tell us something about the rotation of the outer layers of a star. How the angular velocity varies as a function of radius inside the star is uncertain.

The surface chemical abundance in a number of stars indicates that there must be processes able to produce mixing and dispersion in stellar properties (such as the mass-liminosity relation). It is however difficult to distinguish whether this is due to rotation, the process of star formation (section 5.1.1) or the effects of binary evolution (chapter III). Furthermore, as far as the surface chemistry of massive stars is concerned, the presence of microturbulence in the atmospheres of massive supergiants makes the interpretation of the observations ambiguous (section 2.3).

Critical remark

During the evolution, the stellar core is assumed to rotate rigidly. Let us assume for the moment that there is no transport of angular momentum from core to external layers and vice versa. The convective CHB core has an angular momentum

$$J = cM_{core}R^2_{core}\frac{2\pi}{P_{rot}} \tag{6.4}$$

with c of the order 1. We assume that c is constant in time. This is certainly not true but for the present purposes it is sufficient. A mass shell leaving the retreating core has an angular momentum

$$dJ = \frac{2}{3}dMR^2_{core}\frac{2\pi}{P_{rot}}. \tag{6.5}$$

If M_{cc} = mass of the final FeNi core (≈ 1.4 M_o) and $M_{cc,o}$ = the mass of the convective core on the ZAMS, it follows that

$$\frac{J_{cc}}{J_0} \approx \left(\frac{M_{cc}}{M_{cc,0}}\right)^{\alpha} \tag{6.6}$$

J_{cc} = the angular momentum of the FeNi core, J_o = the angular momentum of the convective core on the ZAMS, $\alpha = \frac{2}{3c}$. A generous upperlimit is $\alpha = 4$.

If it can be assumed that the angular momentum of the FeNi core does not change as a consequence of the instantaneous SN explosion, J_{cc} can be replaced by the angular momentum of a neutron star. Most of the periods of ~600 known pulsars range between 0.1 s and 1 s. We consider the underlimit as representative for the rotational period of a NS at birth. Adopting a radius of 10 km and a mass = 1.4 M_o, it follows that $J_{cc} \approx 2.6\ 10^{47}$gr.cm^2/s. A 15 M_o ZAMS star has a core mass $\approx$ 7 M_o, i.e. $J_o \approx 1.6\ 10^{50}$ gr.cm^2/s ($\alpha = 4$). Since the ZAMS core radius $\approx$ 4 R_o, the foregoing values of J_o correspond to very large rotational periods, i.e. the ZAMS core should rotate extremely slow. We therefore conclude

the rotational periods of pulsars indicate that either convective cores of massive core hydrogen burning stars rotate extremely slow (i.e. the angular velocity inside a star increases outwards) or that there must be a very efficient mechanism to transport angular momentum from the CHB core towards the outer layers.

The foregoing criterion is not met by the evolutionary models with rotation discussed in this section.

6.4. Summary.

The evolution of massive single stars hardly depends on the SW mass loss during the O-type phase but it depends critically on the SW mass loss during the RSG, LBV, WR phases and on whether or not the mass loss rates depend on the metallicity.

A comparison between the theoretically predicted and observed HR diagram of massive stars in the SMC seems to indicate that semi-convection is a very ineffecient mixing process. Rotation (at least 30% of the massive stars have an equatorial velocity $\geq$ 200-300 km/s, section 2.4) and rotational diffusion may overtake the role of semi-convection. When it is fast enough, the effect of rotation on the evolution of the mass-luminosity relation is similar to the effect of large(r) convective core overshooting.

CHAPTER III

MASSIVE CLOSE BINARIES

7. General.

A binary is characterized by the masses M_1 and M_2 of both components, the orbital period (P) and eccentricity (e), the spins of the components. The semi-major axis A is determined by Kepler's law

$$\left(\frac{2\pi}{P}\right)^2 A^3 = G(M_1 + M_2) \tag{7.1}$$

and the total orbital angular momentum is given by

$$J = \frac{M_1 M_2}{M_1 + M_2} \frac{2\pi A^2 \sqrt{1-e^2}}{P} \tag{7.2}$$

A binary is called a massive binary when at least one of its components will collapse to form a NS (accompanied by a SN explosion) or a BH. A massive binary will be further abbreviated as MB.

We will always define the primary as the component which was originally the most massive component; the originally less massive star is called the secondary. This evolutionary definition differs sometimes from the observational one where primary means the visually most luminous component, e.g. in WR+OB binaries, the observational primary is the OB star however the evolutionary primary is the WR star since it is this star which was originally the most massive component. Throughout this work, the mass ratio q equals the mass of the secondary divided by the mass of the primary.

8. The Roche model

The evolution of a star which is a component of a binary differs from the evolution of a single star with the same mass and chemical composition, because of the existence of a point in between the two components where the effective gravity becomes zero and where matter from one component can flow freely towards the other.

8.1. Circularization and synchronization

In most of the shorter period (period of the order of days) binaries with intermediate mass components where one or two components have sizes comparable to the orbital separation, the eccentricity is small compared to the eccentricities of binaries with larger separation (Koch, 1976; Wilson, 1989; Torres et al., 1992) whereas the spins of the components appear to be synchronized with respect to their orbital revolution (Plavec, 1970; Levato, 1976). The observed eccentricities in MBs are summarized in the tables of section 13 for resp. OB-type binaries, WR+OB type binaries, the HMXBs and Be-type X-ray binaries. Also here most of the shorter period systems seem to be circularized.

The observed spin periods are less certain. The projected rotational velocity of the optical star of the HMXB Vela X-1 (HD 77581) ≈ 125 km/s and its most probable radius ≈ 28 - 35 R_o (section 14.4.1). This means that the spin period is ~ 11 - 14 days. When compared to the orbital period P = 8.9 days, we conclude that the binary is not far from synchronization. Penny (1996b) studied the rotational velocities of a number of OB+OB binaries. She concluded that the binary components of HD 215835 (O6n + O6n, P = 2.1 days) seem to rotate at ~75 % of the synchronous rotation whereas the components of HD 165052 (O6.5V + O, P = 6.14 days) and of HD 149404 (O8.5I + O7III, P = 9.8 days) rotate at more than 3 times (resp. 2 times) the synchroneous rate. The binaries HD 93403 (O5f + O7.5, P = 15.1 days) and HD 152248 (O7f + O, P = 5.97 days) have almost reached synchronism.

Classical mechanics explains why circularization and synchronization in shorter period binaries should prevail: for given component masses and fixed angular momentum, a binary has least mechanical energy when both stars have circular orbits and synchronized spins (Lynden-Bell and Pringle, 1974). Since all physical systems tend to such a stable equilibrium, as a consequence of the action of torques and/or the exchange of energy, binaries will tend to achieve circularization and synchronization. Typical time scales are given by Zahn (1977) for the case the eccentricity is small:

$$t_{sync} = \frac{1}{6q^2k_2}\frac{I}{MR^2}\left(\frac{A}{R}\right)^6 t_F \tag{8.1}$$

$$t_{circ} = \frac{4}{63q(1+q)k_2}\left(\frac{A}{R}\right)^8 t_F$$

k_2 and I are respectively the apsidal motion constant and the spin moment of inertia of the star with mass M. The most uncertain parameter in the equations (8.1) is the time scale for frictional dissipation t_F. Despite this uncertainty, both equations yield two important effects: first it follows from Kepler's law that the time scales depend strongly on the binary period (the smaller the period, the smaller the time scales) and on the stellar radius (the larger the radius, the smaller the time scales). For massive stars with radiative envelopes, Zahn (1975) investigated the radiative damping of dynamical tides and estimated the time scales to be less than the nuclear lifetime of the primary, for orbital periods of the order of a few days.

However due to the strong dependence of the time scales on the orbital period, synchroneous rotation and circularization may never be achieved by the binary if it has a larger period. Even more, if the eccentricity is large (roughly $e > 0.2$) relations (8.1) are no longer valid. For such binaries it is very difficult to reach their equilibrium configuration and the orbit remains eccentric. Remark however that asynchroneous spins may cause turbulence in stellar envelopes, which enhances considerably viscosity, hence the rate of synchronization (Tassoul, 1987, 1988, 1990; Tassoul and Tassoul, 1992).

All in all we conclude:

a binary with an initial period of the order of days and with a modest initial eccentricity will reach its equilibrium configuration (synchronized and circularized) within a time scale shorter than its evolutionary time scale; if on the contrary the initial period is larger, and/or the initial eccentricity is too large, it is presently unclear whether or not the binary will become circularized.

8.2 Roche equipotentials

We consider a binary with a circular orbit, where the primary has synchronized spin. Evolutionary computations of massive single stars reveal that quite early during CHB, they are centrally condensed so that their gravitational field can be approximated by that of a point source. In a Cartesian coordinate system, with the origin at the center of mass of the binary, the z-axis along the spin axis and the x-axis along the line joining the two stellar centers, the equipotential surfaces take the form:

$$\phi = -\frac{GM_1}{\sqrt{[x-\mu A]^2 + y^2 + z^2}} - \frac{GM_2}{\sqrt{[x-(1-\mu)A]^2 + y^2 + z^2}} - \frac{1}{2}\Omega_B^2\left[x^2 + y^2\right] = \text{const} \tag{8.2}$$

$$\text{with } \mu = \frac{M_2}{M_1 + M_2} \text{ and } \Omega_B = \frac{2\pi}{P}.$$

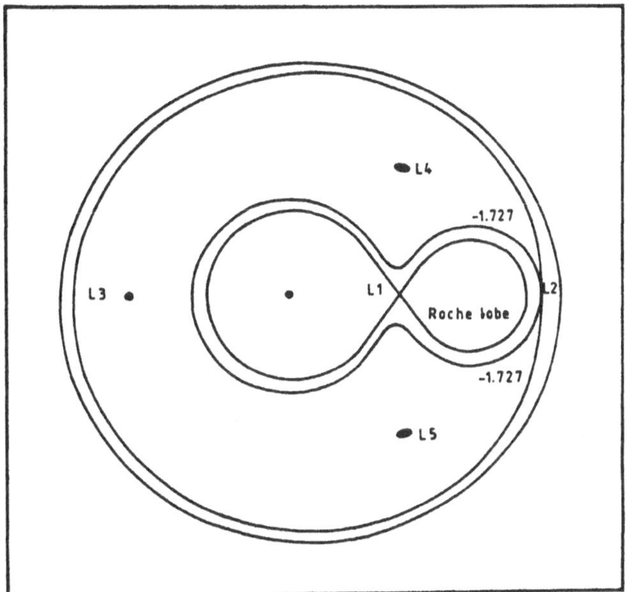

Figure 8.1: The intersection of the equipotential surfaces with the orbital plane.

Figure 8.1 illustrates the intersection of the equipotential surfaces with the orbital plane (z = 0). The function ϕ has five saddle points (Lagrangian points L) where centrifugal and gravitational forces modified by the Coriolis acceleration are in balance. They are shown in figure 8.1. The second and third Lagrangian points L_2 and L_3 are potential minima, located on the x-axis beyond the two binary components. The most important saddle point is the first Lagrangian point L_1, the point of gravitational balance on the x-axis between the two components. The equipotential surface crossing L_1 is called the critical or Roche surface (= Roche lobe). One defines the Roche radius R_c as the radius of a sphere that has the same volume as the Roche lobe. An interpolation formula as a function of mass ratio has been presented by Eggleton (1983):

$$\frac{R_c}{A} = \frac{0.49}{0.6 + q^{2/3}\ln\left(1 + q^{-1/3}\right)}. \tag{8.3}$$

8.3. The Roche lobe overflow process

If the rotation of the envelope of the primary is synchronous with the binary orbital motion, hydrostatic equilibrium applies in the co-rotating frame and the pressure equation reads

$$\nabla P = -\rho \nabla \phi \tag{8.4}$$

meaning that surfaces of equal pressure and density coincide with equipotential surfaces. Hence, the shape of a binary component will be given by an equipotential.

As the star expands, the outer layers may reach the critical Roche lobe, thus also L_1. Since in L_1 we have that $\nabla\phi=0$, hydrostatic equilibrium no longer holds there. When, for simplicity, we consider an isothermal gas near L_1, the one-dimensional (stationary) equation of motion can be written as:

$$(v^2 - c_s^2)\frac{dv}{dx} = \nabla\phi \tag{8.5}$$

with v the velocity of the gas and c_s the local isothermal sound speed.

Therefore, in order to retain a monotonic increasing velocity field near and across L_1 (where $\nabla\phi=0$), $v \approx c_s$. This means that matter funnels near L_1 into a narrow stream and flows hypersonically towards the companion.

The process in which a star fills (and overflows) its Roche lobe and starts losing mass through L_1 is called the Roche lobe overflow (RLOF). It may be accompanied by 'mass transfer' from one component to the other. The component losing matter due to RLOF is sometimes called the mass loser or mass donor, and similarly, the component that accepts (accretes) the mass is sometimes designated as the mass gainer or the mass accretor.

8.4. Limitations of the Roche model

The Roche model discussed in sections 8.2 and 8.3 is a good approximation provided that tidal interaction stabilizes the orbit, i.e. provided that circularization and synchronization are maintained, even during the RLOF process.

However, also in the case where the envelope of the primary rotates asynchroneously with the orbit, a saddle point (with zero effective gravity) exists between the two components (Kruszewski, 1963) and it is possible to define a critical surface from where funneled mass loss occurs. Even more, if the spin period of the star differs not too much from its orbital period, the value of the critical radius is not very different from the Roche radius which is computed as if the star was synchronized.

Even if the binary has an eccentric orbit, a saddle point between the two components exists. Since its exact place depends on the orbital phase, a situation may be considered, where mass loss due to critical lobe overflow has on and off phases during one single revolution.

There are circumstances in which synchronism can never be achieved. As a primary in a close binary evolves, it expands, first on the nuclear time scale during CHB, then on the Kelvin-Helmholtz time scale during hydrogen shell burning [equation (4.2)]. As a consequence of this expansion the star will spin down and rotation becomes asynchronized with the orbital motion. Due to tidal interaction, it will try to absorb orbital angular momentum of the component in order to spin up. It can readily be checked that, for small mass ratios ($q \leq 0.1$), the available orbital angular momentum is not sufficient to meet the need of the primary and the low mass companion is swallowed by the primary (Darwin, 1908; Kopal, 1972; Counselman, 1973; Sparks and Stecher,1974).

8.5. The mass loss rate of the mass loser during the RLOF

As a star in a binary evolves, it may become sufficiently large so that the photosphere attains the first Lagrangian point from where funneled mass loss will occur. In general, the following method is applied in evolutionary computations in order to determine the mass loss rate:

- *assume that the interior structure of the primary can be computed using the set of equations (3.25) derived for single stars*

The latter holds if the structure of a binary component can be assumed to be spherically symmetric. This does not apply for the outer layers of the mass loser. However, in most cases, these outer layers constitute merely a small fraction of the total stellar mass and thus hardly affect the overall stellar interior structure.

Asynchroneous rotation may initiate mixing (Tassoul, 1987, 1988, 1990; Kruszewski, 1963). However this mixing process acts mainly in the outer layers of the stars and since these outer layers are homogeneous, it is clear that it will hardly have any influence on the overall stellar structure.

- *compare the radius of the star to the Roche radius determined by equation (8.3)*
- *if the stellar radius is larger than the Roche radius, determine the mass loss rate by imposing the condition that the stellar radius equals the Roche radius.*

When the Roche radius of a binary component is determined by equation (8.3), one assumes that the binary is circularized, whereas the star rotates synchronously. This may not be correct. However, anticipating, detailed evolutionary computations reveal that

knowing that there exists a critical radius is much more important than knowing its precise value.

Therefore, equalling the stellar radius to the canonical Roche radius derived from the binary period and mass ratio, gives very reasonable mass loss rates of the mass loser and leads to fairly good evolutionary tracks of the binary.

The numerical procedure discussed above works well when the following conditions are satisfied:

- the response of the star to mass loss implies a decrease of its radius,
- due to mass loss, the radius of the star decreases faster than the Roche radius.

This is always the case when the massive star has a (mainly) radiative envelope and when the initial mass ratio of the binary is not too small (roughly $q > 0.2$). Binaries with $q < 0.1$ meat the Darwin instability (previous subsection). The evolutionary computations discussed in section 12 indicate that although the star with a radiative envelope shrinks in response to mass loss, it may not be able to shrink fast enough to keep up with the rapid shrinkage of the Roche lobe when initially the binary has a mass ratio $q \leq 0.2$. It is therefore conceivable that in most of the binaries with initial mass ratio $q \leq$

0.2, where the primary has a radiative envelope, the secondary is engulfed by the primary during RLOF. The further evolution will then be governed by the spiral-in process (sections 10.2.1.2 and 10.2.1.3).

The procedure discussed above does never work when the star has a deep convective envelope (red supergiants). To illustrate what happens, let us approximate the envelope convection in a massive star by adiabatic convection. Since most of the envelope of a massive star is fully ionized, its equation of state then satisfies the relation

$$P \propto \rho^{5/3}. \tag{8.6}$$

When this is used in combination with the hydrostatic pressure equation (3.3), one finds a relation between the stellar radius and the stellar mass, i.e. roughly

$$R \propto M^{-\frac{1}{3}} \tag{8.7}$$

This means that when the star is subjected to mass loss [on a time scale which is shorter than its thermal time scale given by equation (4.2)], its adiabatic response will be an increase of the radius. Therefore, it may be expected that the star will experience a very violent mass loss phase, on a dynamical time scale which is of the order of hours. Due to the increase of the star's radius, the secondary star will be engulfed by the primary; this phase is known as the common envelope phase of a binary.

8.6. The different types of unevolved massive close binaries

A massive binary is called a massive *close* binary (MCB) if its period is small enough that during the evolution one or both components will fill the Roche lobe. Using the single star evolutionary results of section 4 and the Roche model of the previous subsections, figure 8.2 shows the minimum period required for a galactic binary so that the radius of the primary remains smaller than the Roche radius during its entire life (= non-interacting binary). We have chosen an initial mass ratio q = 0.7, although the minimum period only marginally depends on q. We conclude that

- *in binaries with primary mass smaller than 40-50 M_o, RLOF occurs in MCBs with periods up to 3000 days*
- *the LBV phase of a primary with mass larger than 40-50 M_o may significantly reduce the importance of the RLOF as mass loss process. MCBs with a primary*

more massive than 40-50 M_o will be called 'Very Massive Close Binaries', abbreviated as VMCBs.

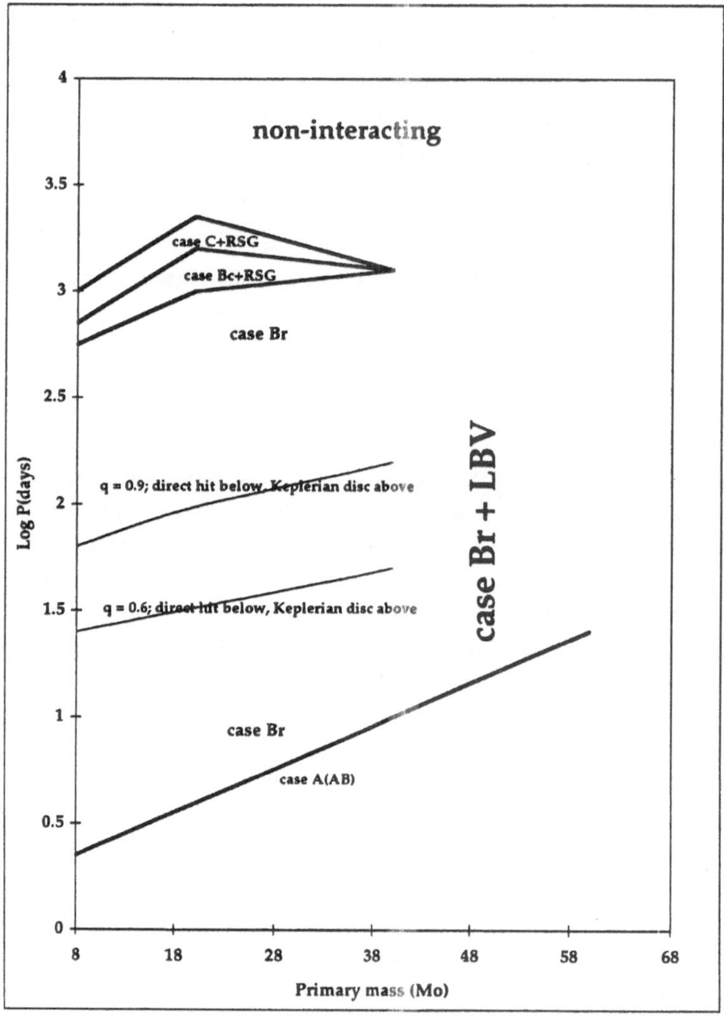

Figure 8.2: The borderline in the Mass-Period plane between case A(AB) and case B_r MCBs, between case B_r and case B_c, between case B_c and case C (where the primary may lose a large amount of its mass by RSG SW mass loss), between case C and non-interacting massive binaries where the period is too large. All these borderlines hold for MBs with initial mass ratio q = 0.7 but they are very similar for binaries with other mass ratios. We also show the borderlines between the two modes of mass transfer i.e. either by direct hit, either through the formation of a Keplerian disc. Since they depend on the adopted q, we consider binaries with q = 0.6 and binaries with q = 0.9. Except for the borderline between case A and case Br, we did not consider the others for binaries with primary mass ≥ 40 M_0 since they may be largely affected by LBV SW mass loss.

A massive star is characterized by one, two or three major expansion phases: the CHB phase, the hydrogen shell burning phase and the He shell burning phase. If this star is the primary of a close binary, depending on the initial orbital period it fills its Roche lobe during one of these three expansion phases. Following the original suggestion of Kippenhahn and Weigert (1967) and Lauterborn (1969), the class of unevolved MCBs can be divided into

case A: the primary fills the Roche lobe during its CHB phase;
case B: the primary fills the Roche lobe during its H shell burning phase,
case C: the primary fills the Roche lobe during its He shell burning phase;

Figure 8.2 also shows the limiting period between case A, case B and case C as function of primary mass for galactic binaries. We have adopted an initial mass ratio q = 0.7; however these borderlines do not very much depend on q. The figure illustrates that:

most MCBs with $P \leq 2\text{-}4$ days will evolve according to case A
most MCBs with $4 < P \leq \sim 1000$ days will evolve according to case B,
most MCBs with $\sim 1000 < P \leq 3000$ days will evolve according to case C.

If the observed period distribution $\Pi(P)$ of binaries where RLOF has not started yet, follows the relation

$$\Pi(P)dP \propto \frac{dP}{P} \tag{8.8}$$

(Popova et al., 1982)

case B binaries are by far the most frequent class of observed binaries.

Due to RSG mass loss, a star with initial mass between M_{min} and 40-50 M_o does avoid the He shell burning expansion and therefore case C will not occur. The value of M_{min} obviously depends on the RSG stellar wind mass loss rates. With the formalism discussed in section 2.6 (see also section 4.1.2), $M_{min} \approx 15\text{-}20$ M_o in the Galaxy and the LMC. When RSG mass loss depends on the metallicity as predicted by the radiatively driven wind theory [equation (2.12)], $M_{min} \geq 40$ M_o in the SMC.

As outlined in the previous subsection, one expects a different evolutionary behaviour in binaries where RLOF starts when the mass loser has a mostly radiative envelope (case B_r) compared to binaries where RLOF starts when the mass loser has a deep convective envelope (case B_c). When we define the phase in which the star has a deep convective envelope as the phase where the envelope convection extends down to layers that previously belonged to the hydrogen burning core (corresponding to the phase where

CNO processed material is dredged-up to the surface), using the evolutionary computations of section 4), figure 8.2 gives the limiting period between case B_r and case B_c as well. It may be concluded that among all case B types, case B_r binaries are the most frequent ones.

The borderlines between case A and case B_r, between case B_r and case B_c, for binaries with primary mass $\leq$ 40-50 M_o, are very similar for the Magellanic Clouds. The separation between case B_c and case C which is affected by the RSG stellar wind mass loss and the effect of LBV mass loss when the primary mass > 40 M_o obviously depends on the relation between the metallicity and the SW rates.

9. The mass transfer process during RLOF

9.1. Direct hit versus the formation of a Keplerian disc

When a stream of matter leaves the primary through L_1 and is trapped into the potential well of the secondary, two situations can be distinguished:

a. the gas-stream hits the secondary directly
b. the gas-stream first forms a disc around the secondary.

This has been studied hydrodynamically by Lubow and Shu (1975) assuming that the forces exerted by the radiation field of both components on the gas-stream can be neglected. Given the binary parameters M_1, M_2 (considered as point masses) and the period P, a detailed computation of the particle trajectories reveals the formation of a disc, provided that the radius R_2 of the secondary is smaller than R_{disc}, of which the value can be calculated from the following best fit formula:

$$\ln(\frac{R_{disk}}{A}) = 0.367\ln(q) - 2.93 \qquad (9.1)$$

Assuming that both components of a binary evolve in the same way as normal single stars (i.e. we use the evolutionary computations of section 4), figure 8.2 gives the limiting period between both modes at the onset of RLOF.

Is it possible that radiation forces in MCBs change the picture sketched above? It is quite clear that only detailed hydrodynamic computations, similar to the case of stellar winds, but adapted for a non-spherical gas-stream, can give an answer but this has not been done until now. The effect can be estimated as follows.

Typical RLOF mass loss rates in MCBs are in the range 10^{-4} - 10^{-3} M_o/yr. Assuming a (cylindric symmetric) gas-stream across L_1 with a typical radius of the order of a solar radius and with a temperature equal to the surface temperature of the mass loser (~ a few 10^4 K), since the velocity of the matter in the stream is of the order of the sonic velocity, straightforward calculations show that the densities in the gas-stream should be ~ 10^{-7} - 10^{-6} gr.cm^{-3}. Under these circumstances the main opacity is electron scattering. It is quite easy to demonstrate that electron scattering radiation forces do not significantly modify the streamlines in between the two components and they are certainly unable to drive the gas-stream out of the binary.

9.2. The spin-up of the mass gainer due to mass transfer

When the gas-stream first forms a Keplerian disc around the mass gainer, it can readily be checked that when matter with mass dM falls onto the mass gainer with mass M_2, it has a specific angular momentum dJ given by

$$dJ = dM\sqrt{GM_2R} \tag{9.2}$$

with R the radius of the gainer. This value has to be compared to the spin angular momentum J_{rot} of the star itself, i.e. assuming rigid rotation

$$J_{rot} = I\omega_{rot} \tag{9.3}$$

with I the moment of inertia of the star and ω_{rot} its rotational angular velocity. Using the evolutionary computations of single stars of section 4, it is possible to calculate the total amount of mass that has to be accreted in this way to force the whole star to rotate at its critical break-up velocity

$$\omega_{crit} = \sqrt{\frac{G(M_2 + dM)}{R^3}} \tag{9.4}$$

where we make the simplifying assumption that during the accretion the radius R does not change. Detailed computations of the accretion process (section 12.3) reveal that in reality this is not true, unless the amount of accreted matter is small, and fortunately this is precisely the case here, i.e. a 15 M_o (resp. a 25 M_o) mass gainer has only to accrete ~2 (resp. 3) M_o through a Keplerian disc in order to rotate at its critical break up velocity.

The latter values were computed assuming that the whole star is spinning up. It is obvious that when only the envelope is affected by the spin-up process, much less matter has to be accreted to make these outer layers rotate at break-up velocity.

We may conclude:

when a Keplerian disc is formed around the secondary during the RLOF of the primary, a very efficient mechanism to remove angular momentum from the disc is necessary to produce efficient mass transfer.

It is not the purpose of the present work to discuss the very complicated physics of accretion discs. An excellent review has been presented by Pringle (1981). The idea is that in the disc, viscosity will transport angular momentum outwards and matter inwards. Tidal effects may then convert the excess angular momentum from the disc-edge back into orbital angular momentum. Of course, since viscosity is involved, the quantitative results of this process are still quite uncertain.

When the gas-stream hits the mass gainer directly, the computation of the transfer of angular momentum is not as straightforward as in the disc-formation case. However the tangential component of the stream velocity is obviously smaller than in the disc-formation case. Even more, the detailed computations of Lubow and Shu (1975) reveal that the larger the radius of the gainer, the smaller the tangential component will be. The spin-up rate depends on this tangential component, hence one may expect that more mass can be transferred before the gainer reaches its critical break up velocity as in the disc-formation case. On the other hand, the case in which the gas-stream hits the gainer directly applies in general to binaries with smaller periods (figure 8.2). Smaller periods imply a more efficient process of synchronization by tidal effects, which may remove the superfluous rotational angular momentum of the gainer. All in all one can conclude that

the smaller the binary period, the less efficient the spin-up of the mass gainer due to mass transfer during the RLOF of the mass loser.

9.3. Computation of the evolutionary behaviour of a mass gainer

Since mass transfer proceeds either by direct hit or by formation of a Keplerian disc, the fully consistent physical and mathematical formalism should be hydrodynamic and 3-dimensional. A few remarks are appropriate.

- The velocity of matter falling onto a star is close to the free-fall velocity given by

$$v_{f-f} = \sqrt{\frac{2GM}{R}} \tag{9.5}$$

The dynamical pressure exerted by this matter is then

$$P_{dyn} = \rho v_{f-f}^2 = \frac{\dot{M}}{4\pi R^2} v_{f-f} \tag{9.6}$$

with $\dot{M}$ the mass accretion rate. It can easily be checked that for accretion rates typical for MCBs (section 12), this dynamical pressure is always much smaller than the hydrostatic pressure and can safely be neglected. We thus assume hydrostatic equilibrium to investigate the reaction of a star accreting matter.

- When gas flows into a star, a shock front is formed close to the stellar surface (Ulrich and Burger, 1976). The gravitational energy released by this inflowing matter is converted into thermal energy and is radiated away as X-rays and UV radiation. Assuming that all the gravitational energy is transformed into radiation, the radiated luminosity (referred to as accretion luminosity L_{acc}) can be expressed as

$$L_{acc} = \frac{GM}{R}\dot{M} \tag{9.7}$$

In MCBs, this luminosity only exceeds the critical Eddington luminosity when $\dot{M}$ is larger than 10^{-2} M_o/yr, which is one or two orders of magnitude larger than the mass transfer rates during a typical RLOF. We therefore conclude that the accretion luminosity does not alter the hydrodynamics of the in-falling matter.

- Accretion of matter in a binary is always local (hence non-spherical). However it is assumed that the redistribution of the mass all over the stellar surface occurs very rapidly so that the effect of accretion may be estimated by adopting a spherical symmetric formalism.

The main effect of mass accretion is then the compression of all mass layers in the stellar interior due to the increase of gravity leading to the release of gravitational energy of all these layers. This release [the entropy term in equation (3.14)] can be computed in a most convenient way by using equation (3.16). We consider two cases:

9.3.1. THE STANDARD ACCRETION MODEL

When dealing with MCBs, it can be assumed that the envelope of the mass gainer is in radiative equilibrium and has a positive entropy gradient. In the 'standard accretion model' it is assumed that the radiative equilibrium of the outer layers is not affected by accretion. This applies when the entropy of the in-falling matter is equal or larger than the entropy of the envelope of the mass gainer and when the effect of rotation (thus of rotational mixing) is neglected.

The overluminosity

The entropy gradient remains positive during accretion and therefore the homologous term in equation (3.16) causes an increase of the luminosity: the star becomes overluminous with respect to its mass. This overluminosity becomes significant when the star does not have sufficient time to radiate away this excess energy, i.e. roughly when the accretion time scale ($\sim M/\dot{M}$) becomes shorter than the thermal time scale [equation (4.2)], i.e. when

$$\dot{M} \geq 3.2 \cdot 10^{-8} \frac{RL}{M} \qquad (9.8)$$

($\dot{M}$ in M_o/yr, R, L and M in solar units). When the mass gainer is a 15 M_o star at the end of CHB, it follows that the star becomes significantly overluminous when the accretion rate is larger than 10^{-3} M_o/yr.

As a consequence of a significant overluminosity, the outer layers of the star may start expanding and a situation may occur in which also the mass gainer fills its critical equipotential: when this happens the binary is classified as a contact binary (they will be discussed in more detail in section 12.4)..

The effect of accretion on the evolution of the convective core

Of particular importance is the reaction of the convective core of a mass accreting star when the standard accretion model applies. A secondary in an interacting close binary may have already performed a significant part of its own CHB at the onset of the RLOF of the primary and thus at the onset of the mass transfer phase. This means that it has built up a molecular weight gradient in its interior (figure 9.1). Increasing the mass of the star results into an increase of the stellar convective core. However as illustrated in figure 9.1, this immediately leads to the formation of a molecular weight barrier at its border which may prohibit a fast increase of the convective core when the stellar mass increases.

The way convection and semi-convection are treated is of fundamental importance here. Let us first remember that (see also section 3.4.2)

layers where $\nabla_{Led} < \nabla_{rad}$ *are convectively unstable; semi-convection applies in layers where* $\nabla_{ad} < \nabla_{rad} < \nabla_{Led}$ *and the timescale of mixing depends on the adopted diffusion coefficient.*

Just as for single stars, we can distinguish two limiting cases:

case 1: the diffusion process in the semi-convective layers on top of the convective core is very fast. In this case, independently of the existence of a molecular weight gradient, the convective core boundary can be determined by the Schwarzschild criterion and the chemical profile of the elements in the semi-convective region will obey the condition $\nabla_{rad} = \nabla_{ad}$.

case 2: the diffusion process in semi-convective regions is slow (small diffusion coefficient). Compared to case 1, we obviously expect a much slower increase of the convective core when the star accretes mass and a situation can arise where the convective core does not increase at all.

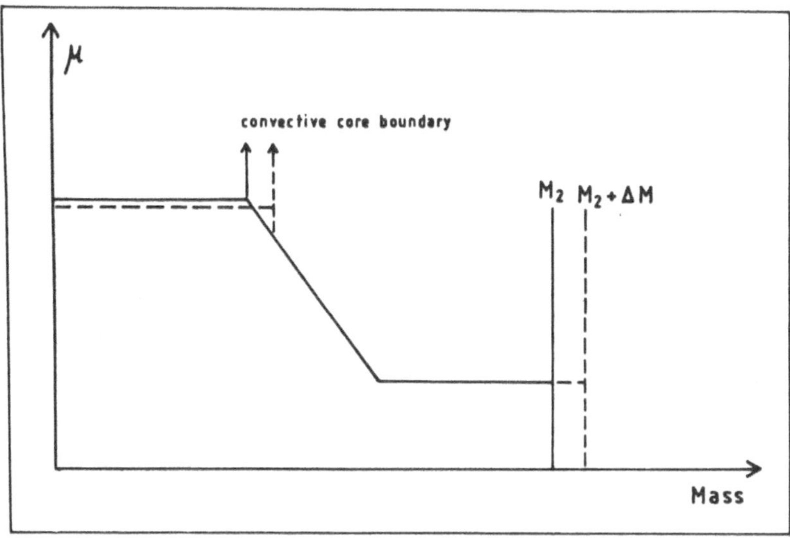

Figure 9.1: The molecular weight gradient in the interior of the mass gainer with mass M_2 and the effect of accretion of matter (DM) with the same molecular weight as the outer layers of the gainer.

Thermohaline mixing

During its RLOF, a primary may lose mass layers which have been nuclearly processed during an earlier evolutionary phase (the CHB phase). This mass has a molecular weight which is larger than the molecular weight of the outer layers of the mass gainer. This means that if at least part of the nuclearly processed matter is accreted by the gainer, it acquires an inverted μ-gradient. This situation is unstable and initiates mixing. The process which is commonly known as 'thermohaline convection', has been studied by e.g. Zahn (1983) and by Kippenhahn et al. (1980) who conclude that it is a very fast process which allows us to treat it as an instantaneous one. The numerical procedure to determine the consequences of this fast mixing is then straightforward (figure 9.2): mix the accreted mass with the outer layers of the gainer until a situation occurs in which the molecular weight of the mixed layers is equal to the molecular weight of the layer just below the mixed region. It is clear that this process will produce stars with increased N and depleted C and O at their surface.

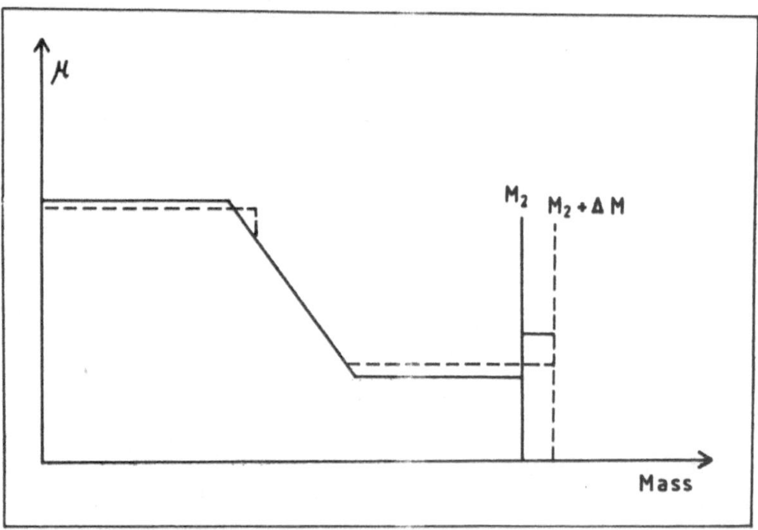

Figure 9.2: The same as figure 9.1 but the accreted mass has a molecular weight that is larger than that of the outer layers of the gainer.

9.3.2. THE ACCRETION INDUCED FULL MIXING MODEL

Mass accretion implies accretion of angular momentum; consequently a mass gainer spins up (section 9.2). Rotation induces mixing, i.e. radiative equilibrium is destroyed. Furthermore, if unlike the assumption made in the standard accretion model, the entropy of the accreted matter is significantly lower than the entropy of the envelope of the gainer, convection will start smearing out the entropy profile (the situation is illustrated in figure 9.3). This convection zone will develop inwards as long as its specific entropy exceeds that of the matter which is accreted on top of it. The homologous term in equation (3.16) is obviously zero in this convection zone and therefore the overluminosity will be smaller than in the standard accretion model. The limiting situation is of course a complete mixing of the whole star. We will call this the 'accretion induced full mixing model'.

It is quite obvious that in the fully mixed case considered here, the way how semi-convection is treated is much less important than in the standard accretion model.

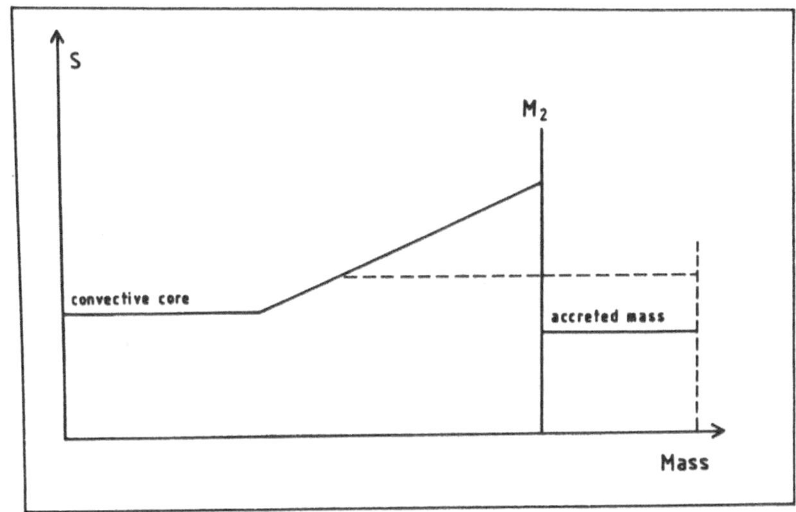

Figure 9.3: The entropy gradient in the interior of the mass gainer and the effect when mass is accreted with entropy significantly lower than the entropy of the envelope of the gainer.

The existing hydrodynamic studies of accretion and accretion discs do not allow a distinction to be made between both accretion models. Therefore, one can study the evolution of mass gainers by adopting both models and by using different values for the diffusion coefficient in semi-convective layers. A comparison with observations may then hopefully favor one of these models.

10. The evolution of the binary period in MCBs

We distinguish the case where both components do not fill their critical Roche lobe but are losing mass by stellar wind, and the case where one of the components loses matter due to RLOF.

10.1. One or both components are losing mass by stellar wind only

We assume that

- the orbit is circular and remains circular during the mass loss phase. The mathematics in the non-circular case are somewhat more complicated but the general behaviour of the period is very similar to the one we will describe here,
- the stellar wind mass loss is spherically symmetric.

If ω = the orbital angular velocity, r_1 (resp. r_2) the orbital radius (with respect to the center of mass of the binary) of the star with mass M_1 (resp. M_2), dM_1 (resp. dM_2) the mass of an infinitesimal shell leaving the star with mass M_1 (resp. M_2) it follows that

$$dJ = \omega r_1^2 dM_1 + \omega r_2^2 dM_2 \tag{10.1}$$

$$= \frac{2\pi}{P}\frac{M_2^2 A^2}{(M_1+M_2)^2} dM_1 + \frac{2\pi}{P}\frac{M_1^2 A^2}{(M_1+M_2)^2} dM_2 \tag{10.2}$$

$$= J\left[\frac{M_2}{M_1+M_2}\frac{dM_1}{M_1} + \frac{M_1}{M_1+M_2}\frac{dM_2}{M_2}\right] \tag{10.3}$$

From equation (7.2) with e = 0

$$\frac{dJ}{J} = \frac{dM_1}{M_1} + \frac{dM_2}{M_2} - \frac{1}{2}\frac{d(M_1+M_2)}{M_1+M_2} + \frac{1}{2}\frac{dA}{A} \tag{10.4}$$

Combining equations (10.3) and (10.4) leads to

$$\frac{dA}{A} = -\frac{d(M_1 + M_2)}{M_1 + M_2} = \frac{1}{2}\frac{dP}{P} \tag{10.5}$$

and thus

$$\frac{P}{P_0} = \left(\frac{M_{1_0} + M_{2_0}}{M_1 + M_2}\right)^2 \tag{10.6}$$

where the subscript 'o' stands for values at the beginning of the stellar wind mass loss phase. We thus conclude that

the period of a binary increases with the square of the sum of the masses of both components when one or both components lose mass by stellar wind.

In binaries where the primary is more massive than ~40-50 M_o, the LBV type stellar wind mass loss rate can be so large that the RLOF phase never starts. Using the single star evolutionary results of section 4, equation (10.6) then predicts a period increase by a factor 2 to 3.

10.2. The variation of the period during RLOF

If we were able to solve the 3-dimensional magneto-hydrodynamical equations -including the effect of radiation pressure, viscosity, intrinsic and orbital rotation, describing the mass loss process in a binary- the variation of the system period could be determined from first principles. However even 'something simple like a spherical stellar wind' of a star is not fully understood. Hence, a solution for the binary problem is certainly not for tomorrow.

One can try to estimate the variation of the binary period in MCBs by accepting the following statement:

radiation pressure hardly affects a gas-stream caused by the RLOF process (section 9.1) however it is the main driver of a more or less spherical symmetric stellar wind.

We therefore consider two cases:

10.2.1. THE RLOF MASS LOSS DOMINATES OVER A POSSIBLE SW MASS LOSS

In this subsection, only binaries are considered where RLOF occurs and where the mass loss due to RLOF is much larger than possible SW mass loss, i.e. the effect of SW mass loss on the variation of the binary period during RLOF can be neglected. Anticipating and accounting for typical values of a RLOF mass loss rate (section 12), only the most luminous LBVs have SW mass loss rates that may be comparable. For the moment, we therefore exclude the VMCBs.

Definitions:

Conservative RLOF: all mass leaving the primary (mass loser) due to RLOF is accreted by the secondary (mass gainer).

Non-conservative RLOF: at least part of the mass lost by the primary during its RLOF leaves the binary system.

It is customary to define a parameter β describing the fraction of the mass lost by the mass loser that is accreted by the mass gainer, i.e. if $\dot{M}_1$ (resp. $\dot{M}_2$) denotes the mass loss (resp. mass gain) rate of the mass loser (resp. mass gainer), it follows that

$$\dot{M}_2 = -\beta \dot{M}_1 \quad , 0 \leq \beta \leq 1 \tag{10.7}$$

10.2.1.1. Conservative RLOF

Since the spin angular momentum of a binary component is always much smaller than the orbital angular momentum (i.e. efficient mass transfer must be accompanied by an efficient mechanism to remove superfluous angular momentum of the transferred matter back into orbital angular momentum of the components, section 9.2), conservation of mass implies conservation of total orbital angular momentum and this means [using equations (7.1) and (7.2) in the circularized case] that

$$\frac{M_1^2 M_2^2}{M_1 + M_2} A = \text{const} \tag{10.8}$$

or using Kepler's law

$$\frac{P}{P_0} = \left(\frac{M_{1_0} M_{2_0}}{M_1 M_2} \right)^3 \qquad (10.9)$$

where the subscript 'o' denotes values at the beginning of the RLOF phase.

Anticipating, when we use the results of MCB evolutionary computations, we conclude that

as a consequence of conservative RLOF, the period of the binary after RLOF is always larger than the period before RLOF.

10.2.1.2. Non-conservative RLOF

a. Non-conservative RLOF in case A/case B_r binaries with $q > 0.2$

As will be discussed in more detail in section 12 dealing with evolutionary computations of close binaries, in most of the MCBs (except the VMCBs), SW mass loss is not sufficiently large to allow the primary to escape RLOF. Due to RLOF, the primary loses matter at very high rates ($\geq 10^{-3}$ M_o/yr). When this is accreted by the secondary, the latter starts expanding and both stars may enter a contact phase. A situation can be encountered such as the one sketched in figure 10.1. A possibility (may be the only one) by which matter can leave a case A/case B_r MCB with synchronized period in an efficient way at rates similar to the mass loss rate of the primary due to RLOF, is mass loss through L_2; in this case it can be expected that a ring is formed around the binary with a radius L_2C with C the center of mass.

Let us assume circular orbits and a circular ring which revolves in the same sense as the binary at a distance A_{ring} from the center of mass of the binary. The angular momentum per unit mass in the ring is given by

$$j_{ring} = \sqrt{G(M_1 + M_2)A_{ring}} \qquad (10.10)$$

and thus the variation of orbital angular momentum of the binary when $-(1-\beta)dM_1$ is lost through L_2, is given by

$$dJ = j_{ring}(1 - \beta)dM_1 \qquad (10.11)$$

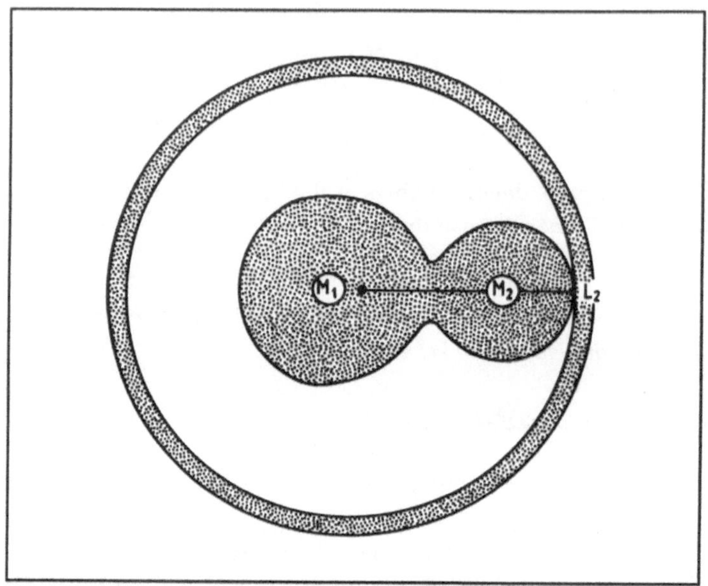

Figure 10.1: Mass loss from the system through L_2 and the formation of a ring around the binary, during the RLOF of the primary in a case A/case Br binary.

Using equation (7.2) with e = 0 and Kepler's law, equation (10.11) can be rewritten as

$$\frac{dP}{P} = (1+3\alpha)\frac{(1-\beta)dM_1}{M_1+M_2} - 3\frac{dM_1}{M_1} - 3\frac{dM_2}{M_2} \tag{10.12}$$

$$\text{with } \alpha = \frac{(M_1+M_2)^2}{M_1M_2}\sqrt{\frac{A_{ring}}{A}} \tag{10.13}$$

If matter leaves the binary through L_2, $A_{ring} \approx L_2C$. For a wide range of binary mass ratios, L_2C ~1.3A. As a consequence, assuming that the ratio $A_{ring}/A = \eta$ = constant, equation (10.12) can be integrated.

When $\beta \neq 0$, one obtains

$$\frac{P}{P_0} = \left[\frac{M_1+M_2}{M_{1_0}+M_{2_0}}\right]\left[\frac{M_1}{M_{1_0}}\right]^{3\left[\sqrt{\eta}(1-\beta)-1\right]}\left[\frac{M_2}{M_{2_0}}\right]^{-3\left[\sqrt{\eta}\frac{1-\beta}{\beta}+1\right]} \tag{10.14}$$

When $\beta = 0$, with $\eta = 1.3$, equation (10.12) reduces to

$$\frac{P}{P_0} = \left(\frac{M_1 + M_2}{M_{1_0} + M_{2_0}}\right)\left(\frac{M_1}{M_{1_0}}\right)^{0.42} e^{3.42\frac{M_1 - M_{1_0}}{M_2}} \tag{10.15}$$

where the subscript 'o' denotes values at the onset of the RLOF. It can readily be checked that the formalism predicts an orbital period decrease if mass leaves the binary, i.e.

a significant fraction of the mass lost by a primary due to RLOF in a case A or case B_r binary, can be removed from the binary at the expense of a reduction of the available orbital energy, leading to a decrease in the orbital period.

b. The common envelope phase in case B_c/C binaries and the spiral-in process in binaries with small mass ratio.

The RLOF in a case B_c/case C MCB is very violent; since mass loss implies an even faster expansion of the primary when the envelope is convective, very soon after the onset of the RLOF, the secondary will be engulfed by the primary: the common envelope phase of a case B_c/case C binary.

When the binary consists of a massive primary and a low mass secondary (binary mass ratio $q \leq 0.2$), RLOF will not occur. Instead, the latter will be dragged into the envelope of the primary (sections 8.4 and 8.5). This means that also here it makes sense to designate the process as a common envelope, although it is not restricted to case B_c/case C binaries.

We know of two energy sources in a massive binary which are able to remove matter from the common envelope: stellar nuclear radiation energy and orbital energy.

Stellar nuclear radiation is able to drive a SW from the common envelope. It can reasonably be expected that the rate is similar to that of supergiants, of the order of a few times 10^{-6}-10^{-5} M_o/yr.

An efficient use of orbital energy may result into larger mass loss rates. The idea: due to viscosity, orbital energy of the secondary is transformed into thermal energy of the common envelope and the secondary spirals-in into the envelope (a common envelope phase is thus always accompanied by a spiral-in process). Part of the thermal energy is radiated away, part is used to drive the matter of the common envelope out of the binary.

Since viscosity is involved, we can only try to answer the following general question:

if viscosity between the secondary and the envelope of the primary is able to transform orbital energy into thermal energy of the envelope in an efficient way, how much orbital energy is required to remove this envelope?

We follow a similar prescription as proposed by Webbink (1984). When $\Delta\varepsilon_b$ is the binding energy of the envelope, the total orbital energy (potential + kinetic) $\Delta\varepsilon_{tot}$ needed to remove this envelope, can be determined from the relation

$$\Delta\varepsilon_b = \alpha\Delta\varepsilon_{tot} \tag{10.16}$$

where the parameter α expresses the fraction of the orbital energy (which is first transformed into thermal energy) that is not radiated away, and hence is effectively used to drive mass from the star ($0< \alpha \leq 1$).

Generally, a star stops expanding (i.e. the RLOF or common envelope phase stops) when most of the hydrogen rich layers (with $X \geq 0.2$-0.3) have been removed (section 12). It is then costumary to introduce a parameter λ so that

$$\Delta\varepsilon_b = G\frac{M_{1_0}(M_{1_0} - M_{1_e})}{\lambda R_0} \tag{10.17}$$

with M_{1_e} = the final mass of the star after the removal of the hydrogen rich layers, and R_o = the stellar radius at the onset of the common envelope/spiral-in phase.

When the primary is a RSG and has a convective envelope, accounting for the adiabatic reaction of the star when it loses mass at a high rate (section 8.5), let us assume that the radius R_o of the common envelope is constant during the whole mass loss process (of the order of the orbital separation at the onset of the process). A rough estimate of the total binding energy of the common envelope at the moment it leaves the binary is then given by

$$\Delta\varepsilon_b = \int_{M_{1_e}}^{M_{1_0}} G\frac{M + M_2}{R_o}dM = G\frac{M_{1_0} + M_{1_e} + 2M_2}{2R_o}(M_{1_0} - M_{1_e}) \tag{10.18}$$

Using the evolutionary computations of section 12, it follows that $\lambda = 0.7$ - 1.5.

When the primary is a blue star with a radiative envelope and when spiral-in is a consequence of the fact that the secondary is a low mass companion, $\Delta\varepsilon_b$ in equation (10.16) can be approximated by the binding energy of the hydrogen rich layers prior to the spiral-in phase. We therefore need the mass concentration M(r) in the stellar interior at the onset of the mass loss process and thus (neglecting the effect of the low mass companion)

$$\Delta\varepsilon_b = \int_{M_{1_e}}^{M_{1_0}} G\frac{M(r)}{r}dM \tag{10.19}$$

Using the evolutionary computations of massive stars (section 4), it follows that the value of the parameter λ ranges between 0.3 and 0.5.

Since the total orbital energy of the binary is given by

$$\varepsilon_{tot} = \varepsilon_{pot} + \varepsilon_{kin} = -G\frac{M_1 M_2}{2A} \tag{10.20}$$

using expression (10.18) [resp. (10.19)] when the binary is a case B_c (resp. when the primary is a blue star and the binary mass ratio $q \leq 0.2$), equation (10.16) predicts the reduction of the orbital period that is needed to remove the whole hydrogen rich envelope. Since the effect of the secondary mass is neglected, the predicted period reduction is an underlimit.

In order to illustrate the consequences of this common envelope/spiral-in process, consider a 20 M_o OB type star ($\lambda \approx 0.4$) with a 2 M_o companion and assume maximum efficiency for the transformation of orbital energy into escape energy, i.e. $\alpha = 1$. Since $M_{1_e} \approx 7$ M_o (section 12), it follows from equations (10.16), (10.19), (8.3) and Kepler's law that the removal of 13 M_o implies a binary period decrease by a factor 2000. We therefore conclude that the spiral-in process in a binary with very small mass ratio is able to remove a significant fraction of the mass of the mass loser at the expense of a very large reduction of the orbital separation (orbital period).

It can easily be checked that whatever the mass of the secondary, a binary with a 20 M_o RSG evolving through a case B_c common envelope also experiences a very large period reduction.

10.2.1.3. Mergers

When the mass ratio of the binary is smaller than ~0.2, the low mass companion is dragged into the envelope of the massive component. As outlined in the previous section, the resulting spiral-in process implies a very large period reduction. Therefore, when the low mass component is a normal CHB star, both stars will probably merge. The low mass secondary is almost a homogeneous star at the onset of the spiral-in and therefore the evolution of the massive component after merging may resemble the evolution of a star that accreted an amount of mass equal to the mass of the low mass star.

Coalescence may also happen in binaries with initial mass ratio > 0.2, where the RLOF is non-conservative implying a reduction of the orbital period as predicted by equation (10.14) or (10.15). How these systems will further evolve is uncertain. A possibility may be similar to the one proposed above for binaries with small mass ratio, i.e. the evolution of the merger resembles the evolution of a star that accreted an amount of mass equal to the mass of the secondary. However in this case, the chemical composition of the secondary may be significantly different from a homogeneous star, and therefore the effect of merging is not easy to predict and remains to be investigated.

A general condition for merging will be derived in section 12.1.5.

10.2.2. THE RLOF AND THE SW MASS LOSS RATES ARE COMPARABLE

In VMCBs where the primary mass > 40-50 M_o, the RLOF competes with a violent LBV stellar wind mass loss process.

When the SW is large enough, the RLOF process can be avoided. In this case the period variation is entirely determined by equation (10.6), implying a significant binary period increase.

When the SW and the RLOF mass loss rates are comparable, one can obtain a first order estimate of the period variation by combining equation (10.6) (to determine the period variation caused by the SW mass loss) and either equation (10.9) (conservative RLOF), equation (10.14) or (10.15) (non-conservative RLOF), or equations (10.16) and (10.18) or (10.19) if the SW mass loss and the common envelope/spiral-in act together.

11. The effect of the supernova explosion of one of the components

When a massive star does not end its life as a BH, it encounters a SN event. In this section we discuss the effect on binary parameters. We assume an instantaneous explosion of a star with mass M_1 in a circularized binary with secondary component mass M_2. The mass of the NS which remains after the SN explosion equals M_{NS}.

The effects of the SN shell on the companion star will generally be small (Fryxell and Arnett, 1981). They will therefore not be considered here.

If the SN explosion is not entirely spherically symmetric, the remaining star with mass M_{NS} will receive a kick velocity $\vec{v}_{kick}$ of which the magnitude depends on the degree of anisotropy of the SN and of which the direction is defined by the angles θ and φ defined in figure 11.1.

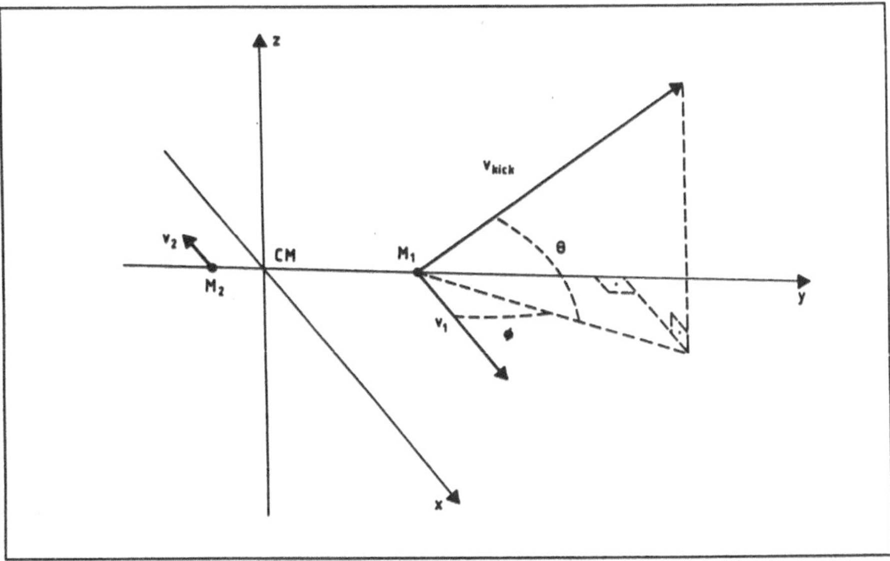

Figure 11.1: The XYZ coordinate system with the center of mass of the binary in the origin and the parameters θ and φ used in order to describe the effect of the SN explosion on the orbital parameters of a MB.

It is convenient to use the following parameters:

$$m = \frac{M_1 + M_2}{M_{NS} + M_2} \tag{11.1}$$

$$\mu = \frac{M_1 M_2}{M_1 + M_2} \tag{11.2}$$

$$\mu_{NS} = \frac{M_{NS} M_2}{M_{NS} + M_2} \tag{11.3}$$

$\vec{v}_{orbit}$ = the relative orbital velocity of the two stars prior to the SN explosion, i.e.

$$\left\|\vec{v}_{orbit}\right\|^2 = \frac{G(M_1 + M_2)}{A} \tag{11.4}$$

$$v_{rel} = \frac{\left\|\vec{v}_{kick}\right\|}{\left\|\vec{v}_{orbit}\right\|} \tag{11.5}$$

Since the explosion is assumed to be instantaneous, the semi-major axis A and the relative orbital velocity $\vec{v}_{orbit}$ are the same just before and just after the SN explosion. Therefore, in the new center of mass frame just after the instantaneous SN explosion, the energy ε_{tot} of the remaining binary with component masses M_{NS} and M_2, is given by

$$\varepsilon_{tot} = \varepsilon_{pot} + \varepsilon_{kin} = -G\frac{M_{NS}M_2}{A} + \frac{1}{2}\mu_{NS}\left\|\vec{v}\right\|^2 \tag{11.6}$$

where $\vec{v}$ is the relative velocity of the two stars *immediately* after the explosion, i.e.

$$\vec{v} = \vec{v}_{orbit} + \vec{v}_{kick} \tag{11.7}$$

and thus

$$\left\|\vec{v}\right\|^2 = \left\|\vec{v}_{orbit}\right\|^2 + \left\|\vec{v}_{kick}\right\|^2 + 2\left\|\vec{v}_{orbit}\right\|\left\|\vec{v}_{kick}\right\|\cos\theta\cos\phi \tag{11.8}$$

Using equation (11.8), equation (11.6) becomes

$$\varepsilon_{tot} = -\frac{GM_{NS}M_2}{2A}\left\{2 - m\left[1 + 2v_{rel}\cos\theta\cos\phi + v_{rel}^2\right]\right\} \tag{11.9}$$

When the binary remains bound, ε_{tot} can also be written as

$$\varepsilon_{tot} = -\frac{GM_{NS}M_2}{2A_{ps}} \tag{11.10}$$

where A_{ps} is the post-SN semi-major axis (the SN-shell at infinity). Consequently, using Kepler's law we obtain the post-SN binary period P_{ps} as a function of the pre-SN period P, i.e.

$$\frac{P_{ps}}{P} = \sqrt{m}\left(2 - m\left[1 + 2v_{rel}\cos\theta\cos\phi + v_{rel}^2\right]\right)^{-\frac{3}{2}} \tag{11.11}$$

Since the total energy of a (bound) binary is negative, equation (11.9) immediately gives us the conditions for a binary to remain bound after the SN explosion, i.e.

- if the SN explosion is spherically symmetric ($\vec{v}_{kick} = 0$), the binary remains bound if the SN shell has a mass less than half the total mass of the pre-SN binary,
- if the SN explosion is not spherically symmetric, the binary will always be disrupted when $v_{rel} \geq 1 + \sqrt{\frac{2}{m}}$
- if the SN explosion is not spherically symmetric, the general requirement for a bound post-SN binary is

$$\cos\theta\cos\phi < \gamma \tag{11.12}$$

with

$$\gamma = \frac{1}{2v_{rel}}\left[\frac{2}{m} - 1 - v_{rel}^2\right] \tag{11.13}$$

- if the SN explosion is not spherically symmetric, assuming an arbitrary direction of the kick velocity, the foregoing implies that the probability p for a binary to remain bound after the SN explosion is given by

$$\begin{array}{ll} p = 0 & \text{if } \gamma \leq -1 \\ p = 0.5(1+\gamma) & \text{if } -1 < \gamma < 1 \\ p = 1 & \text{if } \gamma \geq 1 \end{array} \tag{11.14}$$

When the binary remains bound after the SN explosion, it will receive a 'runaway' velocity $\vec{v}_{rw}$. Since at the moment of explosion, the velocity of the SN shell with mass ΔM is the velocity of M_1, conservation of momentum gives

$$\left(M_{NS}+M_2\right)\vec{v}_{rw}+M_{NS}\vec{v}_{kick}=\Delta M\frac{M_2}{M_1+M_2}\vec{v}_{orbit} \tag{11.15}$$

or

$$\|\vec{v}_{rw}\|=\frac{\|\vec{v}_{orbit}\|}{M_{NS}+M_2}\left\{\left(\frac{M_2\Delta M}{M_1+M_2}\right)^2-2\frac{M_2M_{NS}\Delta M}{M_1+M_2}v_{rel}\cos\theta\cos\phi+\left(M_{NS}v_{rel}\right)^2\right\}^{\frac{1}{2}} \tag{11.16}$$

When the binary is disrupted, the 'runaway' velocity of the remaining optical component equals its orbital velocity with respect to the center of mass just before the SN explosion, i.e.

$$\|\vec{v}_{rw}\|=\frac{M_1}{M_1+M_2}\|\vec{v}_{orbit}\| \tag{11.17}$$

When the binary is not disrupted, the eccentricity e of the orbit after the SN explosion results directly from the angular momentum equation, which states that

$$A^2v_t^2=A_p^2v_{orb_{ps}}^2=A_p^2G\frac{M_{NS}+M_2}{A_{ps}} \tag{11.18}$$

with v_t the tangential (relative to the orbit) component of the velocity of the exploding star, $v_{orb_{ps}}$ the post-SN orbital velocity of the compact star and A_p the periastron distance of the new orbit, i.e.

$$A_p^2=A_{ps}^2(1-e^2) \tag{11.19}$$

It follows then that e can be determined from

$$e^2=1-m\left[2-m\left(1+2v_{rel}\cos\varphi\cos\theta+v_{rel}^2\right)\right]\left[\left(1+v_{rel}\cos\varphi\cos\theta\right)^2+\left(v_{rel}\sin\theta\right)^2\right] \tag{11.20}$$

Of particular importance is the value when the SN explosion is symmetric ($v_{rel}=0$), i.e.

$$e = \frac{M_1 - M_{NS}}{M_{NS} + M_2} \tag{11.21}$$

Using the qualitative evolutionary computations discussed in section 12, formula (11.21) predicts very small eccentricities ($e \leq 0.1$) in all high mass X-ray binaries (section 13.3). However some of them have very high e-values and this can be considered as indirect evidence that the SN explosion is asymmetric.

12. Evolutionary computations of massive close binaries

In figure 12.1 we show the qualitative evolutionary scenario of MCBs first introduced by Paczynski (1967) and completed later on by Van den Heuvel and Heise (1972). The primary is a massive star and possibly also the secondary. Both components first evolve as single stars, independent from each other. As the most massive star expands faster than its companion, it first reaches its critical radius and the RLOF starts. When the mass loser has lost most of its hydrogen rich envelope, an overall contraction phase sets in and the RLOF stops. The system then consists of a hydrogen deficient CHeB star and an OB-type star that has accreted part or all the mass lost by the loser; it was first suggested by Paczynski (1967) that this phase corresponds to the observed WR+OB binaries. Shortly after core helium exhaustion, the He star explodes. If the SN explosion does not disrupt the binary, the post SN-binary resembles an OB+cc binary; Van den Heuvel and Heise (1972) linked this phase to the observed massive X-ray binaries. As the OB-type mass gainer further evolves, it expands and its radius may also reach a critical value: a second RLOF starts but due to the extreme mass ratio of the binary, the low mass compact star will be dragged into the envelope of the OB star and it will start spiralling in (section 10.2.1.2). The result of this process may be a very close binary (period of a few hours) consisting of a helium star and a cc. At the end of the evolution of the helium star, a second SN occurs. In most of the cases, the system will be disrupted leaving behind two single runaway pulsars. There is, however, a non-zero probability that the binary remains bound, in this way forming a binary pulsar like PSR 1855-09 (section 13.4.1).

The evolution of MCBs has been studied quantitatively during the last 30 years by a large number of scientists and science groups, primarily in Europe, i.e. Warsaw (Paczynski, Ziolkowski, ...), Göttingen (Kippenhahn, Weigert, Thomas, Langer, ...), Amsterdam (Van den Heuvel, Heise, Savonije, Van der Linden...), Moscow (Tutukov, Yungelson, Massevitch, ...), Cambridge

(Eggleton, Podsiadlowski ...), Brussels (De Loore, De Grève, Vanbeveren, Packet, Doom, Hellings, ...).

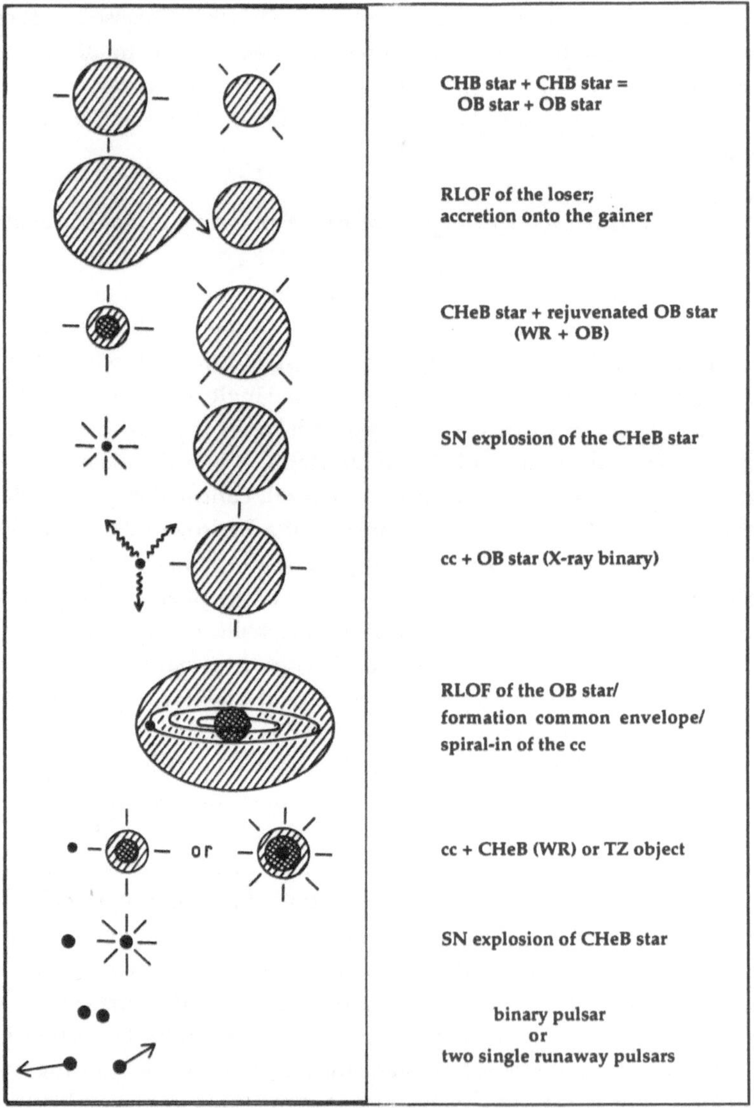

Figure 12.1: The qualitative MCB scenario as it was introduced by Paczynski (1967) and completed by Van den Heuvel and Heise (1972).

For a more detailed description of the computations we refer to the original papers which have been listed in a number of extended reviews: Paczynski (1971), Thomas (1977), Van den Heuvel (1978, 1980, 1993), De Loore (1980), Vanbeveren (1991, 1994, 1996)

In the present work we will summarize the results of these computations, update where necessary and give some general conclusions.

12.1. The evolution of the primary before and during its RLOF

Prior to the onset of the RLOF, both components in a MCB evolve as rotating single stars with the same mass. However, similarly to single stars, most of the evolutionary computations of MCBs use the structure equations holding for a non-rotating spherical symmetric plasma [equation set (3.25)].

When we compare the available observed rotational properties of MCB components with the v_e-distribution of massive single stars (section 2.4), accounting for the effect of synchronization in binaries (i.e. at least for evolved supergiant components, one expects that the equatorial velocity closely satisfies the relation $v_e = \frac{2\pi}{P}R$, with P the orbital period and R the stars' radius), it is reasonable to conclude that

> *components of non-evolved MCBs have rotational velocities which are (on average) smaller than or equal to those of single stars.*

We will first discuss evolutionary computations of MCB components with small convective core overshooting during CHB, similarly as in massive single stars, and neglecting rotation. In section 12.6 we summarize the expected effect of rotation on the evolutionary results of MCBs based on the computations of rotating single stars discussed in section 6.

12.1.1. MASSIVE PRIMARIES WITH INITIAL MASS ≤ 40-50 M_o: CASE B_r WITH $q > 0.2$

First bear in mind that case B_r means that the RLOF starts when the primary is a hydrogen shell burning star with an envelope which is mainly in radiative equilibrium (typical periods are given in figure 8.2; they range between a few days and ~1000 days).

The evolution of the primary can be studied from a phenomenological point of view without computing explicitly the evolution of the secondary component but using its mass in order to determine the critical Roche radius.

Galactic computations

Figure 12.2 gives typical evolutionary tracks of massive primaries in a case B_r binary with initial mass ratio q > 0.2, assuming conservative RLOF.

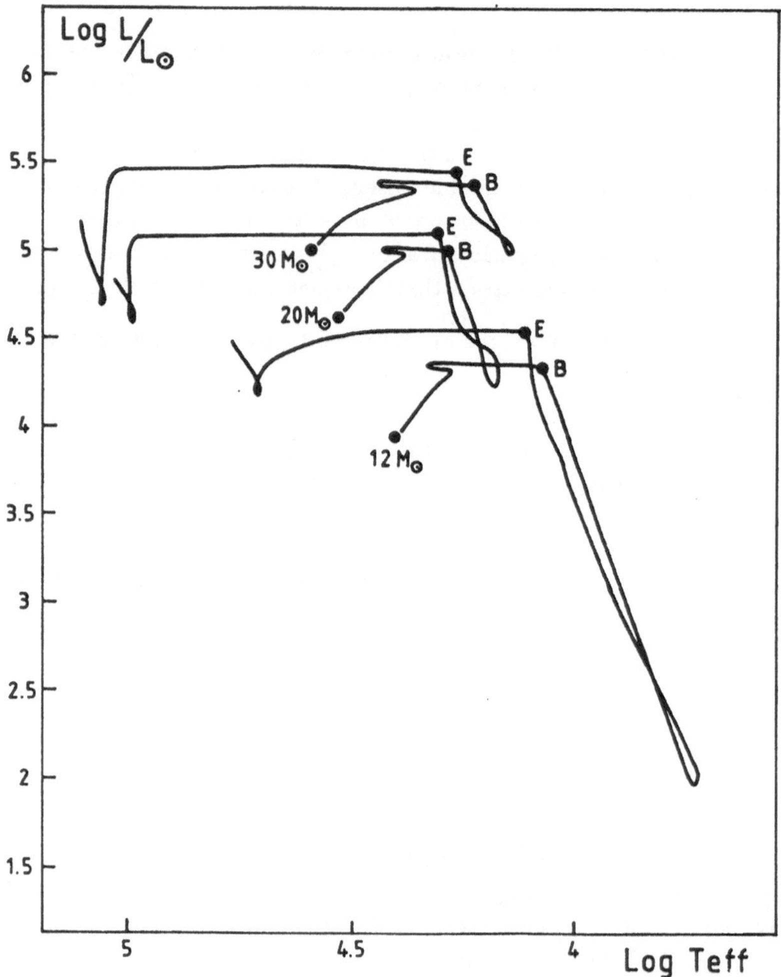

Figure 12.2: Evolutionary tracks of massive primaries in a case B_r binary assuming conservative RLOF; the points B and E indicate resp. the beginning and the end of the RLOF phase (from De Loore and Vanbeveren, 1995).

The timescale of the RLOF process and the behaviour of the primary during RLOF are determined by the tendency of a hydrogen shell burning star to expand (the rate at which the stellar radius increases during hydrogen shell burning). During the first ~10% of this phase, the star evolves on the Kelvin-Helmholtz timescale [equation (4.2)] and it loses mass at very high rates. Roughly, these rates satisfy

$$\dot{M} \approx \frac{M}{\tau_{th}} \approx 3 \cdot 10^{-8} \frac{RL}{M} \tag{12.1}$$

($\dot{M}$ in M_o/yr, M, R and L in solar units).

with maximum values of a few times 10^{-3} M_o/yr: the star is in the rapid phase of the RLOF. Due to the entropy term in the equation of energy [equation (3.16)] this rapid mass loss causes a rapid decrease of the stellar luminosity.

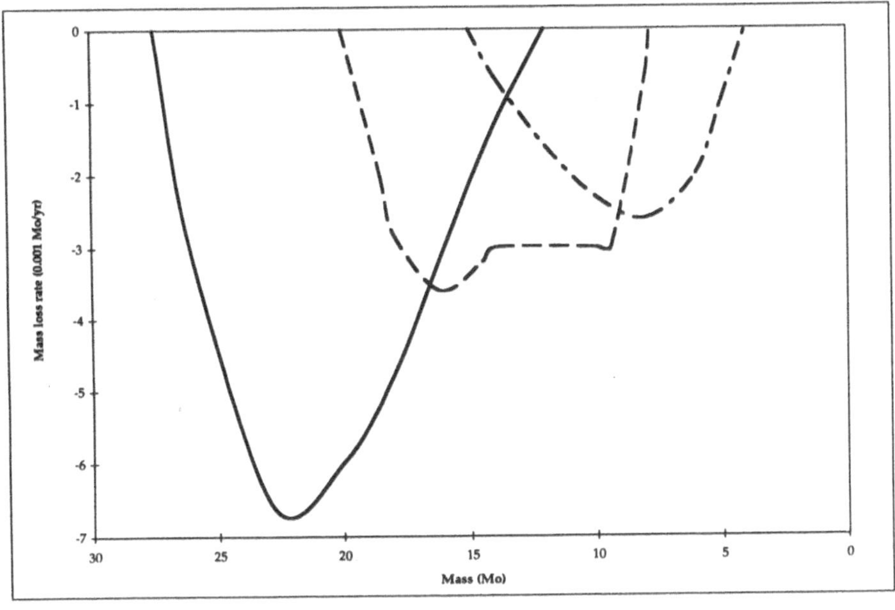

Figure 12.3: The mass loss rate during RLOF of a primary with initial mass = 30 M_o (full line), 20 M_o (dashed line) and 15 M_o (dashed dotted line) (from De Loore and Vanbeveren, 1995).

When the central temperature and density are high enough to start the CHeB process, the expansion rate of a hydrogen shell burning star decreases, i.e. the mass loss rate which is necessary to keep the star inside its Roche radius decreases: the star is in the slow phase of RLOF. Figure 12.3 shows the time variation of the mass loss rate in a few typical case B_r MCBs. During the slow phase the stellar luminosity increases again, and the star tries to regain a state of thermal equilibrium. This equilibrium is attained when most of its hydrogen rich layers have been removed. Detailed evolutionary computations show that, independently of the initial period or initial mass ratio of the binary, a massive hydrogen shell burning star loses its tendency to expand when He is burning in the core and when the atmospheric hydrogen abundance $X_{atm} \approx 0.2$-0.3. The star is almost in thermal equilibrium and starts contracting: the RLOF process has come to an end.

The final structure of a hydrogen shell burning star at the beginning of CHeB, that has reached thermal equilibrium, is largely independent from the details of the RLOF process. As a consequence

a fairly unique relation exists between the mass of the star prior to RLOF (M_b) and at the end of the RLOF (M_a), which is almost independent from the initial mass ratio and period of the binary and from the treatment of the RLOF process, in a conservative or non-conservative way.

It turns out that a different treatment of semi-convection has only a small effect on the evolutionary computations of massive primaries in interacting binaries.

Magellanic Cloud computations.

When we consider only models in which semi-convection is treated as a slow diffusion process, or where semi-convection is treated according to the Ledoux criterion, (both implying a rapid transition from the blue to the red part of the HRD during the hydrogen shell burning of a massive star, corresponding to observations of the MCs, section 5.1.8), it follows that except for relatively small quantitative differences, the evolutionary properties of MCBs for the Galaxy discussed above also apply to the Magellanic Clouds, i.e. they are largely independent of the initial metallicity.

Overall.

In figure 12.4 a comparison between the overall evolutionary behaviour of primaries of case B_r close binaries (small convective core overshooting) with galactic and SMC initial chemical composition is made, the LMC being intermediate. Since the evolution of a primary after RLOF is almost independent from the details during the RLOF phase, we have replaced the RLOF by a black box: for a given input, the output does only marginally depend on what really happens inside the box.

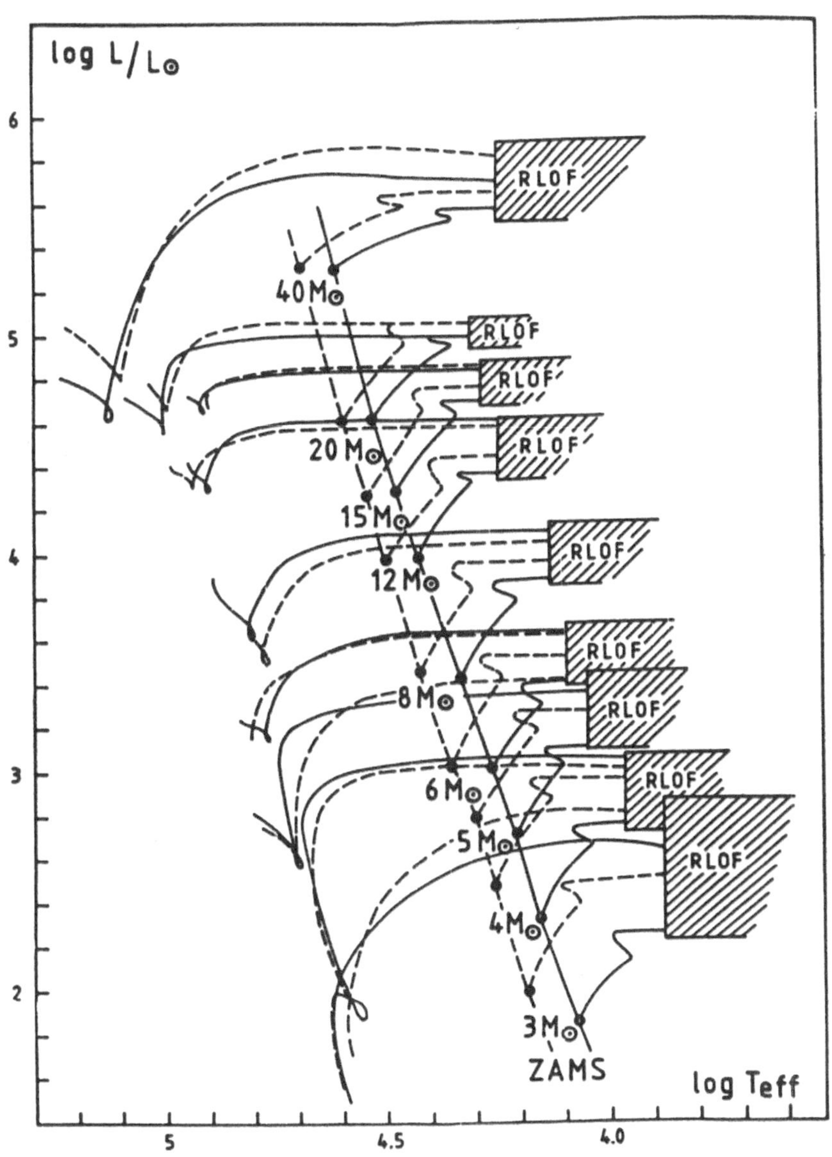

Figure 12.4: The overall evolutionary behaviour of primaries of case B_r close binaries with galactic (full lines) and SMC (dashed lines) initial chemical composition (De Loore and Vanbeveren, 1995).

The masses at the end of CHB and after RLOF are given in table 12.1.

TABLE 12.1: The mass of a primary in a massive case B_r MCB at the end of CHB (M_{eCHB}) and at the end of the RLOF (M_{eRLOF}) as a function of initial ZAMS mass for the Galaxy/SMC.

M_{ZAMS}	M_{eCHB}	M_{eRLOF}
9	9	2.2/2.2
12	12/12	3.3/3.6
15	15/15	4.5/4.5
20	19.5/20	6.7/7.9
25	24/25	9.3/11.1
30	28/29.5	12.3/16.3
40	36/39	18.6/25.3

The following relations between the pre-RLOF and post-RLOF mass can be deduced:

during CHB, SW mass loss according to equation (2.11) [the results are very similar when relation (2.10) is used], small convective core overshooting during CHB

for the Galaxy: $M_{post} = 0.093M_{pre}^{1.44}$

for the LMC: $M_{post} = 0.085M_{pre}^{1.52}$ (12.2)

for the SMC: $M_{post} = 0.048M_{pre}^{1.7}$

with the same SW mechanism as above but with large convective core overshooting (section 4.2), the relation for the Galaxy becomes

for the Galaxy: $M_{post} = 0.055M_{pre}^{1.79}$ (12.3)

A hydrogen stripped star with mass M at the onset of CHeB is in thermal equilibrium and has an equilibrium radius R_e which is very well described by the following relation

$$R_e = 0.62 \cdot 10^{-4} M^3 - 0.49 \cdot 10^{-2} M^2 + 0.18M + 0.17 \qquad (12.4)$$

R_e in R_o, M in M_o. Relation (12.4) hardly depends on the metallicity.

Cooking recipe

The mass relations (12.2) and (12.3) proposed above obviously depend on the treatment of overshooting and on the SW mass loss rate formalism. If one is only interested in the mass of binary components after RLOF (the further evolution can be deduced using the available evolutionary computations of hydrogen poor CHeB stars, section 12.2) it is not always necessary to compute the whole binary evolution. A case B_r RLOF in a MCB stops when the primary has lost most of its hydrogen rich layers, i.e. when $X_{atm} \sim 0.2 - 0.3$ and therefore the final mass of the RLOF remnant must be equal to the mass of the CHB core of the corresponding progenitor when the central hydrogen content $X_c \approx 0.2$-0.3. This means that whatever assumptions are used concerning convective core overshooting or SW mass loss, it is sufficient to calculate the CHB evolution in order to predict the whole evolution of the primary.

12.1.2. MASSIVE PRIMARIES WITH INITIAL MASS $\leq$ 40-50 M_o: CASE B_r WITH $q \leq 0.2$, CASE B_c AND CASE C

As may be seen from equation (4.2), the larger the radius of the hydrogen shell burning star the smaller the thermal time scale. This means that the larger the initial period of the binary, the more violent is the rapid phase of the RLOF and the larger the $\dot{M}$. This $\dot{M}$ becomes critical when the period is large enough for RLOF to start when the primary has a convective envelope: case B_c MCBs. In this case the mass loss due to RLOF will be a very fast process (section 8.5), i.e. the star will lose its mass on the dynamical time scale which is of the order of hours. The secondary is unable to accrete this mass and very soon after the onset of the RLOF a common envelope forms.

Due to the rapid shrinkage of the Roche radius during the RLOF process in a case B_r binary with initial mass ratio $q \leq 0.2$, one can expect that also in this case a common envelope will be formed.

Once a common envelope situation is encountered, the further evolution of the binary depends on the effect of viscosity when the secondary moves in this common envelope. Although detailed evolutionary computations have never been carried out where this process was followed till the end of it, a possible model to predict the evolution of a case B_r binary with $q \leq 0.2$ or a case B_c binary is the following (see also Tutukov and Yungelson, 1979):

when the primary starts filling its Roche lobe, very soon after the onset of the RLOF process, a common envelope phase sets in. The common

envelope will be removed from the binary due to the deposition of orbital energy of the secondary by viscous drag (section 10.2.1.2).
From the discussion in the previous subsection, it can reasonably be assumed that the mass loss process will stop when the primary has lost its tendency to expand, i.e. when most of its hydrogen rich layers have been removed and this means that relations (12.2) or (12.3) should also be valid here.
Knowing the mass of the common envelope that has to be removed, one can use equations (10.16), (10.18), (10.19) and (10.20) to calculate the period variation. Remind that the removal of the whole common envelope implies a very large orbital period decrease.

A particular case of case B_c/case C MCB evolution may occur in galactic MCBs with primary mass between ~15-20 M_o and 40-50 M_o. As shown in section 4 stars in this mass range may lose their hydrogen rich layers by SW mass loss during the RSG phase. Therefore, if the period of the binary is large enough that the RLOF would start during (or even after) this RSG mass loss phase, the RLOF mass loss rate may be largely reduced or the RLOF phase can be avoided. We will refer to this scenario as the RSG scenario of MCBs. Obviously due to SW mass loss during the RSG phase of the primary, the orbital period will increase as predicted by equation (10.6). To illustrate the effect, consider a 30 M_o + 15 M_o case B_c binary. When the hydrogen envelope of the primary is removed through the common envelope phase, using equations (10.16) and (10.18) with maximum efficiency ($\alpha = 1$) we obtain $P/P_o = 0.04$. When 50% of the hydrogen envelope of the primary is removed by SW before the star starts its RLOF and forms a common envelope [where again equations (10.16) and (10.18) apply] $P/P_o = 0.24$, and when all the hydrogen rich layers of the primary are removed by SW, $P/P_o = 2.8$. From the latter we conclude that the RSG scenario predicts the existence of binaries with periods of a few years consisting of a hydrogen stripped star and a normal OB type companion. Some WR+OB binaries with large period (section 13.2) may have followed this evolutionary path.

Note that when, owing to SW mass loss, a significant part of the hydrogen rich envelope of the primary has been removed, an ensuing RLOF may be much less violent and a situation can arise in which matter lost by the primary can be accreted by the secondary.

The mass range of the foregoing RSG scenario obviously depends on the SW mass loss rate during the RSG phase and in particular on how this mass loss formalism depends on the metallicity. If relation (2.12) also applies for RSGs, then the RSG scenario is much less efficient in the SMC.

Detailed computations of case C binaries do not exist, although most probably the primary not only loses all its hydrogen, but also part of the He rich layers outside the He burning shell.

12.1.3. MASSIVE PRIMARIES WITH INITIAL MASS ≤ 40-50 M_o: CASE A

To illustrate the evolution of MCBs evolving according to case A, figure 12.5 shows evolutionary tracks of a 15 M_o and a 30 M_o primary in a galactic case A binary with initial mass ratio q = 0.9 and with initial period = 2 days, evolving in a conservative way.

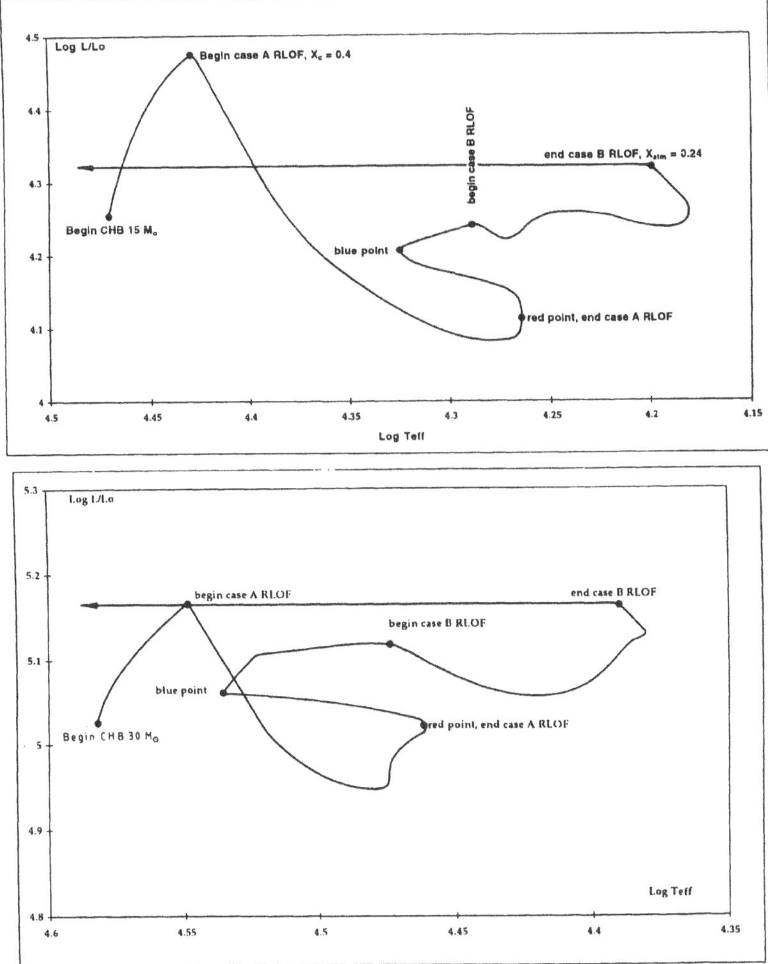

Figure 12.5: Evolutionary tracks of a 15 M_O and a 30 M_O primary in a galactic early case A binary.

The primaries start their RLOF early during their CHB phase (early case A). Table 12.2 summarizes the general evolution of these systems.

TABLE 12.2: The evolution of the early case A binaries 15 M_o + 13.5 M_o and 30 M_o + 27 M_o. The phase marked as Max $\dot{M}$ is the phase in which the mass loss rate during the RLOF is largest.

Phase	Time (10^6 yrs)	M_1 (M_o)	$\dot{M}$ 10^{-6} M_o/yr	Log T_{eff}	Log L/Lo	X_c
ZAMS	0	15	0	4.47	4.26	0.7
Begin RLOF Case A	8.75	15	0.086	4.43	4.48	0.39
Max $\dot{M}$	8.88	13	92.2	4.39	4.3	0.38
End RLOF Case A	16.6	6.6	0	4.27	4.16	0.01
Begin RLOF Case B	16.8	6.6	2.42	4.29	4.24	0
Max $\dot{M}$	16.9	4.3	86.8	4.19	4.24	0
End RLOF Case B	16.9	3.1	0	4.2	4.32	0
ZAMS	0	30	0	4.58	5.03	0.7
Begin RLOF Case A	3.27	29.5	1	4.55	5.17	0.46
Max $\dot{M}$	3.34	26.6	211	4.53	5.07	0.45
End RLOF Case A	7.42	14.1	0	4.46	5.02	0.03
Begin RLOF Case B	7.67	14	10	4.48	5.12	0
Max $\dot{M}$	7.67	12.1	825	4.41	5.06	0
End RLOF case B	7.68	9.52	0	4.39	5.16	0

Just as for case B_r binaries, the RLOF process is characterized by a rapid and a slow phase. The time scale of the rapid phase corresponds to the thermal time scale (as in the case B_r systems) which is considerably longer for CHB stars than for hydrogen shell burning stars. This means that the mass loss rates encountered during the rapid RLOF phase in case A binaries are considerably smaller than in corresponding case B_r binaries. Figure 12.6 shows the variation of the mass loss rate as a function of primary mass. This can be compared to figure 12.3 where the mass loss rates during the RLOF of case B_r binaries are given.

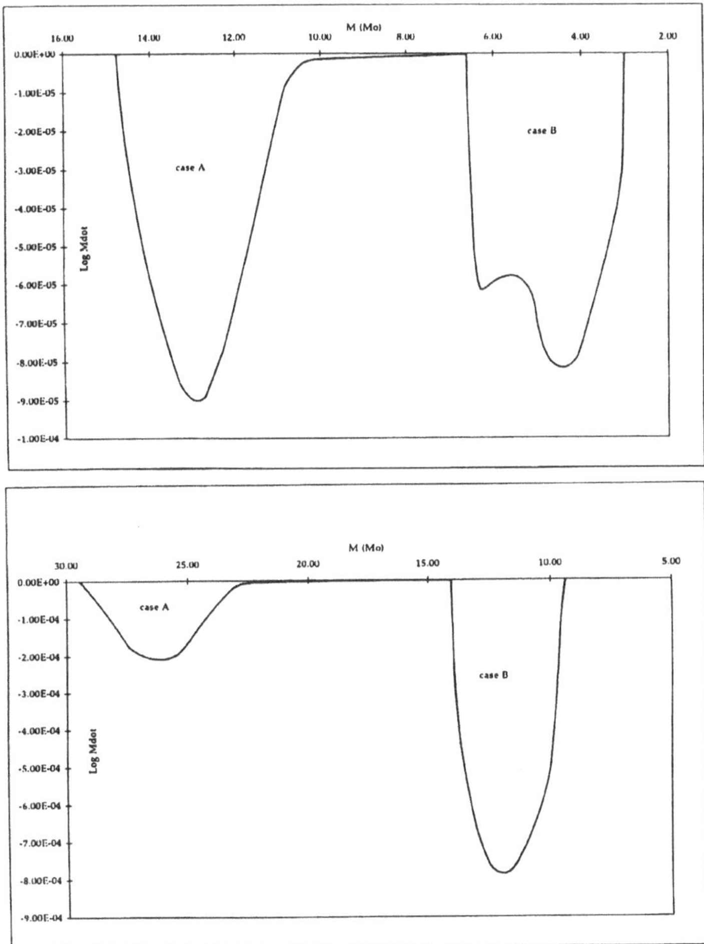

Figure 12.6: The variation of the mass loss rate during RLOF of the primary as a function of its mass for the early case A binaries 15 M_O + 13.5 M_O and 30 M_O + 27 M_O.

The rapid phase is followed by a slow phase which lasts almost the entire remaining CHB lifetime. By the end of this slow phase, layers which were in the hydrogen burning core at the beginning of CHB appear at the surface.

Since these layers all have CN equilibrium abundances, we expect to observe these stars as OBN stars. Note however that the atmospheric hydrogen abundance is still very close to normal.

At the end of CHB, the star contracts and thus the RLOF stops. When hydrogen starts burning in a shell just outside the newly formed helium core, the star starts expanding again, just like a normal single star, i.e. the primary overflows its Roche again and a case B_r type of evolution follows (one refers to this phase as case AB_r RLOF). However since the primary has already lost a lot of matter during the previous case A RLOF, the case AB_r RLOF is much less violent compared to a pure case B_r.

The RLOF stops when He starts burning in the core and when $X_{atm} \sim 0.2$-0.3, i.e. similarly to a case B_r, the remnant mass after case AB_r equals the mass of the hydrogen burning convective core of the progenitor when the central hydrogen content $X_c = 0.2$-0.3. As a consequence

- due to the large mass loss during RLOF in an early case A binary, the mass of the convective core of the mass loser during CHB is smaller than in a case B_r binary. This means that the remnant mass after case AB_r RLOF in a binary that experienced early case A evolution, is smaller than the remnant mass of the same primary but in a pure case B binary,

- when the period is large enough that case A RLOF starts when the central hydrogen content of the primary is smaller than 0.2-0.3 (we propose to call this a late case A MCB), its final mass after the case AB_r RLOF equals the final remnant mass after a pure case B_r RLOF.

Obviously the evolutionary behaviour of a primary in a case A (case AB) binary does not depend on the way semi-convection is treated and therefore does not depend on the metallicity, i.e. the evolution of a case A (case AB) binary in the Magellanic Clouds should be completely similar to the evolution in the Galaxy.

12.1.4. VERY MASSIVE PRIMARIES

When its initial mass is larger than 40-50 M_o, the star is subject to a large LBV like SW mass loss. As a consequence of this SW, the matter lost by a primary during its RLOF could obviously be largely reduced, or, when the SW is large enough, the RLOF will not even start.

When most of the hydrogen rich layers have been removed by SW and RLOF (if the latter occurs) the primary becomes a WR star; SW mass loss continues at the appropriate rates [equation (3.24)]. In table 12.3 we summarize some evolutionary properties of the orbital elements of galactic binaries where this type of SW dominated evolution applies. We call ΔM_{SW} the total mass lost by the primary due to SW, and ΔM_{RLOF} the mass lost due to RLOF (i.e. $\Delta M_{SW} + \Delta M_{RLOF}$ = the total mass of the hydrogen rich envelope of the primary). The period variation caused by the spherical symmetric SW is given by equation (10.6). When RLOF occurs, the loss/transfer of ΔM_{RLOF} implies a period variation predicted by the formalism outlined in section 10.2; for the 60 M_o + 5 M_o binary, we used the spiral-in prescription [equation (10.16) and (10.19), $\alpha = 1$]. Note that, even when the spiral-in process is responsible for the removal from the 60 M_o star of only a small amount of mass, the orbital period reduction is very significant. In section 14.6 we will show that this reduction is necessary in order to explain the low mass X-ray binaries with a BH component, i.e. low mass X-ray binaries with a BH component provide indirect evidence that LBV mass loss is not large enough for a very massive primary to escape the RLOF.

TABLE 12.3: The period evolution in a binary with a 60 M_o primary where due a LBV SW phase the star loses 10 M_o followed by the RLOF where the star loses an additional 6 M_o; the results are compared to the case where the whole 16 M_o are lost by LBV SW; P_o is the initial period of the binary, bWR = beginning of the WR phase, bWC = beginning of the WC phase, eCHeB = end CHeB; the corresponding masses are given in table 12.5. $\Delta M_{SW\text{-}O}$, $\Delta M_{SW\text{-}LBV}$, ΔM_{RLOF} are resp. the total mass lost by the primary during the O-type phase, the LBV phase and during the RLOF/spiral-in phase.

System	$\Delta M_{SW\text{-}O}$	$\Delta M_{SW\text{-}LBV}$	ΔM_{RLOF}		$(P/P_o)_{bWR}$	$(P/P_o)_{bWC}$	$(P/P_o)_{eCHeB}$
60+30	12	16	0		2.2	3	5.2
	12	10	6	β=0.5	1.1	1.5	2.5
				β=1	1.8	2.4	4.1
60+10	12	16	0		2.8	4.5	11.5
	12	10	6	β=0	0.2	0.3	0.8
60+5	12	16	0		3.1	5.4	17.4
	12	10	6	spiral-in	0.03	0.05	0.2

12.1.5. THE BINARY AFTER RLOF: A CHeB+OB BINARY OR A MERGER?

In order to decide upon the remnant system after the RLOF, one may proceed as follows:

- knowing the mass that will be lost by the primary before it becomes a hydrogen stripped CHeB star, the period (orbital separation) of the binary after the RLOF, and thus the final Roche radii of the binary, can be estimated:
    - for binaries with mass ratio $q \leq 0.2$: the spiral-in prescription, and we use equations (10.16), (10.19) and (10.20) when the primary is an OBA type star with the appropriate value of λ (case B_r), or equations (10.16), (10.18) and (10.20) when the primary has a convective envelope (case B_c)
    - for case A/case B_r binaries evolving non-conservatively: possible mass loss from the system through L_2, and we use equation (10.14) or (10.15),
    - for case B_c binaries with $q > 0.2$: the common envelope prescription, and we use equations (10.16), (10.18) and (10.20).
- when the final Roche radii are larger than the equilibrium radii of both stars [for the hydrogen stripped primary, one can use equation (12.4)], we are left with a binary consisting of a hydrogen deficient CHeB star and an OB component,
- when one of the Roche radii is smaller than the equilibrium radius of the corresponding component, it is plausible that both stars merged before the end of the RLOF.

TABLE 12.4: The minimum period P_o (in days) for a binary with mass ratio $q \leq 0.2$ and primary mass M_{10} (in M_o) in order to avoid merging for resp. efficiency 1 and 0.5; the period P (in days) after spiral-in is given as well.

M_{10}	q	$P_{o,min}$	P
10	0.2	200/550	0.3
	0.1	600/900	0.35
20	0.2	180/450	0.38
	0.1	350/850	0.35
30	0.2	200/500	0.65
	0.1	320/750	0.52

It can readily be checked that case B_c binaries rarely merge unless the efficiency factor α in equation (10.16) is quite small. Merging can happen in case B_r binaries evolving in a non-conservative way when mass leaves the binary through L_2, and it is common among binaries with mass ratio $q \leq 0.2$. This is illustrated in table 12.4 where the minimum initial period required to avoid merging is given. Only the binaries with a very large initial period may survive and leave as remnant a very short period system consisting of a hydrogen stripped CHeB star and a low mass companion.

12.2. The evolution of the primary after RLOF: CHeB+[OB or low mass companion] binaries

It is obvious that when a case C binary survives the common envelope phase, the remaining lifetime between the end of RLOF phase and the SN explosion is very short. These binaries will therefore not contribute to the population of post-RLOF and pre-SN MCBs (the WR+OB binaries).

When merging can be avoided, the final system after the RLOF of a case A or a case B MCB is a binary with an OB type or a low mass companion and a component at the beginning of CHeB with X_{atm} ~ 0.2-0.3: a CHeB+[OB or low mass companion] binary. The further evolution of the CHeB binary component depends on the SW mass loss. When the CHeB star is a WR star, we can use equation (3.24). However, when does a CHeB MCB component resembles a WR star? WR stars consist of a core with a surface temperature of about 100000K and a dense stellar mantle which redistributes the energy and emits it at considerably lower temperatures (figure 3.2). It is the stellar mantle which is responsible for the typical emission line spectrum of the WR. Therefore, the following idea seems plausible: a post-RLOF CHeB star will lose its WR character when the SW mass loss has become too small to produce a mantle sufficiently thick and extended, to form a typical WR spectrum.

We account for the following:

a. the majority of the observed WR stars seem to have luminosities Log $L/L_o \geq$ 4.5 (figure 2.10); from the theoretically predicted mass-luminosity relation of massive hydrogen poor CHeB stars (section 12.2.1), this means that the majority of the observed WR stars must have masses exceeding 5 M_o,

b. the semi-empirical mass of most of the observed WR stars in galactic binaries (table 14.1) is larger than 8-9 M_o.

As a working hypothesis, we will adopt the statement:

- *a hydrogen deficient CHeB star will be observed as a WR star when its mass is larger than 5 M_o.*

A possible exception: the WR like component in the binary V Sagittae (section 13.2) indicating that $M \geq 5\ M_o$ is not a condition sine qua non for a CHeB star to be a WR-type.

We discuss separately the evolution of CHeB+OB binaries with CHeB mass larger and smaller than $5\ M_o$.

12.2.1. THE MASS OF THE CHeB COMPONENT $\geq 5\ M_o$.

After the RLOF phase, the remnant of the star very rapidly regains its thermal equilibrium, it moves to the blue part of the HR diagram ($\log T_{eff} \sim 5$) (figure 12.4) and stays there for most of its CHeB lifetime. Its structure is very similar to the structure of single hydrogen deficient CHeB stars (figure 4.4).

TABLE 12.5: The masses M_{bWNL} at the end of the RLOF = beginning of the WNL phase (= WN with atmospheric hydrogen), the masses M_{bWNE} at the beginning of the WNE phase (= WN without atmospheric hydrogen), the masses M_{bWC} at the beginning of the WC phase, the masses M_{eCHeB} at the end of CHeB and the corresponding final CO-core masses for the Galaxy/SMC, for two stellar wind scenarios during CHeB (see also the text):
scenario 1: WR like SW stops when the mass of the CHeB star $\leq 5\ M_o$
scenario 2: WR SW continues without mass limitation.

$M1_o$	M_{bWNL}	M_{bWNE}	M_{bWC}	M_{eCHeB}		M_{CO}	
				scenario 1	scenario 2	scenario 1	scenario 2
20	6.7/7.9	5.4/5.35	-	5/5	4/4	2.6	2.4
25	9.3/11.1	7.6/7.4	5.2/5.3	5/5	4.2/4.3	2.8	2.6
30	12.3/16.3	9.7/10.1	7.1/7.4	5/5	4.3/4.4	3.0	2.7
40	18.6/25.3	14.6/15.1	11.7/12.4	5/5	4.6/5.0	3.8	3.5/3.8
60	32	28	23	10.6	10.6	7.2	

Due to the SW mass loss, first the hydrogen layers which remained after the RLOF are removed. Thus the hydrogen burning shell also dies out and the luminosity of the star decreases. The atmosphere of the star is composed mainly of helium whereas the CNO elements have their equilibrium value (section 3.3.1). Further SW mass loss removes these helium layers and when this removal is efficient enough, layers where He burning occurred appear at the surface. In these layers, N is lacking but C (^{12}C) and O (^{16}O) are very abundant (corresponding to WC stars). Also ^{22}Ne may be two orders of magnitude larger than solar in this type of star. Table 12.5 shows the masses at the end of the RLOF = beginning of the WNL phase (= WN with atmospheric hydrogen), the masses at the beginning of the WNE phase (= WN without hydrogen), the masses at the beginning of the WC phase.

Of particular importance is the existence of a very tight mass-luminosity relation for massive hydrogen poor CHeB remnants after RLOF losing matter by SW at rates typical for WR stars. Such a M-L relation has been derived by Vanbeveren and Packet (1979). It has been studied over again by Maeder (1983), Langer (1989), Vanbeveren (1991), Schaerer and Maeder (1992). All proposed relations closely match the original one, although different assumptions were made concerning a number of uncertainties during CHeB (e.g. different SW mass loss rate formalisms, different CO reaction rates, etc.). Since massive hydrogen deficient CHeB stars are stars in thermal equilibrium, their M-L relations are largely independent from the evolution of their progenitors, i.e. the way the hydrogen rich envelope was removed from the progenitor. This explains why the M-L relation of hydrogen deficient CHeB binary components is very similar to that for single stars where the mass was lost by SW during the LBV phase and/or the RSG phase (section 4).

As was done in De Loore and Vanbeveren (1994), we consider separately CHeB remnants after RLOF where the atmospheric hydrogen abundance X_{atm} is different from zero (corresponding to WNL types), those where $X_{atm} = 0$ but where the products of the He burning do not yet show up at the surface (corresponding to WNE types), and those where He burning products appear in the surface layers (corresponding to WC types). For all the evolutionary models that we have computed (with galactic, LMC and SMC abundances), we consider the mass and the luminosity at the beginning and at the end of the WNL (resp. WNE and WC) phase. Complementary to the results presented by De Loore and Vanbeveren (1994), we also add here a few test calculations of a binary with a 60 M_o primary evolving according to the LBV scenario. The M-L (in solar units) relations hold for WR masses between 5 M_o and ~30 M_o:

$$\begin{array}{ll} \text{WNL} & \log M = -1.691 + 0.524 \cdot \log L \\ \text{WNE} & \log M = -1.658 + 0.520 \cdot \log L \\ \text{WC} & \log M = -1.822 + 0.542 \cdot \log L \end{array} \qquad (12.5)$$

We have identified the CHeB phase of MCBs with the WR phase. However the observed T_{eff} values of WR stars are significantly smaller than the 100000K predicted by evolution. As argued in section 3.7, this is due to the treatment of the atmosphere in evolutionary codes. How to proceed in order to obtain a T_{eff} which is comparable to the observed values, has been outlined in section 3.7 as well.

12.2.2. THE MASS OF THE CHeB COMPONENT < 5 M_o

Observations of hydrogen deficient CHeB binary components with mass < 5 M_o are scarce. This due to the fact that their $T_{eff} \approx 100000$ K, whereas the SW is probably too small to redistribute the emitted energy towards lower temperatures (V Sagittae may be an exception). Therefore, if the companion is an OB type star, the hot component will be hard to observe. The lack of direct observations therefore forces us to consider two possibilities:

a. <u>the SW in a CHeB binary component with mass < 5 M_o is too small to affect the further evolution</u>

In this case all primaries with initial mass between 15 M_o and 40 M_o have a mass ≈ 5 M_o at the end of their CHeB (table 12.5, scenario 1).

Primaries with initial mass smaller than 15 M_o develop hydrogen deficient CHeB remnants after RLOF with mass < 5 M_o. We use the detailed evolutionary results of Habets (1985, 1986) who computed the evolution of helium stars with $2 \leq M/M_o \leq 4$ up to neon ignition, assuming that the SW mass loss during CHeB is small and does not significantly affect the further evolution of the star. Of particular importance is the variation of the radius as a function of core mass (figure 12.7). It can be concluded that

- the CHeB stars with $2 \leq M/M_o < 2.9$ (corresponding to primaries in a case B/late case A binary with initial mass < 15 M_o) expand significantly during the He shell burning phase, after CHeB. It can be expected that they will overflow their critical Roche lobe for a second time: the process is known as case BB RLOF. As a consequence of case BB RLOF a star will lose the remaining hydrogen and most of the helium layers (those on top of the He burning shell).

The mass loss rate during the case BB phase is obviously much lower than during the previous one, so that most probably case BB RLOF is conservative. Since in that case the period varies according to equation (10.9), even the small mass variation during the case BB RLOF results into a very large period increase as illustrated by the following example:

consider a 10 M_o + 9 M_o case B_r binary; after the first conservative RLOF, we obtain a 2.3 M_o + 16.7 M_o system and after the second case BB RLOF a 1.7 M_o + 17.3 M_o system; during this second RLOF, the period has increased by a factor 2.2.

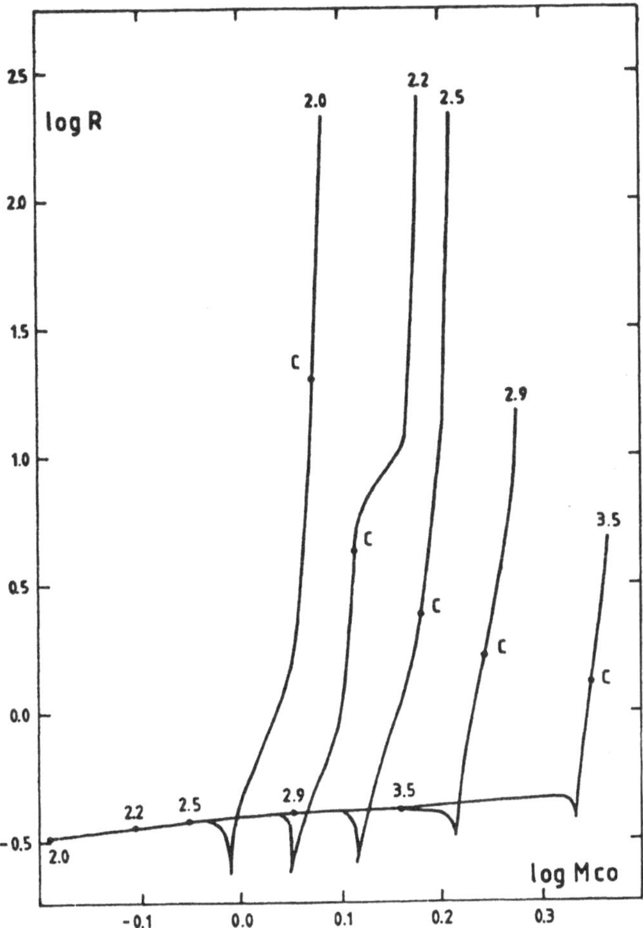

Figure 12.7: The evolution of the radius of helium stars with mass between 2 M_o and 4 M_o as a function of the CO core mass M_{CO} (from Habets, 1986).

b. <u>SW mass loss of CHeB binary components with mass < 5 M_o is large enough to affect the further evolution</u>

In table 12.5 we also give the mass at the end of CHeB if the WR SW mass loss rate formalism [equation (3.24)] is applied without mass restriction. For primaries with initial mass ≥ 20 M_o, the results are almost unaffected.

Since the CHeB lifetime of a hydrogen less CHeB remnant with mass < 5 M_o (corresponding to a binary component with initial mass ≤ 15 M_o) is of the order of 6.10^5-10^6 years, a SW mass loss rate of $1\text{-}2.10^{-6}$ M_o/yr is large enough to affect the evolution of the star. As an example:

> a 15 M_o primary in a case B_r MCB, has a remnant mass after RLOF of 4.5 M_o; using a (constant) SW of 10^{-6} M_o/yr during CHeB, the star ends its CHeB phase with a mass ≈ 3.2 M_o, i.e. a 2.6 CO core and a 0.6 M_o He envelope (no hydrogen left). One can therefore expect that the explosion of the star will be of type Ib.

An important effect of the SW mass loss is a possible suppression of the case BB type of evolution discussed above. The orbital period variation then obviously satisfies relation 10.6. Since the total mass loss is small, the period variation is quite small as well, contrary to the case BB scenario. To illustrate: with the same example as considered above, the period increases by a factor ~ 1.06 (compared to the factor 2.2 with the case BB scenario).

12.2.3. THE FINAL FATE OF PRIMARIES OF MCBs

The final mass of the CO core of the primaries with initial mass ≥ 20 M_o is given in table 12.5 as well. We conclude

- Primaries with initial mass ≤ 40 M_o develop a CO core with mass ≤ 3.8 M_o. Using the computations of Woosley (1986), a CO core mass of 3.8 M_o corresponds to a FeNi core mass of ~1.8 M_o. We therefore conclude that

  evolutionary computations of massive close binaries with small convective core overshooting show that primaries with initial mass up to 40 M_o may end their life as a NS (accompanied by a SN explosion)

  Note the difference with single stars (section 4.1.2).

- Primaries with initial mass > 40 M_o have quite massive CO cores. The corresponding mass of the iron core is large enough to make them potential candidates for BH formation.

For the primaries with initial mass < 15 M_o, one can again use the results of Habets (1986); they allow the conclusion that

- in all helium stars with mass > 2.2 M_o, the heavy element core becomes larger than the Chandrasekhar mass; these stars are therefore expected to run into a SN event leaving behind a compact star. A 2.2 M_o CHeB remnant after RLOF results from a 10 M_o primary in a case B or late case A binary if convective core overshooting during CHB is negligible. The mass limit increases only slightly if scenario 2 applies for the evolution of CHeB remnants after RLOF, i.e. CHeB remnants after RLOF lose mass by SW without mass limitation (section 12.2.2),

12.3. The evolution of the secondary in a MCB

Knowing typical mass loss rates due to RLOF of a primary in a MCB, we can study from a phenomenological point of view the behaviour of a star when it gains mass at these rates. As discussed in section 9.3, we need to distinguish the following cases:

case 1a: accretion with the standard model, very fast diffusion process in semi-convective layers,

case 1b: accretion with the standard model, very slow diffusion process in semi-convective layers

case 2: accretion induced full mixing model.

Accretion case 1a.

In figure 12.8 we show the evolution of an 18 M_o CHB star accreting 8 M_o. We compare the situation in which the mass gain rate is smaller than the critical rate given by equation (9.8) with the one in which it is larger.

We conclude:

- when the accretion time scale is larger than the thermal time scale of the CHB star, the mass gainer always occupies a place in the HR diagram which is very close to the HRD position of a normal CHB single star with the same mass and chemical composition; after the accretion phase, the further evolution of the star is almost entirely the same as the evolution of a normal single star with the same mass,

- when the accretion time scale is smaller than the thermal time scale, the mass gainer becomes overluminous for its mass and its radius increases significantly. When this happens in a binary, also the mass gainer may reach its critical Roche radius and a contact system is formed (contact binaries will be considered in section 12.4).
  At the end of the rapid accretion phase, the star regains its thermal equilibrium very fast and moves in the HRD to the position of a normal

single star with the same mass and chemical composition. Again its further evolution is normal,

- after accretion, the stars are rejuvenated. This means that in the HRD, they are on a time-isochrone with a time scale smaller than their real lifetime.

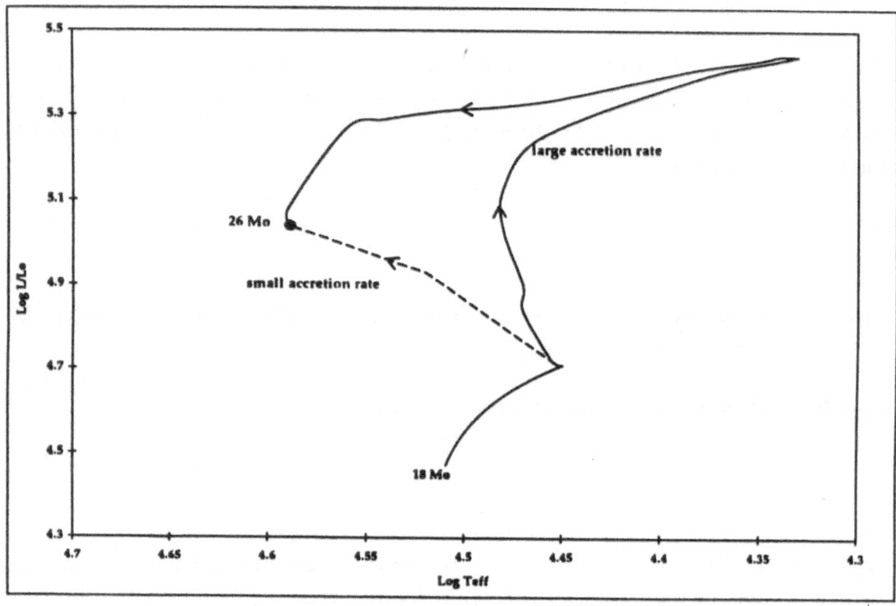

Figure 12.8: The evolution of a 18 M_0 CHB star accreting 8 M_0 at a rate smaller (dashed track) and larger (full track) than the critical accretion rate.

When the mass ratio of a binary is close to one, the RLOF starts when the secondary is also a hydrogen shell burning star. In figure 12.9 we show the evolution of a 12 M_0 hydrogen shell burning star accreting 8 M_0 and we compare the track with that of a 12 M_0 CHB mass gainer.

We conclude:

- since the He core is fixed and somehow protected by the hydrogen burning shell, accretion does not enlarge the core but instead produces an extended fully convective region on top of the hydrogen burning shell. After the accretion phase the star is (slightly) underluminous for its mass,

- the most important difference when compared to a CHB mass gainer is the further evolution of the gainer after the accretion process. The star has a He core which is small for its mass whereas due to the large convection region on top of the burning shell, the shell has a lot of fuel at its disposal and a rapid expansion of the star does not occur during the hydrogen shell burning and CHeB phase. As a consequence the star stays in the blue part of the HRD and does not make an excursion towards the RSG region. Although we have not continued the evolutionary computations after CHeB up to the SN explosion, we can reasonably accept that the star will remain a blue star till the end.

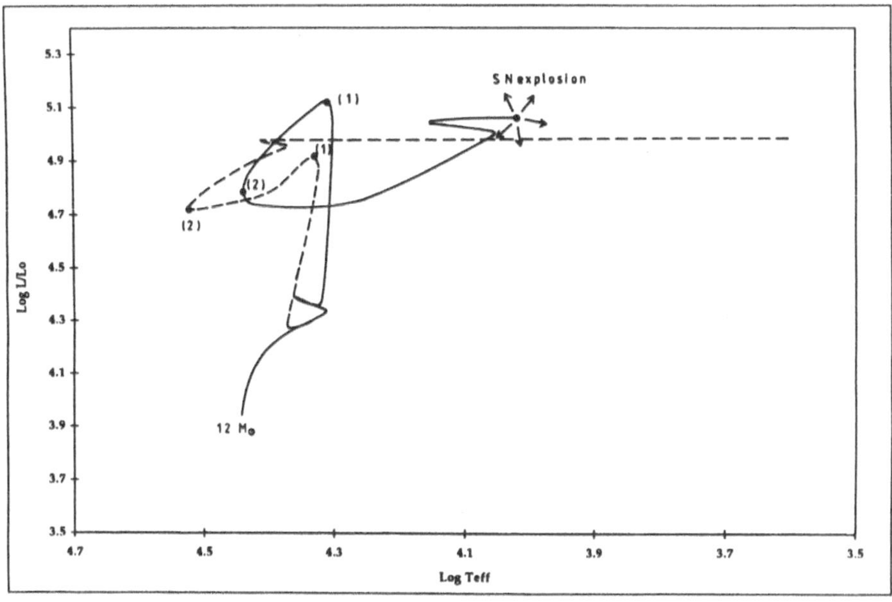

Figure 12.9: The evolution of a 12 M_O star accreting 8 M_O when it is a CHB star (dashed track) and when it is a hydrogen shell burning star (full track); (1) = the end of the accretion phase: the star starts contracting and rapidly regains its thermal equilibrium configuration; (2) = the end of the contraction phase when the star has reached thermal equilibrium and occupies a place in the HR diagram that is normal for its mass and chemical composition. The evolution is continued till the end of CHeB.

Accretion case 1b.

When the diffusion in semi-convective layers is very slow, the formation of a μ-barrier on top of the convective core right from the beginning of the accretion phase prevents a rapid core mass increase. Therefore even when the gainer is a CHB star, a situation can be encountered which is similar to the one described above when the accretion case 1a applies and when the gainer is a hydrogen shell burning star. Detailed computations were performed by Braun and Langer (1995). We conclude:

- after accretion, the stars are underluminous and are less rejuvenated compared to case 1a. Even when the gainer was a CHB star at the onset of mass transfer, it can remain a blue star during its entire further evolution after accretion. Furthermore, although the stellar mass increased considerably, its core mass remained almost the same as before the accretion process. Its further lifetime (and certainly its CHeB lifetime) may therefore be considerably larger than that of a normal single star with the same mass. This scenario explains (at least partly) the large number of AB-type supergiants observed in the BHG in the HR diagram.

Accretion case 2.

As discussed in section 9.3.2, even when the accretion rate is large, the large expansion of the star, which is typical for accretion case 1, is suppressed. Obviously, when the whole star is mixed, the way semi-convection is treated is of no importance.

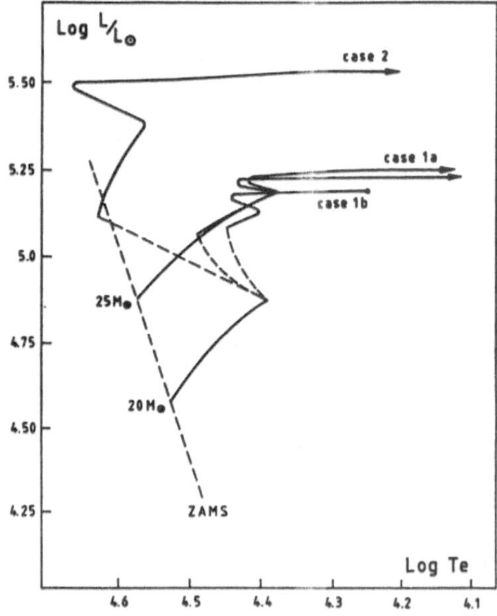

Figure 12.10: Evolutionary tracks of a 20 M_0 star accreting 5 M_0 when it was at the end of CHB, with the three accretion models discussed in the text. The tracks are compared to the evolution of a 25 M_0 single star.

In figure 12.10 evolutionary tracks are shown of a 20 M_o star accreting 5 M_o when it was at the end of CHB. We compare the three accretion cases; the evolution of a normal 25 M_o single star is added as well. In order not to overload the figure, we have replaced the tracks during the accretion phase by a smooth line joining the HRD positions before and after the mass gain phase.

We conclude:

- as expected, the rejuvenation of a mass gainer is very pronounced with the accretion induced mixing model.
- when accretion started during the second half of the CHB phase of the gainer, the post-accretion star is significantly overluminous and remains overluminous during its further evolution. It is obvious that for these mass gainers, the mass calibration of table 5.1 is not valid. Given the spectral type and luminosity class of the gainer, the mass can be 20% - 30% lower than for normal stars of the same class.

The variation of the chemical abundance in the surface of a mass gainer.

- During the second part of the RLOF phase, the primary loses CNO processed matter and thus mass transfer due to RLOF and thermohaline mixing (section 9.3.1) produces stars with surface layers which are significantly N enhanced (more than a factor 2) and CO depleted.
  If, due to accretion, the star is mixed completely (the accretion induced mixing model, section 9.3.2), the effect is obviously much more pronounced.
- The surface hydrogen abundance in mass gainers is hardly affected in the standard case. However with the accretion induced full mixing process, the gainer can have a significantly reduced hydrogen abundance in its outer layers.

12.4. The formation and evolution of contact binaries

The evolutionary computations of MCBs performed in Brussels reveal that

> *if, during mass transfer, accretion is treated in the standard way, the mass gainers in all MCBs expand and most (all) of them reach their own Roche limit during the fast phase of the RLOF, i.e. the majority (all) of the MCBs evolve into a contact system.*

Pols (1994) made a detailed study of the evolution of case A and case B MCBs and concluded that only binaries with large mass ratio may avoid contact if accretion is treated in the standard way. His conclusion was based on evolutionary computations performed with old opacity tables (Cox and Stewart, 1969). With new OPAL tables, also systems with large mass ratio become contact binaries.

Remember (section 9.2) that the star may spin-up during mass transfer. Furthermore, matter may settle down onto the gainer with an entropy that is significantly smaller than the entropy of the gainer's envelope. This means that large scale mixing of the gainer cannot be excluded. In this case, the expansion is not that rapid and contact may be avoided.

The further evolution of contact binaries is not straightforward. One needs an evolutionary code where both components are followed simultaneously. For the RLOF, the following model is a possibility:

- start with a conservative RLOF where the mass loss rate of the primary (and thus the mass gain rate of the secondary) is determined by imposing the condition that the primary radius must be equal to the Roche radius; remember that although the assumptions of circularization and/or synchronization may not be 100 % valid, this approximation will give very reasonable mass loss rate values,

- when due to mass transfer/accretion, both components come into contact, determine the mass loss rate of the primary (= mass gain rate of the secondary) by imposing the condition that both components are filling a common equipotential surface. Since evolutionary models are spherically symmetric whereas common equipotentials are not, approximations have to be used. The best approximation has been proposed by Packet (1988), i.e. the mass loss/gain rate is determined by imposing the condition

$$M_1\left(\frac{1}{R_{1_c}}-\frac{1}{R_1}\right)=M_2\left(\frac{1}{R_{2_c}}-\frac{1}{R_2}\right) \qquad (12.6)$$

  where M_1 (resp. M_2), R_1 (resp. R_2) and R_{1_c} (resp. R_{2_c}) are the mass, the radius and the critical Roche radius of the mass loser (resp. mass gainer),

- when both stars overflow the equipotential surface defined by L_2, calculate the mass loss rate of the primary by imposing the condition that the star must remain inside this surface, compute the mass gain rate by imposing the condition that the mass gainer must remain inside this surface; the difference then determines the amount of mass that has to be removed from the binary through L_2 resulting in a period variation given by equation (10.14) [or (10.15)].

It is clear that these computations are very time-consuming and depend on the initial mass ratio and period of the binary. A few test computations have been performed in Brussels, but a systematic investigation is still lacking. However, from these computations we can already conclude that when the accretion process is treated with the standard model (section 9.3.1), all MCBs with primary mass larger than ~ 20 M_o evolve into deep contact during RLOF and will lose mass through L_2. Our computations reveal that this deep contact is avoided when during the rapid phase of the RLOF 50 % of the mass lost by the primary during its RLOF also leaves the binary. During the slow phase, a conservative assumption is perfectly possibly.

We have also studied the early case B_r evolution of MCBs with primary mass smaller than 20 M_o and mass ratio q close to unity. These systems seem to be able to survive a conservative RLOF.

12.5. The evolution of a MCB after the collapse of the core of the primary

The evolution of the primary comes to an end when its core is composed mainly of iron and nickel, and nuclear processing ceases.

When the mass of the FeNi core is larger than some critical value, it is very probable that the whole star will collapse to form a massive BH (BH). This critical value depends on the details of the equation of state of matter where electrons and nuclei are degenerate, and this is still uncertain (section 3.2.6). Furthermore, also the post-CHeB evolution of a massive star contains uncertainties so that translating the critical core mass value into a critical value for the initial mass of a star is uncertain. The current 'theoretical' idea promotes a minimum initial mass of ~40-50 M_o above which binary components may form BHs (section 12.2.3). As will be discussed in section 14.4.1, the observations of massive X-ray binaries reveal that NSs are remnant of binary components with initial mass up to at least 40 M_o, which is not inconsistent with the theoretical value.

If the SN explosion does not happen when a BH is formed, one can expect that the majority of binaries with primary mass larger than ~40-50 M_o will first evolve through a LBV mass loss phase (possibly followed by a RLOF but, due to the LBV SW, with a largely reduced mass loss/mass transfer) and then form an OB+BH binary.

When the initial mass of the primary is smaller than ~40-50 M_o, its final FeNi core collapses into a NS accompanied by a SN explosion and the ejection of the mass layers outside the core. The event could disrupt the binary. It is clear that in this case, the further evolution of the OB component is that of a single star but, due to accretion, with a chemical composition that may differ from the chemical composition of a normal single star. We distinguish the following possibilities:

- the OB type mass gainer has a mass ≥ 40-50 M_o. We may expect that in this case the further evolution of the gainer will be similar to the evolution of a single star with the same mass and thus will be governed by the LBV/RSG mass loss processes,

- the OB type mass gainer has a mass < 40-50 M_o. When the initial mass ratio of the binary is not too close to one, the further evolution of the gainer after the mass transfer phase is comparable to the evolution of a single star with the same mass. However when the initial mass ratio is close to one (the exact value depends on the treatment of semi-convection, section 12.3) the gainer can remain a blue star during its entire life and thus populate the BHG. These stars will then explode as blue supergiants, producing events like SN 1987A.

When the binary is not disrupted as a consequence of the SN explosion, a OB+NS binary remains.

The evolution of an OB+BH or an OB+NS binary (both are designated as OB+cc binaries) depends on the orbital period, on the mass of the OB component but also on the initial mass ratio of the system, since this determines the evolutionary phase of the mass gainer at the onset of mass transfer (section 12.3).

When the mass of the OB type component is larger than the minimum mass of LBVs (~40-50 M_o for the Galaxy), the OB star will lose most of its hydrogen rich layers first by a LBV type SW possibly followed by a spiral-in phase. The total mass lost by the OB star as a consequence of the latter process is largely reduced due to the preceding SW mass loss. It may be expected that the binary will evolve into a CHeB+cc (WR+cc) binary. It is obvious that such systems are rare. As an illustration, let us consider a 60 M_o +50 M_o binary. After two LBV phases (possibly followed by a RLOF or a spiral-in with reduced total mass loss, section 12.1.4), the system evolves into a 28 M_o (WNL) + 10 M_o (BH) binary with a period of the order of days to decades.

When the mass of the OB type component is smaller than ~40-50 M_o, we can distinguish three subclasses:

- the period is large enough to allow the OB star to become a RSG before it fills the critical Roche volume; in this case the OB star can lose most of its hydrogen rich layers by RSG stellar wind and the binary evolves into a WR+cc, obviously with a very large orbital period; it can readily be understood that the frequency of such large period systems is expected to be very low,

- accretion onto the mass gainer in the progenitor binary started late enough (how late depends on the treatment of semi-convection, section 12.3), so that after the accretion it stays in the blue part of the HR diagram during its remaining lifetime. We then expect that spiral-in of the cc will not happen and the OB+cc binary will remain an OB+cc binary till the SN explosion of the OB-type star. This explosion will be a type II SN event but from a blue progenitor, much like SN 1987A (section 2.9). Note that, since the OB+cc binaries are HMXB candidates (thus also the class of OB+cc binaries discussed here), the X-ray lifetime of this class could be very long, possibly the whole CHeB lifetime of the OB type star,

- after accretion the OB star evolves more or less as a normal star and the period of the OB+cc binary is small enough that the OB star fills the critical Roche volume before it becomes a RSG. The further evolution of the binary is governed by the spiral-in process described in section 10.2.1.2. Let us just remember that since viscosity is involved, we can at most answer the following question:

  if viscosity between the cc and the envelope of the OB-type star is able to transform efficiently orbital energy into thermal energy of the envelope, how much orbital energy is needed to remove this envelope?

  What the final system after the spiral-in phase will look like can then be estimated as follows (we only consider the most frequent case which corresponds to the one where the spiral-in starts when the OB type star is a hydrogen shell burning star = case B spiral-in):

  - use the general property that a hydrogen shell burning (massive) star loses its tendency to expand at the beginning of CHeB when most of its hydrogen rich layers have been removed (section 12.1); furthermore the equilibrium radius R_e of a hydrogen stripped star is given by the relation (12.4),

  - starting from an initial OB+cc binary, we determine the Roche radius of the remnant after the removal of the hydrogen rich layers of the OB type star using equations (10.16), (10.19) and (10.20) ($\lambda = 0.4$),

- when the Roche radius is larger than the equilibrium radius, we conclude that initially there was sufficient orbital energy available and that the formation of a binary with a hydrogen poor CHeB star and a compact companion (when the mass of the CHeB star is large enough, a WR + cc) is an acceptable possibility,

- when the Roche radius is smaller than R_e, both stars merged before the removal of the whole hydrogen rich envelope, i.e. the system consists of a massive (hydrogen shell burning) star with a compact star in its He core: a Thorne-Zytkow object (TZO) (Thorne and Zytkow, 1977). From the detailed structure models of Biehle (1991) and Cannon et al. (1992) it follows that TZOs have the structure of a RSG. Due to SW mass loss, first the hydrogen rich layers will be stripped off; when its mass and SW are large enough, the star may be observed as a 'weird' WR star (WR_{TZ}). Most probably this mass loss dominated evolution continues until the whole stellar mass has been blown away and the cc becomes visible again.

TABLE 12.6: The minimum period $P_{o,min}$ (in days) an OB+cc (NS or BH) must have in order to avoid merging for efficiency factors $\alpha = 1$ and 0.5; the minimum period P (in days) after spiral-in is given as well.

M_{OB}	NS	BH	$P_{o,min}$	P
10 +	1.4		400/900	0.2
10 +		5	100/300	0.2
10 +		10	55/150	0.2
20 +	1.4		220/600	0.12
20 +		5	60/170	0.13
20 +		10	30/90	0.15
30 +	1.4		200/540	0.08
30 +		5	50/150	0.1
30 +		10	25/75	0.1

In table 12.6 we calculate the initial conditions a pre-spiral-in binary must have in order to avoid merging using the previous prescription. We conclude:

- when the cc is a NS with mass ~1.4 M_o, complete merging can be avoided provided that the pre-spiral-in period is of the order of 100 days or larger. As will be discussed in chapter IV, this means that very few WR stars are expected to have a NS companion.

When merging due to the spiral-in of the compact star is avoided and a CHeB + (NS or BH) is formed, depending on its mass the CHeB star becomes a WD, a NS (with SN explosion) or a BH, i.e. the binary ends its life as

WD + (NS or BH) binary

or

NS + (NS or BH) binary if the SN explosion did not disrupt the system

or

two single compact stars if the SN explosion disrupted the binary

or

BH + (NS or BH) binary

12.6. The effect of rotation on the evolution of a MCB

12.6.1. THE PRIMARY

First notice that if the primary in a MCB is initially a rapid rotator, accounting for the orbital periods of most of the binaries, tidal torques will slow down most of these primaries, i.e. the effect of rotation on the structure and evolution of a star is expected to be smaller when this star is a close binary member compared to the case where it is a single star.

Probably the most important consequence of fast rotation for primaries in MCBs, is the enlargement of the convective core. This will not affect the qualitative evolutionary behaviour of the primary discussed in the previous subsections, although the final mass after RLOF/LBV/spiral-in/common envelope will be larger compared to the non-rotating case. This means that similarly to single stars

the effect of fast rotation on the evolution of primaries in MCBs ≈ the effect of large(r) convective core overshooting.

A few conclusions:

- the total mass lost by RLOF (and transferred) of a fast rotating primary is smaller than of a slowly rotating one,

- given the mass of the hydrogen deficient post-RLOF CHeB remnant (a WR when its mass is large enough), the mass of the pre-RLOF progenitor is smaller when it was a rapid rotator compared to the case where it was a slow rotator,

- similarly as for single stars, the minimum mass of a binary component that will collapse into a NS will be smaller when the star was a rapid rotator during its CHB phase compared to the case where it was a slow rotator.

12.6.2. THE SECONDARY

Accretion of mass lost by the primary during its RLOF may spin-up the outer layers of the secondary. When the whole star spins up, it is obvious that the further evolution of the secondary will be affected by rotational mixing. Accounting for the discussion in section 6 we can expect that

- a rapidly rotating secondary will evolve at a larger luminosity compared to a slow rotator producing stars that are overluminous with respect to their mass; in the case of the accretion induced full mixing model, this overluminosity will obviously be very large,

- due to mass transfer, the secondary may already have an altered surface abundance; dredge-up of CNO processed material continues when the star is a rapid rotator.

13. Massive binaries: observations

General observations of massive stars which are important for binaries as well, were discussed in section 2. Here we focus more specifically on observations of binaries.

The data of extra-Galactic MCBs are scarce, so the dependence on metallicity of binary properties such as binary frequency, mass ratio distribution, period distribution etc., is unknown. Even more, MCB data are reasonably complete for the solar neighbourhood only (≤ 3 kpc). Whether or not the statistics apply for the whole Galaxy (for the whole cosmos?) is uncertain.

We distinguish the following classes of massive binaries (MBs):

- MBs where both components are OBA-type stars (OBA+OBA),

- MBs with a WR star component and an OBA-type component (WR+OBA),
- the X-ray binaries for which we separately consider the high mass X-ray binaries (HMXB) where the optical component is an OB-type star, a Be type star or a WR star and the companion is a NS or BH, and the low mass X-ray binaries (LMXB) with a massive binary evolutionary history,
- binary pulsars where at least one component is a NS or a BH; the other component is either a low mass white dwarf or a NS (BH).

Due to the SN explosion of one of the components in a binary, the remaining component may acquire a large space velocity (section 11): the star may be classified as a 'runaway' star. Therefore, when discussing observations in function of MCB evolutionary computations, we have to add

- the runaways.

13.1. Massive OBA+OBA binaries

The class of massive OBA+OBA binaries can be subdivided into:

detached binaries:	both components are smaller than their critical Roche lobes,
semi-detached binaries:	one component fills its Roche lobe
contact binaries:	both components fill their Roche lobe.

We further distinguish pre-RLOF binaries from the others.

13.1.1. PRE-RLOF OBA+OBA MBs

Pre-RLOF binaries are obviously detached binaries. If we make the very reasonable assumption that prior to RLOF both components in a binary evolve as single stars do, a fundamental property of a pre-RLOF binary is that the most luminous component is also the most massive one.

Table 13.1 is a fairly complete list of known pre-RLOF binaries where at least one component is either an O-type or an early B-type star (B3 or earlier). In theory, pre-RLOF MBs may exist with later spectral type and luminosity class III to I provided that the orbital period is large enough. However, the few cases that have been observed all show evidence that mass transfer is going on or occurred. Using the spectral type - luminosity class - M_{bol} - T_{eff} calibration of table 2.1, figure 13.1 shows the HR diagram of these binary primaries. The O-type systems HD 215835 and HD 93205 are included in the table although there are indications that at periastron, the radius of the primary is larger than the Roche radius (Penny, 1996b). This means that in both systems RLOF has started but it is expected that significant mass transfer has not occurred yet.

It is important to realize that some orbits are still uncertain (see the DAO8 cat) and confirmation of the data is certainly desirable.

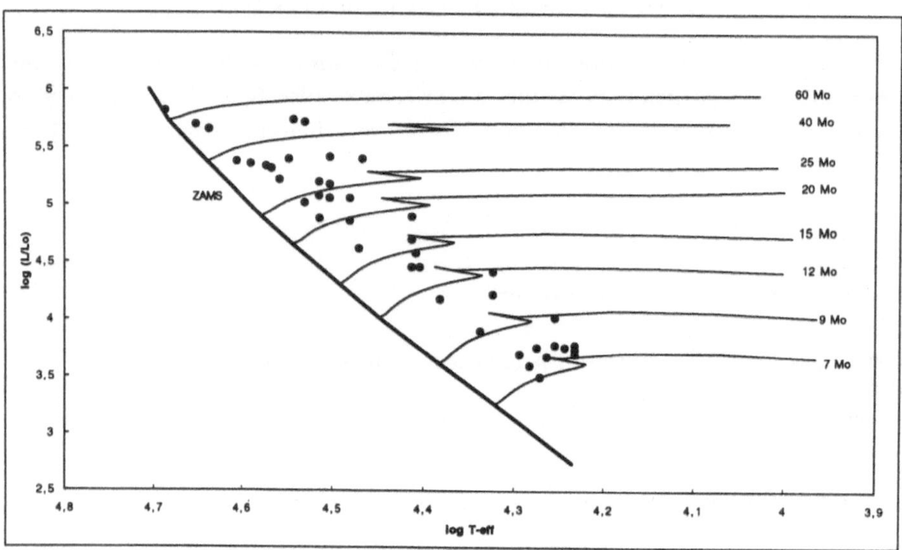

Figure 13.1: The HR diagram of observed primaries of non-evolved MCBs in the solar neighbourhood. Owing to the fact that in some cases the calibration of table 2.1 is used, one point in the diagram may represent different stars. For comparison, we also show evolutionary tracks of galactic massive stars.

Figure 13.2 illustrates the dependence of the eccentricity on the systems period. As expected, the smaller the orbital period the larger the number of binaries with a circularized orbit. Although a majority of systems seem to have (almost) zero eccentricity, some of them still have very large e-values, even if the orbital period is relatively small (~20% among the O-type binaries and ~30% among the early B-type binaries have e > 0.2; ~13% even have e > 0.4). Following the discussion in section 8.1 it is questionable whether these binaries will ever become circularized.

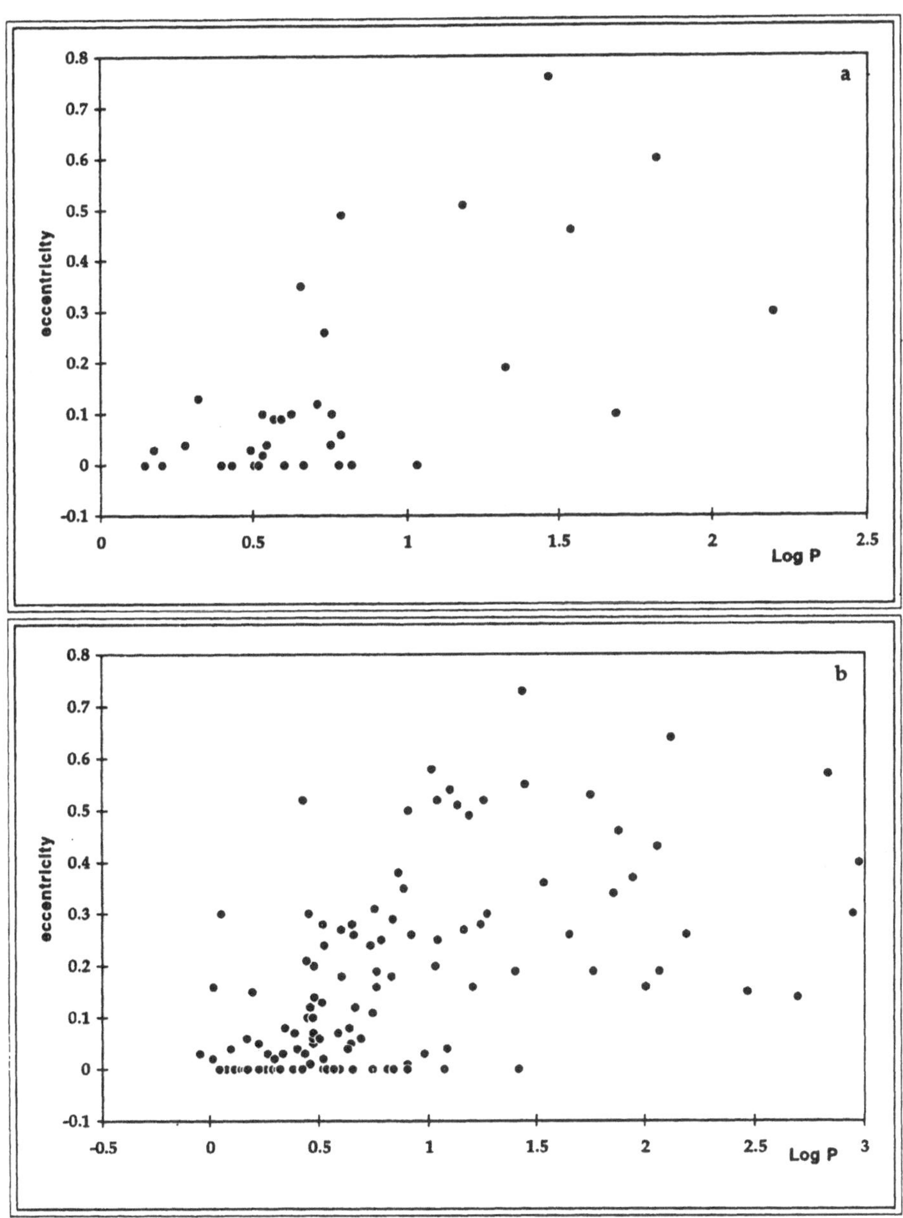

Figure 13.2: The dependence of the eccentricity on the system period for O-type binaries (fig. 13.2a) and for early B-type binaries (fig. 13.2b).

TABLE 13.1a: 30 suspected pre-RLOF binaries with an O-type primary as taken from the Batten catalogue DAO8 listed with ascending periods; 16 binaries are SB2s, whereas 14 binaries are SB1s. (*) The data for HD 152248 have been taken from Penny (1996b).

Star HD	Spectral Type	P days	e	K $km\ s^{-1}$	f(m) M_o	$m \sin^3 i$ M_o
175514	O9Vnn	1.6219	0.0	156		26
	B3V			435		9.2
215835	O6n	2.1110	0.13	261.7		23
	O6n			321.4		19
37737	O9.5III(n)	2.4900	0.0	69.7	0.088	
303503	O8V	3.2316	0.0	202		17
	BOIII			247		14
52533	O9V	3.2934	0.0	33.0	0.012	
328(Batten)	O4f	3.301	0.0	224		41
	O6V			355		26
159176	O6V	3.3668	0.02	218.8		15
	O6V			221.7		14
19820	O9IV	3.369	0.10	141.7		19
	O9IV			291.8		9.2
97484	O7.5	3.4143	0.02	236		23
	O8.5			263		21
206267	O6.5V	3.7098	0.09	103.7		18
	O9			298		6.4
135240	O9V	3.9025	0.09	158.3	1.6	
167771	O7.5III(f)	3.9735	0.0	107.1		2.7
	O9III			122.9		2.3
152219	O9.5IV	4.16	0.10	117.4	0.69	
152590	O8	4.4870	0.35	73.2	0.15	
151564	O9.5IV	4.56	0.0	59.9	0.10	
152218	O9.5IIIn	5.40	0.26	161		13
				202		11
924(Batten)	O8V	5.64	0.04	89.0	0.41	
36486	O9.5II	5.7324	0.10	101.0	0.60	
152248	O7I*	5.816083	0.124	215.5		20.92
	O7I*			202.8		2.23
93206	O9.7Ib(f)	5.9983	0.0	260	11	
93205	O3V	6.0810	0.49	139.1		38
	O8			360		14
165052	O6.5V	6.14	0.06	91.0		2.5
				104.0		2.2

228766	O7.5	10.7424	0.0	150		16
	O5.5f			156		16
93403	O5f	15.093	0.51	80.0		4.9
	O7.5			121		3.2
37041	O9.5V	20.9672	0.19	90.5	1.5	
37043	O9III	29.1338	0.76	102.5	0.89	
166734	O7.5If	34.54	0.46	148		29
	O9I			140		31
199579	O6	48.608	0.10	42.2	0.37	
37020	O7	65.4323	0.60	33.3	0.13	
57061	O9III	154.90	0.30	49.5	1.7	

TABLE 13.1b: 39 suspected pre-RLOF SB2s with an early B-primary. Our sample is taken from the Batten catalogue DAO8 listed with ascending periods. Primaries having CP, n, nn, s or e-characteristics have been excluded.

Star HD	**Spectral Type**	**P days**	**e**	**K km s^{-1}**	**m sin^{3}i M$_o$**
52942	B2.5IV-V	1.2730	0.0	215	5.4
	B2.5IV-V			217	5.3
305543	B1V	1.3278	0.0	261	11
				278	10
36695	B1V	1.4854	0.0	128.6	7.6
	B5V-B9V			285.6	3.4
143018	B1V	1.5701	0.15	121	1.9
	B1V			156	1.5
191567	B0.5V	1.8179	0.0	104.5	2.6
				174.6	1.5
176853	B2V	1.8491	0.03	150.6	7.1
				241.2	4.4
23625	B2V	1.9406	0.0	82	0.88
				114	0.63
185507	B3V	1.9503	0.0	164.2	5.8
	B3V			208	4.6
175544	B3V	1.9858	0.02	149.4	4.5
				190.2	3.6
29376	B3V	2.2075	0.08	124.9	7.0
	B5			235.8	3.7
177624	B3V	2.3741	0.0	81.9	.7
	B6V			196	1.6

35715	B1III	2.5260	0.04	139.1	7.3
	B2V			219	4.7
218066	B0.5IV-V	2.7291	0.03	207.6	12
	B0.5IV-V			221.3	11
193611	B0V	2.8815	0.01	230.2	14
	B0V			226.9	14
29763	B3V	2.9565	0.05	53.6	3.0
				179	0.89
104337	B1.5V	2.9631	0.06	120.5	8.2
				225	4.4
193536	B2V	2.9847	0.07	115.0	2.8
	B2V			141.0	2.3
208947	B2V	2.9899	0.20	116.2	2.5
	B2V			134.6	2.1
198846	B0IV	2.9963	0.14	244.2	16
	B0IV			234.4	17
161783	B2V	3.1691	0.06	150.4	6.2
	B3V			176.6	5.3
147971	B3V	3.2617	0.13	122.5	2.9
				132.9	2.6
139365	B2.5V	3.2907	0.28	75	3.0
				167	1.3
227696	B0.5IV	3.8797	0.07	181.8	17
	B0.5IV			237.4	13
116658	B1V	4.0145	0.18	120	7.2
	B3V			189	4.5
23180	B1III	4.4192	0.05	109.3	5.3
				159.4	3.6
86118	B2V	4.4780	0.28	167.0	9.2
				182.5	8.4
136504	B3IV	4.5598	0.26	56.1	0.40
	B3V			64.8	0.35
3950	B1III	5.5429	0.0	34.9	0.56
				77.0	0.25
90707	B1III	5.5621	0.0	222	20
	B			199	23
37021	B3V	6.4705	0.0	52.8	5.8
	A			171	1.8
144217	B0.5V	6.8282	0.29	125.3	13
				198.0	8.1

77464	B2.5V	6.8895	0	127.0	6.1
	B2.5V			129.2	6.0
218440	B2.5IV	7.2511	0.38	87.9	4.8
				146.7	2.9
39698	B2V	7.9969	0.01	70	8.8
				176	3.5
121263	B2IV	8.024	0.5	110.7	6.3
				159.4	4.4
191201	B0III	8.3343	0.26	157.0	14
	B0III			168.5	13
164852	B3IV	12.4573	0.54	57.6	0.59
	B3IV			57.5	0.59
217312	B0IV	15.2767	0.49	108	27
	B0IV			228	13
37756	B3III	27.1546	0.73	88.5	6.4
				137.9	4.1

TABLE 13.1c: 74 suspected pre-RLOF SB1s with an early B-primary. Our sample is taken from the Batten catalogue DAO8 listed with ascending periods. Primaries having CP, n, nn, s or e-characteristics have been omitted.

Star HD	Spectral Type	P days	e	K km s^{-1}	f(m) M_o
134687	B3III	0.9014	0.03	22	0.0010
178329	B2V	1.0309	0.02	12.1	0.00019
1414(Batten)	B2IV	1.0392	0.16	64.3	0.028
64503	B2.5V	1.1120	0.0	71	0.041
217463	B1.5V	1.1368	0.30	20.7	0.00091
214652	B2V	1.2545	0.04	64.2	0.034
68243	B2III	1.4826	0.06	45	0.014
42933	B0.5III	1.6725	0.05	167.0	0.81
1344(Batten)	B3V	1.6770	0.0	120.8	0.31
189103	B3IV	2.1051	0.0	15.9	0.00088
209961	B2V	2.1727	0.03	121.7	0.41
23466	B3V	2.4079	0.0	23	0.0030
928(Batten)	B1V	2.446	0.07	162.5	1.1
120307	B2V	2.6252	0.0	20.6	0.0024
217061	B1V	2.661	0.0	125	0.54
1400(Batten)	B0III	2.6677	0.52	50.2	0.022
56310	B1V	2.7704	0.21	62.2	0.065

138690	B2IV	2.8081	0.10	28.4	0.0066
65041	B2V	2.826	0.30	34	0.010
35588	B3V	2.8884	0.12	71.2	0.11
46792	B3V	2.9723	0.10	119	0.51
198784	B2V	3.3035	0.02	63.8	0.089
40005	B3V	3.306	0.0	77	0.16
218407	B2V	3.3378	0.24	86.1	0.2
104841	B2IV	3.4280	0.0	51.3	0.048
31237	B2III	3.7005	0.0	57.9	0.075
142669	B2IV-V	4.0033	0.27	16.4	0.0016
191473	B0IV	4.2876	0.04	106.5	0.54
1367(Batten)	B2III	4.3553	0.08	76.0	0.20
87059	B3III	4.515	0.0	61.3	0.11
36954	B3V	4.6239	0.12	48.1	0.052
217224	B3V	4.9088	0.06	70.5	0.18
235807	B0.5IV	5.4150	0.24	87.4	0.34
145502	B2IV-V	5.5521	0.11	26.5	0.011
216711	B1V	5.6556	0.31	47.8	0.055
1374(Batten)	B2III	5.7786	0.16	18.7	0.0038
145482	B2V	5.7805	0.19	31.5	0.018
221253	B3V	6.0663	0.25	56.7	0.10
79351	B2IV	6.7447	0.18	21.5	0.0066
113791	B2IV	7.6497	0.35	38.8	0.038
35411	B1V	7.9841	0.0	145.2	2.5
30836	B2III	9.5191	0.03	25.8	0.017
142883	B3V	10.0535	0.58	64	0.15
25799	B3V	10.67	0.2	20	0.0083
205021	B2III	10.893	0.52	3.1	0.000021
214240	B3IV	10.9188	0.25	82.3	0.57
193924	B3IV	11.753	0.0	7.2	0.00046
216916	B2IV	12.097	0.04	23.0	0.015
3264	B2V	13.504	0.51	23.7	0.012
142096	B2.5V	14.4829	0.27	28.5	0.031
176819	B2IV-V	15.9526	0.16	55.1	0.027
217919	B0.5III	17.4	0.28	42.2	0.12
59543	B2V	17.911	0.52	45.6	0.11
37017	B1.5V	18.65	0.3	30	0.045
217979	B1V	25.249	0.19	32.98	0.088
24190	B2V	26.111	0.0	13.0	0.0060
206672	B3V	26.33	0.0	16.5	0.012

37438	B2V	27.864	0.55	25.5	0.028
147165	B1III	34.23	0.36	34.0	0.11
42560	B3IV	45.10	0.26	22.4	0.047
180163	B2.5IV	56.4	0.53	2.8	0.000078
32990	B2V	58.31	0.19	36.7	0.28
207330	B3III	72.0162	0.34	7.8	0.0030
108248	B1IV	75.7794	0.46	41.7	0.40
175426	B2.5V	88.352	0.37	39.7	0.46
218342	B0IV	101.0	0.16	28.5	0.23
160762	B3IV	113.804	0.43	5.4	0.0014
81188	B2IV	116.65	0.19	46.5	1.2
41753	B3V	131.211	0.64	33.3	0.23
35039	B2IV-V	293.0	0.15	4.1	0.0020
16908	B3V	490	0.14	8.8	0.034
44402	B3V	675	0.57	13.5	0.10
213420	B2IV	880	0.30	9	0.058
3901	B2V	940.2	0.4	11.9	0.13

13.1.2. MASSIVE OBA+OBA BINARIES DURING OR AFTER RLOF

The semi-detached systems, the contact systems and the detached massive OBA+OBA binaries where the less massive component is the most luminous of the two, are post-RLOF OBA+OBA binaries or are OBA+OBA binaries where RLOF is going on. Table 13.2 lists a representative number. We consider three subclasses:

- both components look like normal stars but the primary is filling its Roche lobe and is in some cases the less massive star. Indications of gas streams are present in the spectra of some of them. Both components in HD 209481 have very similar spectral types and magnitudes but very different masses. This suggests that the binary is close to the end of its RLOF,

- binaries where at least one component is classified as an OBN type star, i.e. the star shows strong lines of nitrogen and weak carbon/oxygen lines, similarly to single stars. The OBN star in these confirmed binaries is generally the more luminous component whereas in HD 72754, HD 163181 and HDE 235679 strong evidence exists that it is the less massive star. Note the (very) large eccentricity of HD 12323 and HD 14633,

- the post-RLOF binaries; Φ Per and HD 49798 both contain a hydrogen deficient hot O-type subdwarf and υ Sgr has a very hydrogen deficient A type supergiant component. Both stars in HD 25638 have nearly equal spectral types but the masses are very different. This is considered as (strong) evidence that the B0 star is post-RLOF. Some of these systems will be discussed in more detail in section 14.1.

TABLE 13.2: a representative set of evolved OB+OB binaries; (1) = Batten et al. (1989), (2) = Vreux et al. (1997), (3) = Penny et al. (1994), (4) = Penny et al. (1996), (5) = Bolton and Rogers (1978), (6) = Bozic et al. (1995), (7) = Harries et al. (1997), (8) = Gies et al. (1996), (9) = Hutchings (1975), (10) = Dudley and Jefferey (1990); (11) = Popper (1982); (12) = Popper (1980); (13) = Hilditch (1974); the masses indicated by an asterisk are $M\sin^3 i$ values.

HD (name)	Sp. type	P (days)	e	f(M) (M_o)	M (M_o)	ref
1337(AO Cas)	O8.5I+O8V	3.5	0.04		12.4+18.3	(3)
35921	O9.5III	4.0	0.0		22.0+8.1	(11)
57060(29 CMa)	O7Iabf+O8.7Ib	4.4	0.09		16.0+19.0	(3)
100213	O7/O8	1.4	0.0		21.0+14.0	(12)
149404	O8.5I+O6.5III	8.8	0.0		1.9+2.3*	(4)
190967(V448 Cyg)	B1Ib+O9.5V	6.5	0.04		14.0+25.2	(7)
BD+40°4220	O6f+O7f	6.6	0.0		6.5+24.0*	(2)
209481	O9V+O9V	3.1	0.03		6.2+2.9	(13)
228854(V382 Cyg)	O7 V+O7.5 V	1.9	0.04		26.0+19.3	(7)
BD+66°1521(XZ Cep)	O9.5 V+B1 III	5.1	0.12		15.8+6.4	(7)
12323	ON9V	3.1	0.21	0.0034		(1)
14633	ON8V	15.3	0.68	0.019		(1)
72754	BN2pe	33.7	0.04	8.9		(1)
163181(V453 Sco)	BN0.5Iae+OBN	12	0.08		13.0+22.0	(1,9)
193516	BN0.7IV	4.0	0.06	0.043		(1)
201345	ON9V	sp. binary				
		unresolved				(5)
E235679	BN2.5Ib	225.2	0.1	5.9		(5)
υ Sgr	AI+B4/6	138	0.0		2.5+4*	(10)
Φ Per	sdO6+B0.5Ve	126	0.02		1.7+17.3	(6)
					1.14+9.3	(8)
49798	sdO6	1.5	0.03	0.27		(1)
25638	O9.5V+B0	2.7	0.0		17.0+4.3	(1)

13.2. WR+OB binaries

Table 13.3 summarizes the data of all known Galactic WR+OB binaries (with orbits).

A few properties:

- the majority of the short period WR+OB binaries have very small eccentricities; ~20% have e > 0.2 (0.4) and surprisingly this percentage is not very much different from the percentages of non-circularized binaries found among non-evolved O-type MCBs. This observation is compatible with the statement:

  when a massive binary is born with small e it will remain a binary with small e (or evolve into a state of circularization); when a massive binary originally had a large e, the binary will remain eccentric up to the evolutionary phase where the primary has become a WR star,

- most of the WR+OB binaries have periods smaller than 112 days; 3 WR+OB binaries have a period larger than 1000 days; WR binaries with periods between 112 days and 1000 days are lacking,

- in section 14.2 we compare in detail the observations of WR+OB binaries with MCB evolutionary computations. We conclude that most WR components in WR+OB binaries have masses ≥ 9±1 M_o; the WR component of the binary V Sagittae is an interesting exception:

  A special case: V Sagittae.

  The eclipsing close binary V Sagittae has been studied in detail by Herbig et al. (1965). The primary is identified as a hot star with the spectral characteristics of a WN5 star although the emission lines are narrower than in the WN5 component of the system V444 Cyg. The orbital solution gives $M_{WN5}\sin^3 i = 0.74$ and $M_2\sin^3 i = 2.8$. From the duration of the true primary eclipse (= 0.102 days), the magnitude difference at primary minimum $\Delta mag_p \geq 1.28$ and from the obvious restriction that the radii of both components have to be smaller than their Roche radius, it follows that sin i ≥ 0.776. This means that the mass of the component showing the WN5 characteristics is smaller than 1.6 M_o.

- two binaries contain a WR star showing WN and WC features,

- among the systems with P ≤ 112 days, the number ratio WN/WC ≈ 2,

- the majority of the secondary companions are O-type stars; 2 at the most are early B. Some of the O-type companions have a very early spectral subtype and some of them are classified as a dwarf. We will show in section 14.2 that this is an absolute proof that accretion of mass onto the secondary star has occurred during the RLOF of the primary,

- for γ^2Vel, two quite different mass estimates appeared in the literature. Both are given in table 13.3.
  Hipparcos measured the distance to the spectroscopic binary as d ≈ 260 pc (Van der Hucht et al, 1997) resulting in an angular projected semi-major axis A = 4.3 milli-arc-sec (Hanbury Brown et al., 1970). With a period P = 78.5 days, it follows then that M(WC8+O) ≈ 30 M_o, which supports the solution promoted by Pike et al. (1983), i.e. M(WC8) = 8 M_o and M(O) = 22 M_o.

TABLE 13.3: Data for Galactic WR+OB binaries; the WR nr refers to the catalogue of Van der Hucht et al. (1981). The references: (1) = Moffat et al. (1986), (2) = Niemela (1982), (3) = Niemela and Moffat (1982), (4) = Conti et al. (1979), (5) = Niemela et al. (1985), (6) = Cerruti (1984), (7) = St.-Louis et al. (1988), (8) = Massey (1981), (9) = Cherepaschuck (1975), (10) = Lundstrom and Stenholm (1984), (11) = Lipunova and Cherepaschuck (1982), (12) = Drissen et al. (1986), (13) = Conti (1982), (14) = Niemela (1995), (15) = St.-Louis et al. (1987), (16) = Niemela et al. (1983), (17) = Moffat and Seggewiss (1977), (18) = Eaton et al. (1985), (19) = Annuk (1995), (20) = Annuk (1991), (21) = Herbig et al. (1965), (22) = Pike et al. (1983), (23) = Niemela and Sahade (1980), (24) = Breysacher and Perrier (1991).

WR nr	HD name	P days	e	Sp. type	q	$M\sin^3 i$ M_o	i	ref
3	9974	46.9(?)	0.0	WN3				(1)
12	CD-45°45	23.9	0.0	WN7				(2)
21	90657	8.3	0.04	WN4	1.92	8.4	61°±5	(3)
				O4-6		16.3	eclipse	
22	92740	80.4	0.64	WN7	~2			(4)
31	94546	4.8	0.0	WN4	2.27	2.7±0.4		(5)
				O8V		6.3±0.4		
47	E311884	6.3	0.0	WN6	1.1	40±4	76°9	(6)
				O5V		47±10	eclipse	
97	E320102	11.6	0.0	WN3-4	2	1.8		(14)
				O5-7		5.5		
127	186943	8.6	0.07	WN4	2.0	8-11	56°±8	(7)
				O9/B0V		14-27	eclipse	
133	190918	111.8	0.43	WN4	2.33	0.24±0.04		(8)

				O9Ib		0.56±0.15		
139	193576	4.2	0.0	WN5	2.56	8.5	78°±1	(9)
	V444 Cyg			O6		24.1	eclipse	
141	193928	21.6		WN6	1.82			(10)
151	CX Cep	2.1	0.0	WN5	2.33	5.3		(11)
				O8		11.2		
153	211853	6.7	0.0	WN6	~2		78°±1	(7)
	GP Cep			O6			eclipse	
155	214419	1.6	0.0-0.3	WN7	0.7-1.2		65°-78	(12)
	CQ Cep			O			eclipse	
8	62910	85.4	0.4	WN6				(2)
145	AS 422	22(?)		WNC				(13)
9	63099	14.7	0.0	WC5	1.59	5		(14)
				O7		7		
11	68273	78.5	0.4	WC8	1.89	8	>70°	(22)
	g^2Vel			O		22		
						17±2	>70°	(23)
						32±6		
30	94305	18.8	0.0	WC6	2.13	15±3		(16)
				O6-8		32±5		
42	97152	7.9	0.0	WC7	1.69	3.6±0.3	35-46°	(15)
				O7V		6.1±0.5		
48	113904	18.3	0.0	WC6				(17)
	q Mus			O8.5/BOIab				
79	152270	8.9	0.0	WC7	2.78	1.8±0.3	35°±8	(15)
				O5-8		4.9±1.1	45°±5	
113	168206	28.7	0.0	WC8	2.08	11	76°±4	(18)
	CV Ser			O8-9III-V		22	eclipse	
70	137603	3.29yrs	0.0	WC9	2.1			(14)
				B0Ia				
137	192641	15.56yrs	0.07	WC7				(19)
				O9				
138	193077	4.21yrs	0.1-0.3	WN6				(20)
				O9				
140	193793	7.9yrs	0.85	WC7	2.7	23±3.4		(19)
				O4-5		62±10		
	V Sagittae	0.51	0	WN5(?)	3.8	0.74	eclipse	(21)
				F8V (?)		2.8	≥ 57°	

5980	19.3	0.2-0.4	WN3-5 O7 (?)	(24)

13.3. X-ray binaries[7]

There are about 200 Galactic X-ray sources known with fluxes in the energy range 1-10 keV (hard X-rays). A few strong sources have been found in the Magellanic Clouds. Using a reasonable estimate of the distance of a few sources, the observed fluxes yield X-ray luminosities $L_X \approx 10^{33} - 10^{38}$ erg.s^{-1}. Zeldovich and Guseynov (1966) were the first to suggest that at least some strong X-ray sources are rotating NSs accreting mass lost by a companion star. This picture was confirmed by Schreier et al. (1972) who showed that the source Cen X-3 pulses with a period of 4.84 s (a period characteristic for the rotation period of a NS with a strong magnetic field) while it is a member of an eclipsing binary with a 2.087 day period. One separates the high mass from the low mass X-ray binaries.

13.3.1. HIGH MASS X-RAY BINARIES (HMXB)

A HMXB is a X-ray binary where the optical component is a massive star. Since 1966, ~70 HMXBs have been confirmed. An orbital period is known for 22 systems. They are listed in table 13.4. When the X-ray source is pulsating, the pulse period is given.

One distinguishes three subclasses:

13.3.1.1. HMXBs with an OBe type optical component

The OBe/X-ray binaries are characterized by orbital periods between 15 days and a few hundred days. While 'normal' Be stars have spectral types ranging from early A to late O, the components in OBe/X-ray binaries have spectral types earlier than B2. The X-ray component is pulsating and this indicates that it is a rotating, magnetic NS.

The OBe optical components have radii which are much smaller than their Roche radius. They are losing mass by stellar wind and as a consequence of their high rotation velocities a large part of this stellar wind matter may be

[7] Literature: '*X-Ray Binaries*', eds. W.H. Lewin, J.Van Paradijs, E.P.J. Van den Heuvel, Cambridge University press 1995.

confined within a disc (the wind-compressed disc model of Bjorkman and Cassinelli, 1993). It is the matter in the disc that is accreted by the NS which produces the X-ray radiation.

OBe stars are known not to be persistent e-types whereas a 'normal' OB type star may become an OBe. The reason why a star becomes an OBe star is still unknown although a combination of pulsation, rotation and stellar wind mass loss is very plausible. The on and off stage of the OBe phenomenon is probably the reason why OBe/X-ray binaries turn on and off (these sources are therefore called transient X-ray sources).

The system V0332+53 is an interesting example; it was on for a few months in 1973/74 and in 1983/84 and it was off in between.

13.3.1.2. Standard HMXBs

The optical components are giants or supergiants, most of them have an orbital period smaller than 10 days and very small eccentricity (4U 1223-62 is an exception with period = 41.5 d and e = 0.47) and unlike the OBe/X-ray binaries they are persistent X-ray sources.

One distinguishes two groups of HMXBs: the largest group where the X-ray source is pulsating and thus is most likely a rapidly rotating NS (spin period = pulse period) with a strong magnetic field, and a small group (Cyg X-1 and LMC X-3) with a non-pulsating X-ray source.

Pulsating standard HMXBs.

Quite accurate mass estimates of the binary components of a HMXB can be obtained provided that the optical star is identified, the system is eclipsing and the X-ray source is pulsating. The radial velocities can be determined by Doppler tracking of the pulse period of the pulsar and by measuring the Doppler shifts in the spectra of the optical component taken at different orbital phases. When the optical star is close to filling its Roche lobe, it is pear-shaped (section 8) and one expects photometric variability. Comparison between theoretical light curves of nearly Roche lobe-filling stars and the one observed allows the constraint of the inclination angle i. This has been possible with reasonable success for six standard pulsating HMXBs (table 13.4). Note that the observed average mass value of the six pulsars ~1.4 M_o, the Chandrasekhar limit of NSs (section 3.2.5).

In section 14.4.1 we will discuss in more detail Vela X-1 and Wray 977, which may be of particular importance for MCB evolution.

TABLE 13.4: Orbital data of known HMXBs; M_{cc} and $M_{o.s}$ are resp. the mass (in M_o) of the compact companion and of the optical star. Sources of these data can be found in Lewin et al. (1995); the minimum mass of the optical component of Wray 977 is taken from Kaper et al. (1995).

Source	Sp. Type	P_{orb} (days)	e	M_{cc}	$M_{o.s}$	Pulse period (sec)
HMXBs with an OBe optical component						
A 0538-66	B2 IIIe	16.7	>0.4			104
4U 0115+63	Be	24.3	0.34			3.61
E0236+610	B0e	26.5				-
V 0332+53	Be	34.2	0.31			4.38
EXO 2030+375	Be	46	0.4			41.8
A 0535+26	O8.7 IIIe	111	0.2-0.4			104
GX 304-1	B2 Vne	133				272
4U 1145-679	B1 Vne	188				292
Standard HMXBs						
A 0114+65	B0.5 Ib	11.6				850(?)
X Per	O8.5 III-V	580?				835
4U 1700-37	O6.5f	3.4	0.01	1.8±0.4	52±2	-
Vela X-1	B0.5 Ib	9	0.09	1.9±0.6	24 ± 2.5	283
1E 1145.1-614	B2 Iae	9.8				297
LMC X-1	O7-9III	4.22				-
4U 1538-52	B0 I	3.7		1.1±0.4	21 ± 9	529
Wray 977	B1.5 Ia	41.5	0.47		≥48	696
SMC X-1	B0 I	3.9	≈0	1.2 ± 0.3	17 ±4.5	0.78
Cen X-3	O6-8(f)p	2.1	≈0	1.1 ± 0.6	21 ± 4	4.84
LMC X-4	O7 III-V	1.4	<0.1	1.5±0.4		13.5
Cyg X-1	O8.7 Iab	5.6	≈0			-
LMC X-3	B3 V	1.7	0.13			-
HMXBs with a WR type component						
Cyg X-3	WR?	0.2				-

Non-pulsating HMXBs.

Using the mass - spectral type -luminosity class relationship of table 2.1 for the optical component, it follows that $M_X \approx 6\ M_o$ for the HMXB LMC X-3, a value which is much larger than the upper mass limit of a stable NS (section 3.2.6), and therefore the X-ray source is thought to be a BH.

The most likely BH candidate is the X-ray source in Cyg X-1. An evolutionary model for this HMXB will be proposed in section 14.4.1.

13.3.1.3. <u>HMXB candidates with a WR type component</u>

Cyg X-3 is a X-ray binary discovered in 1966 (Giacconi et al., 1967). The X-ray flux shows a 4.8 hour period modulation and it has been classified as a WR+cc system by Van Kerkwijk et al. (1992). The WR classification is based on I- and K-band spectra where WR-like features have been detected but an unambiguous classification similar to other WR stars is still lacking. This particularly interesting HMXB will be discussed in detail in section 14.5.3.

Prior to 1985, ~13 other WR+cc candidates were listed in literature. However, after extensive photometric and spectroscopic studies of these suspected WR+cc candidates (Vreux, 1985, 1987; Gosset and Vreux, 1987; Lamontagne and Moffat, 1987), it was concluded that the variability in most cases was due to non-radial pulsations or rotation of star spots. The only WR+cc candidates that are left are HD 50896 (period of variability = 3.76 days) and HD 197406 (period of variability = 4.32 days) although the variability of HD 50896 can very well be attributed to intrinsic wind instabilities in the WNE star and/or rotation as well (St.-Louis et al., 1993a). If HD 197406 is a WR+cc, it may be a WR+BH binary (Drissen et al., 1986). Both candidates are located far above the galactic plane whereas for HD 50896 the radial component of the runaway velocity ~100 km/s (Schmutz, 1997).

An important observational constraint is that none of the 13 WR+cc candidates (HD 50896 and HD 197406 included) is a hard X-ray emitter, similar to the standard HMXBs. As will be shown in section 14.5.3, this limits the possible models for the WR+cc candidates.

13.3.2. LOW MASS X-RAY BINARIES (LMXB) WITH A MCB HISTORY

An LMXB is a X-ray binary where the optical component is a low mass star. A majority of the LMXBs are old binaries belonging to Globular Clusters. Most probably they are formed by tidal capture or by exchange collisions of an old NS passing by (Verbunt, 1990; Bhattacharya and Van den Heuvel, 1991).

TABLE 13.8: Observed properties of four LMXB candidates with a MCB history (Van Kerkwijk et al, 1995, for Her X-1; McClintock, 1992, for the three others).

system	L_X (erg/s)	Sp. type	period (days)	f(m) (M_o)	pulse period (sec)
Her X-1	2.10^{37}	A	1.7	0.9	1.24
LMXB black hole candidates					
A0620-00	10^{38}	K5V	0.32	2.91±0.08	-
GS 2023+338	2.10^{38}	K0III-G9V	6.47	6.26±0.31	-
GS 1124-68	10^{38}	K0-4 V	0.433	3.07±0.4	-

Some LMXBs are found in the galactic disc. Since the star densities there are much smaller than in Globular Clusters, the close encounter mechanism is not very likely for their formation. A more plausible formation model is the process called 'accretion-induced collapse' (see a.o. the review of Canal et al., 1990): a white dwarf (a O-Ne-Mg WD) with mass very close to the Chandrasekhar mass limit, accretes mass from a Roche lobe filling low mass blue optical counterpart. When its mass is large enough, the WD collapses and a NS + low mass companion binary is formed. We will not further discuss the evolution of such binaries.

There are, however, a few LMXBs which are in our opinion of fundamental importance in shedding some light onto MCB evolution, i.e. the X-ray pulsar Her X-1 and the LMXB BH candidates. They are listed in table 13.8. The evolutionary model for these systems will be discussed in section 14.6.

13.3.3. THE FORMATION OF X-RAYS WITH ENERGIES ≥ 1 keV

The basic process for the formation of hard X-rays:

> The optical star loses mass at a rate $\dot{M}$; part of this mass (= $\dot{M}_{acc}$) is trapped gravitationally (accreted) by a compact star of mass M_c and radius R. The accreted mass is accelerated and heated up to energies of a few keV, explaining the hardness of the X-ray spectrum.

Matter lost by the optical star, settles into an accretion disc and spirals in towards the cc by gradually losing angular momentum through viscosity. At any radius, this matter rotates at the local Keplerian velocity.

Suppose that the cc is a NS with mass M_{NS} and with a strong magnetic dipole field. When the disc material enters the NS magnetosphere (with radius R_m), due to the strength of the magnetic field ($B \approx 10^{10}$ -10^{13} Gauss is most common for a NS, see also section 13.4) it will be forced to corotate with the NS. Two possibilities:

- when the NS rotates faster than the disc-Keplerian speed at radius R_m, it will use part of the angular momentum in order to accelerate the disc matter: the NS will slow down. Since the matter now rotates too fast to remain in the disc, it will be flung out of the system (the propellor mechanism) and no accretion onto the NS will happen,
- when the NS rotates too slowly, accretion will obviously be possible. In this case it is part of the angular momentum of the disc that will be transferred to the NS via magnetic torques: the NS spins up.

It is obvious that, given enough time, an equilibrium state will be achieved when the spin period P_s of the NS equals an equilibrium value P_{eq} corresponding to the situation where the co-rotation speed at R_m equals the Keplerian speed ω_K at R_m, i.e.

$$P_s = P_{eq} = \frac{2\pi}{\omega_K} = 2\pi\left(\frac{R_m^3}{GM_{NS}}\right)^{1/2} \tag{13.1}$$

A fair estimate of P_{eq} can be obtained by replacing R_m by the Alfvén radius R_A defined as the radius where the magnetic pressure equals the ram pressure of the infalling matter, i.e.

$$\frac{B_m^2}{8\pi} = B_s^2\frac{R^6}{8\pi R_A^6} = \rho v_{infall}^2 \approx 2\rho\frac{GM_{NS}}{R_A} \approx \frac{\dot{M}_{acc}}{2\pi R_A^2}\sqrt{\frac{GM_{NS}}{R_A}} \tag{13.2}$$

where B_m (resp. B_s) is the magnetic field strength at radius R_m (resp. on the NS surface), i.e.

$$R_m \approx R_A \approx \left(\frac{B_s^2 R^6}{4\sqrt{GM_{NS}}\dot{M}_{acc}}\right)^{2/7} \tag{13.3}$$

and thus

$$P_{eq} = (0.7\text{sec})\left(\frac{B_s}{10^{12}}\right)^{6/7}\left(\frac{M_{NS}}{1.4}\right)^{-5/7}\left(\frac{\dot{M}_{acc}}{10^{-8}}\right)^{-3/7}\left(\frac{R}{10^6}\right)^{18/7} \tag{13.4}$$

B_s in Gauss, M_{NS} in M_o, $\dot{M}_{acc}$ in M_o/yr, R in cm.

To summarize:

a binary with a NS component may become a X-ray binary when the spin period P_s of the NS is larger than or equal to the equilibrium period P_{eq}; if P_s is smaller than P_{eq}, no mass accretion is possible and no X-ray radiation will be observed. However, independently of the initial spin period of the NS, the system will always try to reach its equilibrium state and will remain near to it.

Obviously, when the cc is a BH and no magnetic field is present, it can be expected that mass accretion will occur and X-ray radiation will be observed independently of the constraints discussed above.

If the cc is a BH or if the cc is a NS with $P_s \geq P_{eq}$, the maximum accretion luminosity L_{acc} is obtained if it is assumed that the entire accretion rate of potential energy U (in ergs/s) is radiated away,

$$L_{acc} = \left|\frac{dU}{dt}\right| = \frac{GM_c\dot{M}_{acc}}{R} \tag{13.5}$$

Since L_{acc} must be smaller than the Eddington luminosity L_{crit}

$$L_{crit} = \frac{4\pi GM_c}{k_e} = \frac{1.3 \cdot 10^4}{k_e}\frac{M_c}{M_o}L_o \tag{13.6}$$

k_e is the electron scattering coefficient.

With $k_e = 0.19(1+X)$ it follows that

$$\dot{M}_{acc} \leq \frac{3 \cdot 10^{-8}}{1+X}R_6 \tag{13.7}$$

(R_6 is the radius of the cc expressed in 10 km).

For a standard NS $R_6 = 1$ and thus $\dot{M}_{acc}$ must be smaller than $1.76 \cdot 10^{-8}$ M_o/yr (X = 0.7) or $3 \cdot 10^{-8}$ M_o/yr (X = 0).

For a BH

$$R = \frac{2GM_c}{c^2} \tag{13.8}$$

and thus

$$\dot{M}_{acc} \leq \frac{8\pi GM_c}{ck_e} \tag{13.9}$$

There are three ways by which the optical star can lose matter: by RLOF, where the star loses its mass through the first Lagrangian point L_1; by a spherical symmetric stellar wind; or by an OBe type mass loss, i.e. due to rotation through an equatorial disc.

a. RLOF

A typical RLOF mass loss rate of a massive primary (section 12) is several orders of magnitude larger than the maximum accretion rates given above. Therefore, soon after the onset of a RLOF the source will be quenched, i.e. the excess inflowing matter will pile up around the compact star and absorb the X-rays. One does not expect, therefore, that many HMXBs are powered by RLOF.

When the optical component is a low mass star (as is the case in LMXBs), the RLOF mass loss rates are much smaller than those of massive primaries. When the period is small, the RLOF may proceed on the nuclear timescale (e.g. a case A) and the mass transfer rate may be smaller than the critical value given by equation (13.6) or (13.8). This is a fortiori true when the mass of the optical component is smaller than the mass of the compact component since in this case mass transfer causes the orbital period to increase [equation (10.9)].

b. OBe type mass loss

The situation is sketched in figure 13.3. Most of the observed OBe/X-ray binaries are eccentric, i.e. as the NS orbits around the OBe star, there are phases where it does not interact with the disc and no (or only weak) X-ray radiation is emitted. When the NS enters the disc, large mass transfer may be possible and the binary shows up as a strong X-ray source. Since the physics of the discs in OBe stars is still poorly known, it is difficult to estimate how much mass is accreted by the cc from the disc and thus how large the X-ray radiation will be.

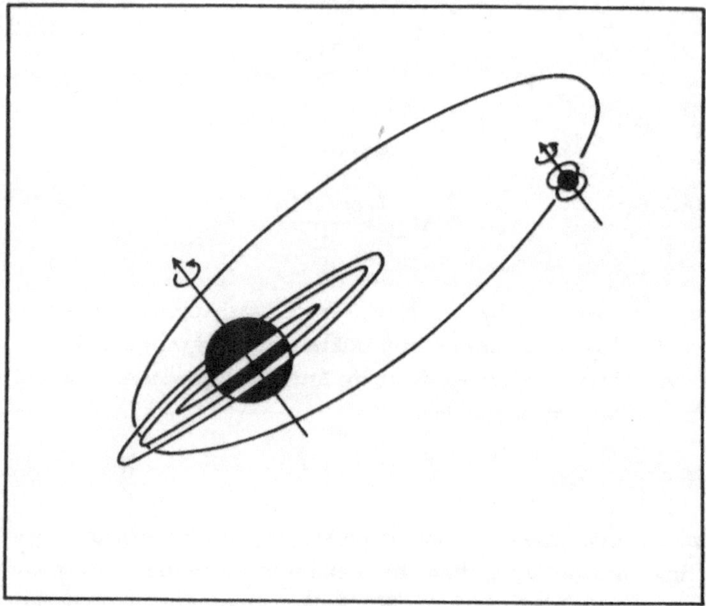

Figure 13.3: The formation of X-rays in an OBe+NS binary with an eccentric orbit.

c. Accretion from a spherically symmetric SW

The situation in which X-ray radiation is a consequence of accretion from a more or less spherical symmetric stellar wind is sketched in figure 13.4.

The total accretion luminosity L_{acc} produced by the capture of matter of a SW by the compact companion can more precisely be described by the model of Davidson and Ostriker (1973), wherein L_{acc} is the root ($< L_{crit}$) of the quadratic equation

$$\frac{L_{acc}}{L_{crit}} = 1.46 \times 10^{13}\, \xi\, \dot{M}_{acc} [1 - \frac{L_{acc}}{L_{crit}}]^2 \frac{L_o}{L_{crit}} \qquad (13.10)$$

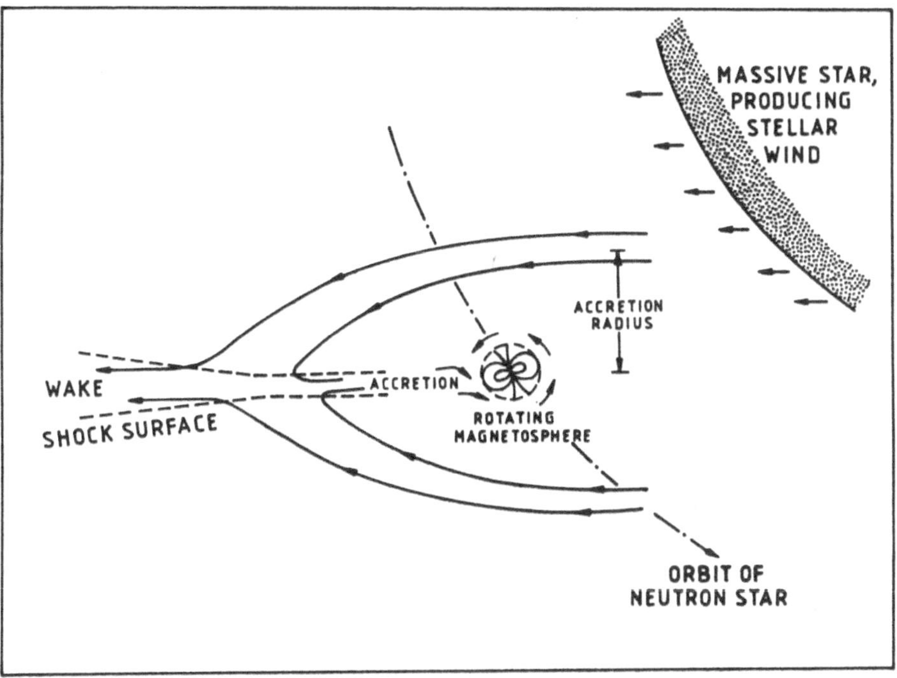

Figure 13.4: The formation of X-rays due to accretion by a compact star of matter from a spherically symmetric stellar wind (from Davidson and Ostriker, 1973).

The parameter ξ is an efficiency parameter for converting gravitational energy into X-ray radiation ($0 \leq \xi \leq 1$) whereas the accretion rate $\dot{M}_{acc}$ is only a small fraction of the stellar mass-loss rate $\dot{M}$ and is given by

$$\dot{M}_{acc} = \frac{3.63 \times 10^{10} \dot{M} M_c^2}{v_{rel}^4 A^2} \tag{13.11}$$

$\dot{M}_{acc}$ and $\dot{M}$ in $M_\odot$/yr , A is the separation between the two stars in $R_\odot$ and v_{rel} is the relative wind velocity (km s^{-1}) with respect to the compact star. In the co-rotating case, v_{rel} can be expressed as

$$v_{rel}^2 = 1.91 \times 10^5 (1 - \frac{R_D^2}{A^2})^2 \frac{M + M_c}{A} + v_{wind}^2(A) \tag{13.12}$$

R_D is the radius of the hydrostatic core of the donor star and the wind velocity $v_{wind}(A)$ (km s^{-1}) is taken at the location of the compact star, M is the mass of the mass losing optical component and M_c is the mass of the compact star (masses in M_o).

The X-ray luminosity L_X is then an appreciable fraction of the total luminosity L_{acc}. If the radiation field has an energy distribution I(E), expressed in Hz, one can write:

$$L_{acc} = \int_0^\infty I(E)\,dE \qquad (13.13)$$

and in the X-ray range between E_0 and E_1

$$L_X = \int_{E_0}^{E_1} I(E)\,dE \qquad (13.14)$$

For I(E) one may consider a bremsstrahlung distribution

$$I(E) = C_1.e^{-E/kT} \qquad (13.15)$$

or a Planck- distribution

$$I(E) = C_2.E^3.\frac{1}{e^{E/kT}-1} \qquad (13.16)$$

With a bremsstrahlung distribution one obtains readily from (13.13) that

$$L_{acc} = C_1.kT \qquad (13.17)$$

and hence a distribution

$$I(E) = \frac{L_{acc}}{kT}.e^{-E/kT} \qquad (13.18)$$

From this expression one can derive the emitted X-ray luminosity L_X in any X-ray interval $[E_0 - E_1]$.

13.4. Radio pulsars

13.4.1. GENERAL

A radio pulsar is a rapidly rotating, strongly magnetized, NS emitting beams of radiation above his magnetic poles. The direction of this radiation is strongly aligned with the magnetic axis. Since the magnetic axis does in general not line up with the rotation axis, the beam sweeps around. When the Earth is located within the cone of a beam, the pulsar can be observed. Depending upon the orientation of the line of sight towards the magnetic and rotation axes this happens once or twice per rotation period.

A rotating magnet emits magnetic dipole radiation E_{em} (Pacini, 1967). When the NS is in vacuum and the magnetic axis is perpendicular to the rotation axis, it follows that

$$\frac{dE_{em}}{dt} = -\left(\frac{2R_s^6}{3c^3}\right)B_s^2\omega^4 \tag{13.19}$$

which obviously equals the rate at which rotational energy E_{rot} of the NS is lost, i.e.

$$\frac{dE_{rot}}{dt} = I\omega\dot{\omega} = \frac{dE_{em}}{dt} \tag{13.20}$$

with I the moment of inertia of the NS (~10^{45} g.cm^2 for a standard NS with M =1.4 M_o and with radius $R_s = 10^6$ cm); or, using $\omega = \frac{2\pi}{P_s}$ (P_s the pulse period in sec)

$$\begin{aligned} B_s &= \left(\frac{3c^3 I}{8\pi^2 R_s^6} P_s\dot{P}_s\right)^{1/2} \\ &= 3.2\cdot 10^{19}\sqrt{P_s\dot{P}_s}\ \text{Gauss for a standard NS.} \end{aligned} \tag{13.21}$$

This means that the magnetic field B_s of a NS can be estimated from the pulse period and its time derivative (see also Ostriker and Gunn, 1969).

The model described above may be over-simplified since the magnetic axis does not need to be perpendicular to the rotation axis. At present, theory is unable to predict the consequences (Bhattacharya and Scrinivasan, 1995). However, for ~100 pulsars, the angle between the magnetic and rotation axis has been measured (Lyne and Manchester, 1988; Rankin, 1990) and it looks as if the distribution of spin-down torques $I\dot{\omega}$ is independent from this angle.

The present state of the (B_s, P_s) relation for ~600 pulsars has been reviewed by Bhattacharya (1995). We recall two important properties:

- ≥ 90% of the pulsars have a magnetic field $B_s \approx 10^{11}$ - 10^{13} G and a pulse period $P_s \approx 0.1$ - 2 sec. Note that this value is significantly smaller than the average pulse period of the X-ray sources in the HMXBs (table 13.4). The reason has been explained in section 13.3.3. When the NS spins too fast, accretion of mass lost by a companion will slow down the NS until an equilibrium period is reached. At that point the NS may become a X-ray source.
  Using $B_s = 10^{12}$ G, $M_{NS} = 1.4\ M_o$, $R = 10^6$ cm and typical values for the velocity and the rate of the SW of a massive star, it follows from equation (13.4) that P_{eq} is of the order of a few 100 sec, corresponding to the observed pulse period of the NS in some HMXBs,

- ≤ 5% of the pulsars have a magnetic field $B_s \leq 10^{10}$ G (down to 10^8 G) and $P_s \leq 0.01$ sec: *the millisecond radio pulsars*. Most of the millisecond pulsars in the galactic disc and at least 50% of the globular cluster members are in binary systems, indicating that binary radio pulsars and millisecond pulsars are closely related.

13.4.2. BINARY RADIO PULSARS

The majority of the known pulsars are single. Presently only 13 binary pulsars are known in the galactic disc, compared to a total of ~600 pulsars (i.e. 2% are binaries). They are listed in table 13.9, with the orbital period and mass of the companion if it is known.

Our present knowledge of the masses of the NSs in these binary pulsars was summarized in figure 5.5. The masses of both components of the systems PSR1913+16 and PSR1534+12 have been determined very accurately from special and general relativistic effects. The values that are obtained suggest that the companions are also NSs in both systems. The same is expected for PSR2303+46 as well as for the globular cluster (M15) binary pulsar PSR 2127+11C (Phinney,1991). The companion in PSR0655+64 has been optically identified as a ~1 M_o white dwarf.

TABLE 13.9: The known binary pulsars in the galactic disc, with their orbital period and companion mass if they are known. The source 1820-11 has been placed among the high mass binary pulsars, although this is still uncertain. References for these data can be found in Lewin et al. (1995).

Binary pulsar	**orbital period (days)**	**mass companion (M_o)**	**pulse period (ms)**
high mass binary pulsar			
1534+12	0.42	1.36	37.9
1913+16	0.32	1.39	59.0
0655+64	1.03	>0.7	195.6
2303+46	13.3	1.5	1066.4
1820-11	357.8	>0.8	297.8
low mass binary pulsar			
1957+20	0.38	0.02	1.6
1831-00	1.81	0.07	521.0
1855+09	13.3	0.23	5.4
1953+29	117.4	0.22	6.1
0820+20	1232.4	0.23	864.9
1259-63			47.8
1937+21			1.6
1257+12			6.2

The mass determinations in these binary radio pulsars are probably the best mass estimates in the whole of astrophysics. They very nicely confirm the theoretically expected value of NS masses (section 3.2.6).

One can distinguish two classes of binary radio pulsars:

- the high-mass binary radio pulsars: the companion is a 1-1.5 M_o WD or NS; their pulse period ranges between ~30 msec and 1 sec which is moderately smaller than the average pulse period of the whole pulsar sample,

- the low mass binary radio pulsars: the companion is a [0.2 - 0.4] M_o WD and they are most likely the descendants of LMXBs (e.g. see Bhattacharya, 1995); 3 of them are millisecond pulsars. An explanation is the following: due to RLOF, accretion of mass onto the NS in a LMXB proceeds at the critical rate of $\sim 10^{-8}$ M_o/yr. When the magnetic field strength $B_s \leq 10^{10}$ G, it follows from equation (13.4) that during the RLOF the NS may achieve a pulse period ≤ 0.01 sec.
  In the table, we have also included the millisecond pulsars PSR 1937+21 and PSR 1257+12. Although they are single, they most probably have had a binary history. The idea is the following (see also Ruderman et al., 1989):

one starts with a normal millisecond binary pulsar with a low mass WD. The radiation and e^-e^+ wind from the pulsar impinges on the companion. As a consequence, this companion evaporates completely and a single millisecond pulsar remains.

13.5. Massive star runaways

13.5.1. OB TYPE RUNAWAYS

Most of the OB type stars have a peculiar velocity v_{pec} of the order of a few km/s. However some of them have significantly larger values. Blaauw (1961) introduced the term 'runaway' OB type star when v_{pec} is larger than 30 km/s.

When the z-component of the space velocity of a star is large enough, its present position can be high above the galactic plane, i.e. also the z-value of a star can be used as a criterion to classify the star as runaway.

Since v_{pec} is the combination of the radial velocity component and of the transverse velocity component, its value is often not known. Therefore the real percentage of observed runaways among OB type stars is still highly uncertain. This is illustrated by the values quoted by different authors: Gies (1987) proposes a percentage not larger than 10 % however the study of Stone (1979, 1985, 1991) promotes 30% as OB type runaway frequency. The criteria used by Gies are very severe and this means that one may be confident that the stars listed in his work as runaway are indeed runaways, but there may be more of them.

There are two processes known in astrophysics which are able to produce OB runaways:

- close encounters in dense OB type clusters. The scenario has been studied in detail by Leonard and Duncan (1988, 1990). In their model a star can be ejected from a dense cluster up to 10 million years after the formation of the cluster provided that the cluster star density is quite high and contains many binaries. As a consequence, when a massive star was ejected from a cluster and when the star's present lifetime is small, one may expect to see at least the remains of the cluster in the neighbourhood.

- the SN explosion in MCBs, introduced originally by Blaauw (1961) (see also section 11). If the SN explosion in a MCB is responsible for the runaway nature of an OB type star, it is clear that the OB star was the mass gainer in an earlier evolutionary phase of the binary. Its properties may therefore teach us something about the accretion process.

A first class of OB type stars which became runaway as a consequence of a SN explosion in a binary are obviously the HMXBs with a NS component discussed in the previous section. Although the radial velocity of the optical star of Vela X-1 is only a few km/s, the annual proper motions taken from Turon et al. (1992) indicates a $v_{pec} \sim 45$ km/s which places the system among the runaways. As shown by Van Rensbergen et al. (1996), it is quite possible that the progenitor of Vela X-1 belonged to the cluster Vela OB1. This means that although a runaway is observed close to a massive star cluster, this does not necessarily prove that it is the close encounter mechanism which is at the origin of the runaway nature of a star.

As will be shown in section 19.2.2, we expect that a significant fraction of the binaries are disrupted during the SN explosion of the originally most massive component. This means that when a runaway star is observed as a single star, it may have had a MCB history. ζ Oph and ζ Pup are two most interesting cases and will be discussed in more detail in section 14.4.2.

13.5.2. RUNAWAYS OTHER THAN OB TYPE RUNAWAYS

An OB type star may evolve into a LBV, a yellow or red supergiant, a hypergiant, or a WR star. If the OB star was a runaway, the descendants will be runaway as well. Obviously, the complicated spectrum of the OB star descendants makes the detection of reliable peculiar velocities very difficult. Still, a few interesting cases were quoted in the literature:

- the hypergiant ρ Cas has a radial velocity of ~ - 47 km/s (Wilson, 1953). Located at ~ 3 kpc (Lobel et al., 1994) and using its proper motions as given by Röser and Bastian (1991) this gives the star a total velocity towards the local standard at rest of ~ 91 km/s.
- the WR star HD 50896 has a peculiar radial velocity of ~100 km/s (Schmutz, 1997),
- the WR star HD 197406 has been classified as a runaway based on it large distance above the galactic plane (799 pc according to Moffat and Seggewiss, 1980).
- the WN8 star 209 BAC has been called the 'fastest WR runaway' (Moffat et al., 1982). It has a peculiar radial velocity component ~156 km/s and is located ~250 pc above the galactic plane.

Apart from the descendants of OB type runaways, there may be intermediate mass and low-mass stars which have acquired a high peculiar space velocity as a consequence of the SN explosion of a massive star:

- the peculiar space velocity of the LMXB Her X-1 was larger than 120 km/s just after the SN explosion (section 14.6). This means that the A-type component of Her X-1 is a runaway. As will be discussed in sections 16.6 and 19.2.2, owing to the expected asymmetry of the SN explosion, most of the binaries will be disrupted during the SN explosion, e.g. most of the binaries similar to the progenitor of Her X-1 will produce single A type runaways with a space velocity larger than 100 km/s. A systematic study of space velocities of A-type stars may therefore be very interesting but has not been done yet.

14. Comparison between the predicted evolution of MCBs and observations

Theoretical evolution can be used either to explain observations of individual stars and binaries, or to compare predictions and observations of the number of different types of stars in regions of continuous star formation or in starburst regions. The latter can obviously be done ONLY by simultaneously considering single stars and close binaries. This will be the topic of chapter IV.

In the present section we will compare the evolutionary computations discussed in section 12 with observations of specially interesting binaries considered in section 13.

14.1. The binaries HD 163181, HD 12323, HD 14633, HD 193516, HD25638, HD 209481, Φ Per and υ Sgr

HD 163181

HD 163181 is an eclipsing binary with a period P = 12 days, with a nitrogen enriched BN0.5Ia primary. Hutchings (1975) derived masses for the components, i.e. $M_1 = 13\ M_o$ (the primary) and $M_2 = 22\ M_o$. The secondary star (M_2) is 1.5-2 mag fainter than the primary and it looks also nitrogen enhanced, indicating that mass transfer has occurred. The primary is largely overluminous for its mass (it has a luminosity matching a normal 30 M_o star). We consider this as evidence that the star is a CHeB star, in the slow phase of the RLOF, close to the end. Its actual mass suggests an initial mass of ~ 30 M_α.

Evolutionary tracks for the components of this system are given in figure 14.1. The overall evolutionary scenario is illustrated in figure 14.2. The system started as a 30 M_o + 16 M_o case B_r binary with a period P ≈ 60 days. We assume that 50% of the mass lost by the primary also leaves the binary through L_2 (forming a ring around the binary). This reduces the period to the presently observed value. Since mass transfer will spin-up the gainer, it may be a rapid rotator.

The computations predict that the BN0.5 Ia star has an atmospheric hydrogen abundance $X_{atm} \leq 0.3$, whereas the ratio N/C should correspond to CNO equilibrium, i.e. the supergiant has all the evolutionary properties to be a WR star, but it is not. We therefore conclude that in order for a star to be a WR type, being a hydrogen deficient massive CHeB star is a necessary condition, but it is not a sufficient one.

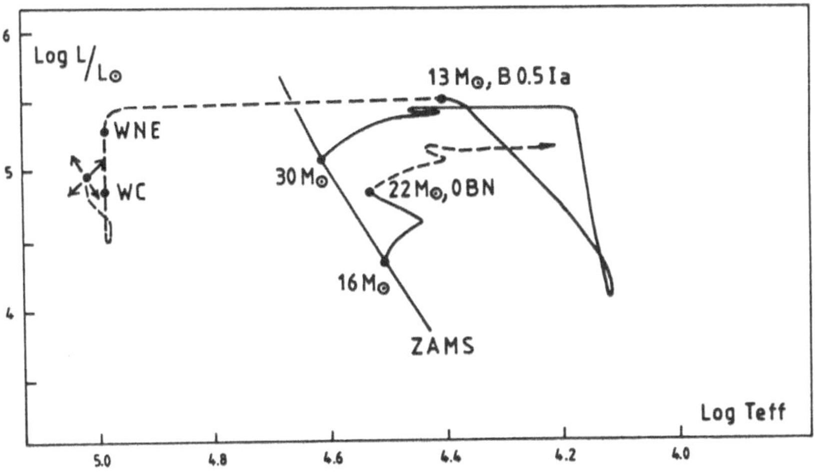

Figure 14.1: The theoretically predicted evolution in the HRD of the binary HD163181.

The binary orbit is eccentric (e = 0.08) although the binary is very close and RLOF/mass transfer has occurred. This means that a binary with a period of the order of a decade where RLOF has occurred will not necessarily be circularized, supporting the general conclusion made in section 8.1.

In figure 14.1 we also show the further evolution of the binary (dashed part). After a WR phase, the primary will explode as a SN leaving a NS as remnant and producing an OBN runaway. If the binary is disrupted (most probable event, see also section 16.6 and 19.2.2), the runaway will evolve as a normal single star.

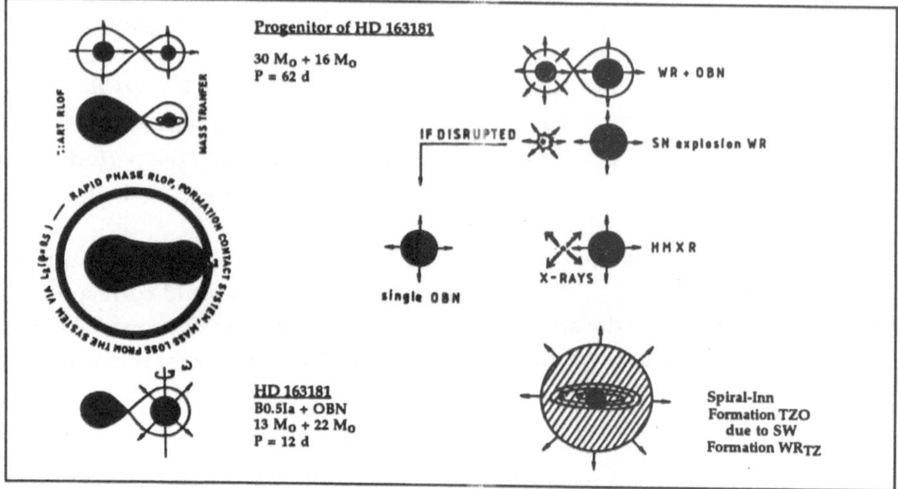

Figure 14.2: The overall evolutionary scenario of the binary HD 163181.

If the binary is not disrupted as a consequence of the SN explosion, its further evolution will be determined by the spiral-in process. The present period of the binary is too small for the possible OBN+cc binary to escape at a complete spiral-in of the cc down to the center of the OBN star. We therefore expect that a TZO will be formed, possibly evolving into a 'weird' WNL star (WR_{TZ}, section 12.5).

Φ Per

The binary *Φ* Per is classified as sdO6+B0.5Ve with a period P = 126 days. In correspondence with the evolution of MCBs, the O6 subdwarf has to be a CHeB remnant after a previous RLOF whereas the B-type companion is a former mass gainer which has acquired a large rotation as a consequence of mass accretion and therefore shows the e-feature. Two sets of orbital masses have been determined for this binary, i.e. 1.7 M_o + 17.3 M_o (Bozic et al., 1995) and 1.14 M_o + 9.3 M_o (Gies et al., 1996) for resp. the subdwarf and the Be component. The evolutionary model is quite different for both sets of masses but since the scenarios are typical of the formation of Be type binaries, we discuss both.

a. Φ Per is a 1.7 + 17.3 M_o binary

When it is assumed that the sdO6 star is at the beginning of CHeB (i.e. its mass equals the remnant mass after the first RLOF), its pre-RLOF mass must have been smaller than 8 M_o and it can readily be checked that there is no binary scenario where the mass gainer can attain a mass of 17.3 M_o.

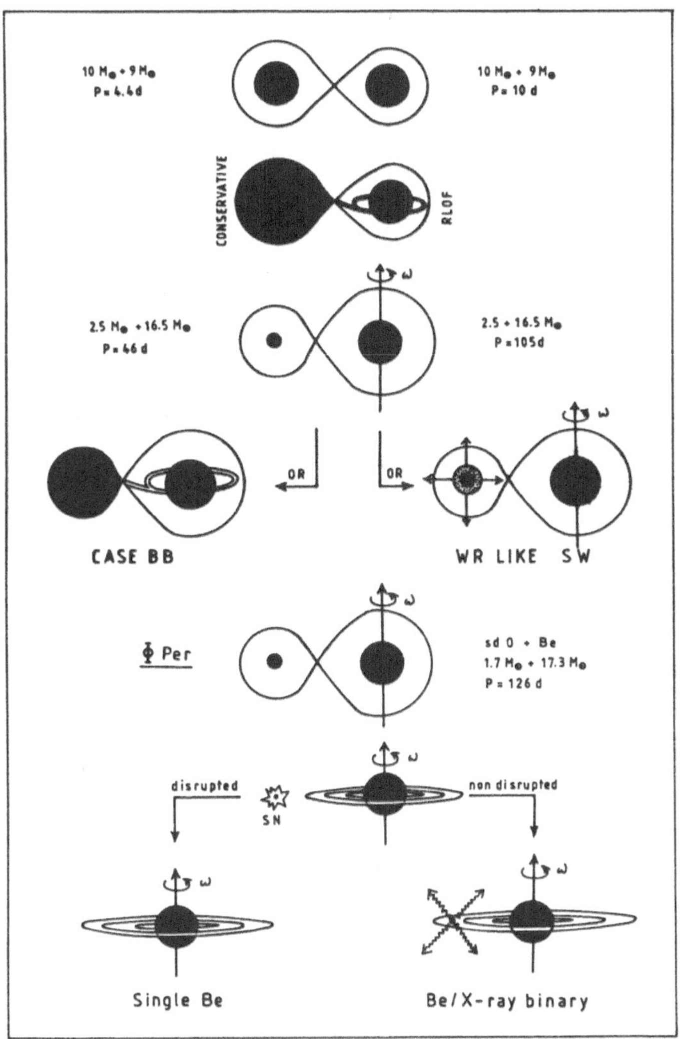

Figure 14.3a: The evolutionary scenario of Φ Per adopting the mass estimates 1.7 + 17.3 M_o.

When the subdwarf is at the end of (or even beyond) the CHeB phase, in addition to the mass lost during the first RLOF, it may have lost matter either due to case BB RLOF, or due to post-RLOF SW mass loss (section 12.2). The evolution is sketched in figure 14.3a. The initial system was a binary with mass ratio close to unity and a primary mass ~ 10-11 M_o. In order to obtain a gainer mass of 17.3 M_o, a conservative RLOF phase is needed.

With the conservative case BB scenario, the initial period must have been 4.4 days, whereas with the SW scenario during CHeB, an initial period ~10 days is more appropriate.

The rapid rotation of the Be component is most likely due to the accretion process, either during the first RLOF or during the second case BB mass transfer (if it occurred). Since during the RLOF phases matter has been transferred which was nuclearly processed during the CHB phase of the primary, the computations predict that the B0.5V mass gainer has enhanced N (factor >3) and depleted CO.

We expect that the sdO6 component is now at the end of CHeB (or even beyond) which means that it is very close to a SN event where a NS will remain. When the SN will not disrupt the binary, *Φ* Per will evolve into a Be/X-ray binary.

The scenario for *Φ* Per discussed here, can be considered as typical for Be/X-ray binaries and their progenitors.

Note that if the masses of the binary used here are correct, CB evolutionary computations with large convective core overshooting during CHB cannot explain the observations.

b. Φ Per is a 1.14 + 9.3 M_o binary

If these masses apply we propose a model in which the sdO6 star is at the beginning of CHeB.

A binary with initial masses 6 M_o + 5 M_o with a period P = 13.5 days, where the RLOF is conservative, reproduces the present system parameters (figure 14.3b). A B0.5V star in general corresponds to a star with mass ~15 M_o, i.e. the star is undermassive (or overluminous) and this may indicate that accretion induced mixing has occurred (sections 9.3.2 and 12.3).

Particularly interesting is the further evolution of the binary. The sdO6 component will evolve into a White Dwarf (with mass ≈ 1 M_o) and the binary may become a low-luminosity X-ray source (X-ray luminosity in the range 10^{29}-10^{33} erg/s) when the WD accretes mass from the Be component, similar to the Be+WD candidates μ^2Cru and HR 4804 (Waters et al., 1989).

As in the case where *Φ* Per is a 1.7 + 17 M_o binary, it is impossible to find an evolutionary model with large convective core overshooting.

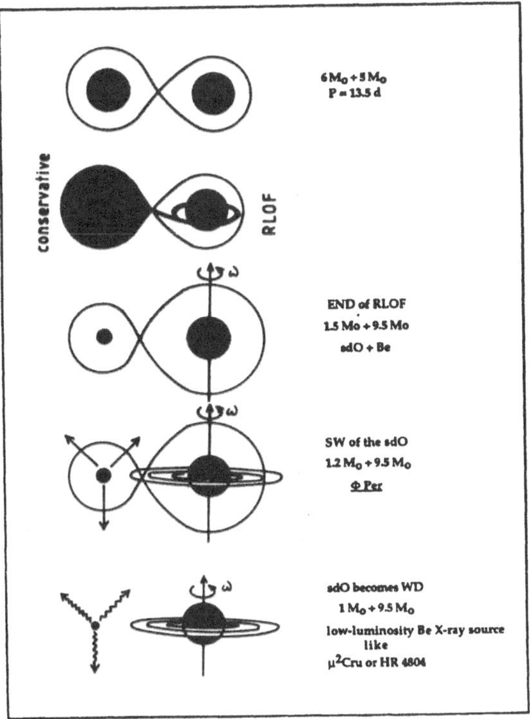

Figure 14.3b: The evolutionary scenario of the Φ Per adopting the mass estimates 1.14 + 9.3 M_o.

HD 12323, HD 14633 and HD 193516

The three systems have an OBN V or IV optical component. Their mass function suggests a lower mass companion. We propose two evolutionary scenarios.

a. The systems are post-common envelope/spiral-in

As an example, consider a 15 M_o + 3 M_o CB with a period of the order of 2-3 years. The spiral-in of the 3 M_o component removes the hydrogen rich envelope of the primary. At the end, the system has a period of the order of days, the secondary has remained a 3 M_o late B-type and the primary has evolved into a ~5 M_o hydrogen deficient ($X_{atm} \approx 0.2$-0.3) CHeB star with CNO equilibrium abundances. It is highly overluminous for its mass ($\log L/L_o \approx 4.5$), it may show up as an ONV and is on its way to become a subdwarf similar to the O6 component in Φ Per but more massive.

b. The systems are comparable to Φ Per

In this case the OBN components are accretion stars in an OBe 'off' phase. The overabundance of N is due either to thermohaline mixing when the standard accretion model applies (section 9.3.1), or to the accretion induced fullmixing (section 9.3.2). In the latter case the stars are expected to be overluminous. The companion may be a subdwarf, a WR-like star just like in the binary V Sagittae (section 13.2), or it can be a compact star.

HD 25638

The orbital parameters of this particularly interesting binary are given in table 13.2. The spectral type and the orbital masses of both components and the magnitude of the B0 star are strong indicators that the system is a post-RLOF binary (the B0 is a post-RLOF remnant). Accounting for the present period, we propose a 12 M_o + 10 M_o case A progenitor. Although the age of the system ~20 million years, the O9.5 component has luminosity class V. This indicates that mass accretion and rejuvenation has played a very important role in its evolution; the star could be a fast rotator.

If the B0 star is post-RLOF, it can be expected that it has a significantly reduced hydrogen abundance. The star may evolve into a sdO or WR type.

HD 209481

Both components are classified as O9V although their masses differ by a factor 2 at least. A most plausible explanation is that the binary went through a mass tranfer phase. The most massive star is the accretor and the lower mass star is the mass loser. Since both stars do not differ much in magnitude, we expect the loser to be close to the end of its RLOF.

The gainer is classified as luminosity class V and therefore, despite accretion, it is close to thermal equilibrium. Adopting a normal mass-spectral type relation, the mass of the gainer equals ~24 M_o and thus the mass of the loser ~11 M_o. Evolution predicts that the latter should be hydrogen deficient, on its way to become a WR star in the near future (probably less than 20000 yrs from now)

υ Sgr

The ≥ 2.5 M_o primary is an A type supergiant that has lost most of its hydrogen rich layers ($n_H/n_{He} < 10^{-4}$, Schönberner and Drilling, 1983). The very small hydrogen abundance, the mass of the secondary and the 138d orbital period indicate that either the system went through a case C common envelope phase, either through a case B common envelope phase where the remnant system experienced further mass loss due to SW or due to a second case BB RLOF.

A 2.5 M_o He core (no hydrogen left) corresponds to a primary with initial mass ~12 M_o (small convective core overshooting during CHB). We propose a 12 M_o + 4 M_o progenitor binary with a period ~2-3 years. When the 12 M_o star becomes a RSG, SW mass loss (equation 2.18) first removes 5-6 M_o (implying a period increase as predicted by equation 10.6) before the common envelope phase starts. The latter process then removes another 3-4 M_o, reducing the period (equations 10.16, 10.18 and 10.20) to a value comparable to the observed one. The details of the period evolution obviously depend on whether a case C or a case B common envelope scenario is considered.

It is important to realise here that, if convective core overshooting during CHB is small, it is very hard to find a binary evolutionary model for υ Sgr where SW during the RSG phase is assumed to be very small, i.e. the existence of this system can be considered as indirect evidence that also for stars with initial mass ~12 M_o SW during the RSG phase is large enough in order to remove a significant fraction of their mass.

Let us finally remark that beside υ Sgr, there are three other candidate binaries with an evolved very hydrogen deficient (probably CHeB or hydrogen shell burning) component, i.e. KS Per (Plavec, 1986), LSS 1922 and LSS 4300 (Drilling et al., 1985; Jeffery et al., 1987). Whether or not these are MCB products is unclear at present.

14.2. WR+OB binaries

For 18 galactic WR+OB binaries, we obtain the WR mass from the orbital properties and a value of the inclination. When sufficient information is not available, we estimate the mass of the OB-type component from a mass-spectral type-luminosity class calibration. For the latter we have first used the overall conclusion that mass gainers are normal stars when the standard accretion model is used (i.e. the calibration of table 5.1 applies). As a second possibility we have considered the results for the gainers when the accretion induced full mixing model applies during the mass transfer. In this case the masses are typically 20%-30% lower than the values of table 5.1. From the evolutionary computations of hydrogen poor CHeB remnants after RLOF (section 12.2) and the mass relation (12.2) (holding for small convective core overshooting during CHB), one can estimate the initial mass of these systems. When the WR star is a WNL (resp. WNE), we first assume that the star is at the beginning of its WNL (resp. WNE) phase; this gives us minimum progenitor masses. A maximum value is obtained when it is assumed that the WNL (resp. WNE) star is at the end of its WNL (resp. WNE) phase. When the WR star is a WC type, we only present the results when it is assumed that the WC star is at the beginning of the WC phase (giving minimum initial masses). The results are shown in table 14.1. We conclude:

- all the observed WC binary components seem to have an initial mass larger than 35 M_o. It therefore cannot be excluded that the majority of them will finally collapse to form a BH and no SN explosion occurs,

TABLE 14.1: The probable mass range of the WR components in galactic WR+OB binaries; M_{aRLOF} = the mass estimate of the WR progenitor just after RLOF and M_{init} = the probable initial mass of the WR component when it is assumed that convective core overshooting during CHB is small. All masses are in M_o.

WR nr	HD (name)	P(days)	Sp. Type	M_{WR}+M_{OB}	M_{aRLOF}	M_{init}
21	90657	8.3	WN4+O4-6	15+29	17-21	35-45
31	94546	4.8	WN4+O8V	(9-13)+(20-30)	11-18	27-40
47	E311884	6.3	WN6+O5V	42+50	42-44	>70
97	E320102	11.6	WN3-4+O5-7	(11-26)+(23-52)	14-29	35-55
127	186943	8.6	WN4+O9/B0V	9+22	11-15	27-35
133	190918	111.8	WN4+O9Ib	(14-18)+(33-42)	16-23	35-50
139	193576	4.2	WN5+O6	9+26	12-15	30-35
151	CX Cep	2.1	WN5+O8	11+21	13-17	33-38
153	211853	6.7	WN6+O6	(13-17)+(26-35)	16-23	35-50
155	214419	1.6	WN7+O	31+26	31-33	57-65
9	63099	14.7	WC5+O7	(16-25)+(25-40)	>22	>45
11	68273	78.5	WC8+O	9+23	>16	>38
30	94305	18.8	WC6+O6/8	(11-16)+(23-35)	>17	>37
42	97152	7.9	WC7+O7V	(12-18)+(20-30)	>18	>39
79	152270	8.9	WC7+O5-8	(8-19)+(22-52)	>15	>35
113	168206	28.7	WC8+O8-9III-V	12+24	>18	>39
140	193793	7.9yrs	WC7+O4-5	23+62	>29	>56
	V Sagittae	0.51	WN5(?)+F8V	1+3.8	>1	>10(?)

- at least 90% (resp. at least 60%) of the observed WN binary components seem to have an initial mass larger than 30 M_o (resp. 35 M_o) and therefore, as with the WC binary components, it is possible that a significant fraction of them will collapse into a BH.

Using evolutionary computations where convective core overshooting or rotation was large during the CHB phase, the initial mass of the WR progenitors may be ~30% smaller on average than the values given in table 14.1, but the conclusions stated above concerning the BH formation obviously remain valid, since overshooting and/or rotation also tend to reduce the minimum mass for BH formation (section 12.2.3).

14.2.1. WR+OB SYSTEMS WHERE RLOF/MASS TRANSFER DID NOT OCCUR

Candidates where SW during a LBV phase dominated the mass loss process

When the initial mass of the primary of a MCB is larger than 40-50 M_o, within the uncertainty of the observed mass loss rates of LBVs, it is possible that the binary reaches the WR+OB phase without mass transfer. Since a spherically symmetric SW increases the orbital period of the binary [equation (10.6)], it can readily be checked that such WR+OB binaries should have a period larger than 15 days when the WR star is a WN type, and larger than 20 days when the WR star is a WC type. Furthermore, since there was no accretion, rejuvenation did not occur. This means that the WR star and the OB type companion have to be on the same time-isochrone, i.e. the OB type companions in these binaries should be preferentially late O-type giants/supergiants or early B type supergiants.

Accounting for the spectral type of the OB type component, the LBV scenario works for the WC binaries HD 94305 and HD 168206 but fine tuning is needed.

The early spectral type of the O type component (if confirmed) in E311884 and HD 193793 cannot be explained if rejuvenation has not played a role.

HD 68273 (γ^2 Vel) and HD 190918 are two 'LBV scenario' candidates. Figure 14.4 summarizes the evolution of both systems if it is assumed that both systems indeed underwent a LBV phase rather than a RLOF.

The figure also illustrates their further evolution. We expect that primaries with initial mass larger than ~40-50 M_o will finally collapse to form a BH and no SN occurs (see also section 14.4.1). This will be the fate of our two candidates here, i.e. both systems will evolve into a BH+OB binary, with a period of the order of 100 days.

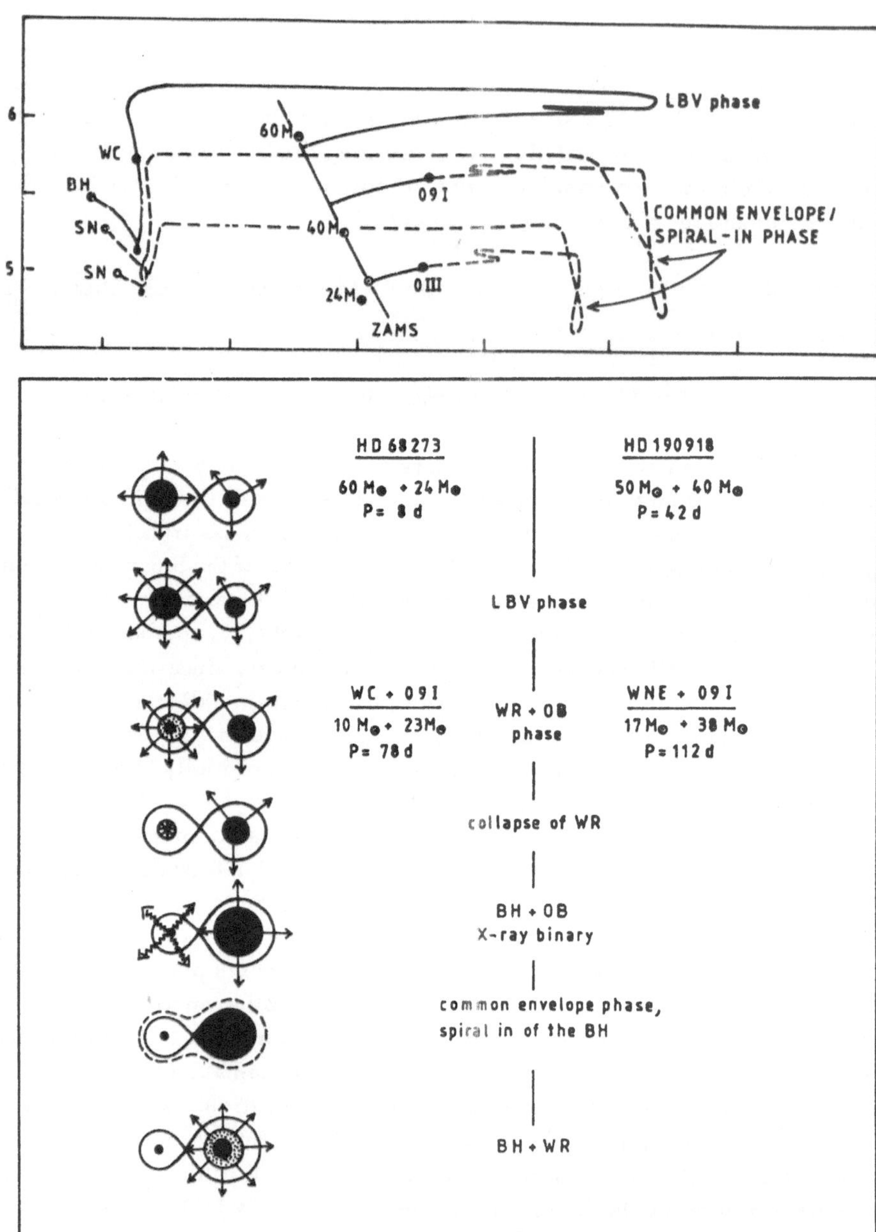

Figure 14.4: The theoretically predicted evolution of the WR binaries γ^2 Vel and HD 190918, two LBV scenario candidates.

When the OB type component reaches its own Roche lobe, the BH will start spiraling-in into the OB star atmosphere (section 12.5). If the orbital energy is efficiently transformed into potential energy of the OB star's outer layers (α = 1 in equation 10.16), the final remnant after spiral-in will be a WR+BH binary with a period of ~5-8 hours.

Many single WR stars seem to have shell like structures around them (section 2.8) which are explained as being due to previous very efficient SW mass loss phases (RSG and/or LBV). If the LBV scenario were at work in both WR+OB binaries considered here, one could expect that these structures would be present there as well. As far as we know this has not (yet?) been observed.

Candidates where SW dominated the mass loss process during the RSG phase

When the primary has a mass that is smaller than 40 M_o, it may lose its hydrogen rich layers by a large SW during the RSG phase (the RSG scenario, section 12.1.2) if the original binary period was large (> 600 days). Since the binary period increases due to SW mass loss, it can be expected that the period of post-RSG WR+OB binaries is larger than 1000 days.

Due to the fact that accretion (and thus rejuvenation) has not occurred in these binaries, as with binaries evolving according to the LBV scenario, the OB type component should be a late O type giant/supergiant or an early B type supergiant. The three best candidates are HD 137603, HD 192641 and HD 193077. Obviously the same argument as for the LBV scenario candidates (discussed in the previous paragraph), concerning the existence of shells around these binaries, also applies here.

14.2.2. WR+OB SYSTEMS WHERE RLOF/MASS TRANSFER DID OCCUR: V444 Cyg

We recall two important evolutionary properties holding for MCBs where the SW of the primary is not large enough to escape the RLOF:

- the removal (out of the system) of a significant fraction of the mass lost by the primary due to RLOF implies a reduction of the orbital separation hence of the orbital period,

- mass transfer during RLOF can rejuvenate the mass gainer; this means that when the OB type star in a WR+OB binary looks younger than the WR star, mass transfer must have taken place.

If we believe the 'observed' spectral types of the OB stars in the WR+OB binaries of table 13.3, we conclude that mass transfer must have played a fundamental role in the binaries HD 90657, HD 94546, E320102, HD 186943, HD 193576 (V444 Cyg), HD 63099, HD 97152, HD 193793 (?). If the O8 component of the CX Cep system is a dwarf, then also here mass transfer must have played an important role. The spectral type and luminosity class of the OB type component in the other systems is not known accurately enough in order to draw definite conclusions. Except HD 193793, all of them are (almost) circularized, which could indicate that tidal effects have played an important role.

As prototype of the class of binaries where mass transfer must have played an important role, we consider in more detail V444 Cyg.

The WR+OB binary V444 Cyg

Facts

- the observed mass of the WNE component ≈ 9 M_o; its progenitor must therefore have had a mass of ~ 30 M_o (convective core overshooting during CHB is assumed to be small) and thus the age of the binary is ~7 million years,

- the OB companion is an O6 star. Since the age of a normal O6 star is ~ 1-2 million years, the OB star must have been rejuvenated, i.e. mass transfer must have occurred,

- the observed mass of the O6 component is ~ 25-26 M_o. However, a normal O6 star has a mass of ~ 37 M_o when it is a class V and ~ 48 M_o when it is a class III (table 5.1). We therefore conclude that either the O-type component is not an O6-type star but rather an O7-O8, either the O6 component is significantly undermassive compared to its spectral type or, in other words, overluminous compared to its mass.

A conservative RLOF scenario

With a conservative RLOF, in order to obtain the observed post-RLOF component masses and the orbital period, the progenitor system must have been a 30 M_o + 12 M_o case A binary with a period of ~3 days. Evolutionary tracks are presented in figure 14.5.

Note that with this scenario, the O-type mass gainer is not expected to be overluminous.

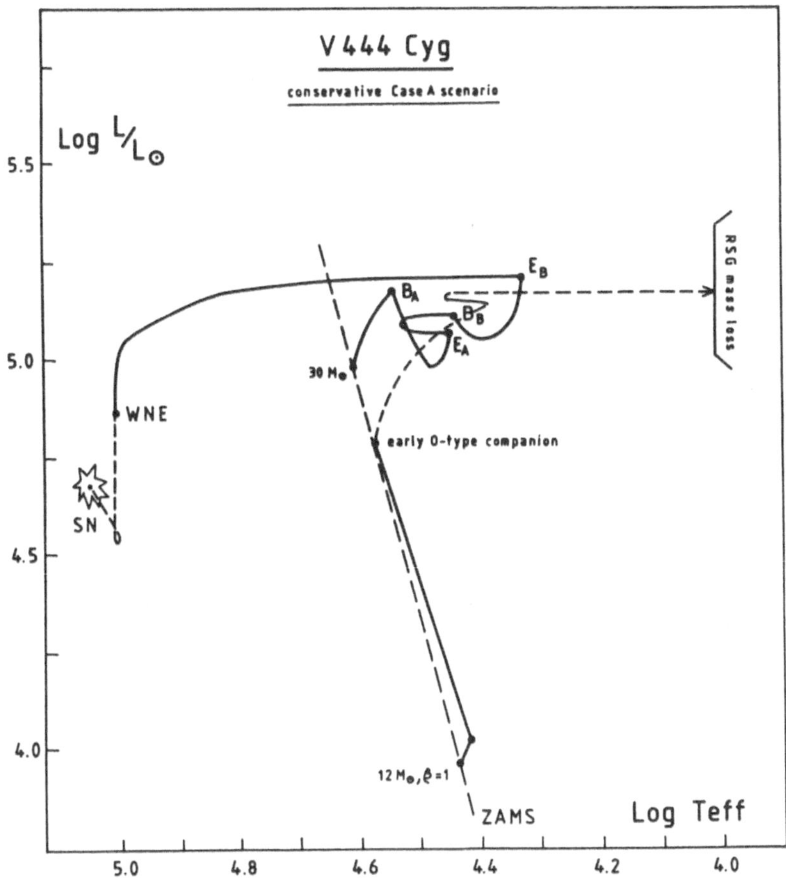

Figure 14.5: The theoretically predicted case A evolution of the WR+OB binary V444 Cyg under the assumption that the RLOF process was conservative; B_A = beginning of the case A RLOF, E_A = end of case A RLOF, B_B = beginning of the case B RLOF, E_B = end of case B RLOF. The dashed part of the tracks is the expected further evolution of the binary.

When accretion is treated in the standard way, very soon after the onset of the mass transfer phase also the mass gainer fills its Roche lobe and a contact binary is formed. We continued our computations as if there was no contact, however, we can conclude that, when mass accretion is treated in the standard way, a conservative scenario for the progenitor binary of V444 Cyg is rather improbable.

A quasi-conservative RLOF scenario.

A probable quasi-conservative MCB evolutionary model is illustrated in figure 14.6 (during CHB, convective core overshooting is small). Evolutionary tracks are given in figure 14.7.

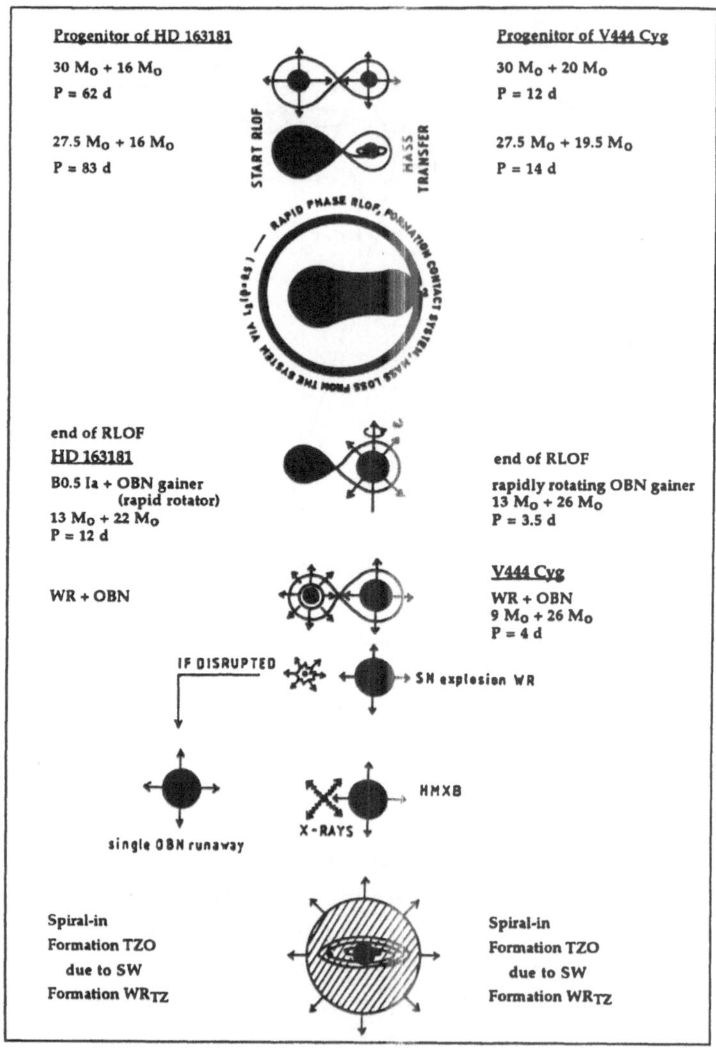

Figure 14.6: The evolutionary scenario of the WR+OB binary V444 Cyg, compared to the one of the OBN+OB binary HD 163181.

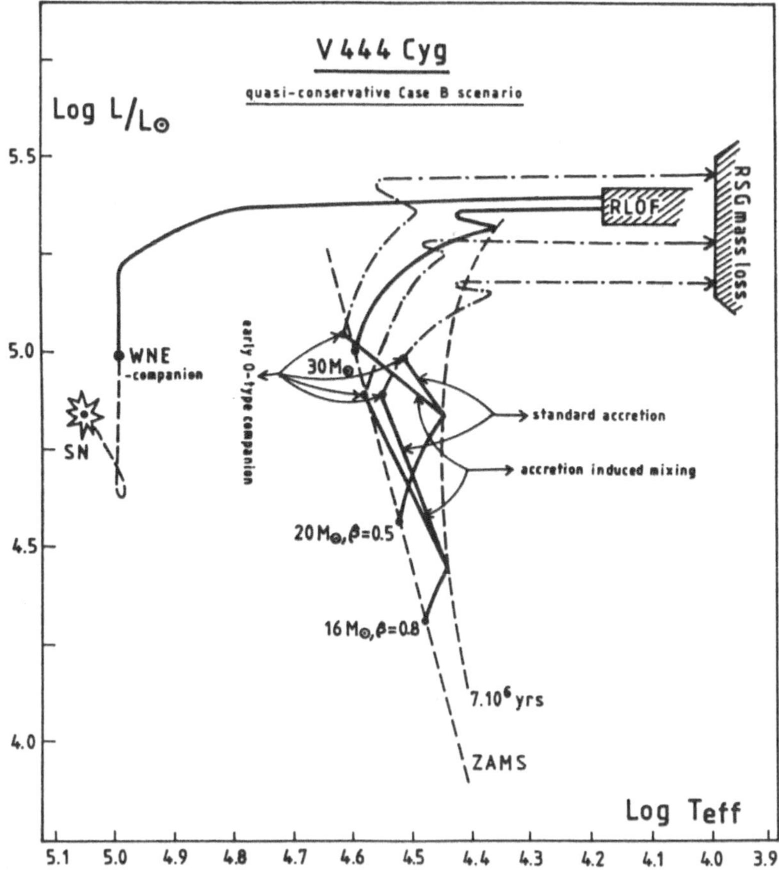

Figure 14.7: The theoretically predicted case B evolution of the WR+OB binary V444 Cyg when it is assumed that the RLOF process was quasi-conservative and convective core overshooting during CHB is small.

We show the evolutionary behaviour of the gainer when the accretion induced mixing process applies and when accretion is treated in the standard way. The progenitor system is a 30 M_o + 20 M_o MCB with initial period P = 12 days. Due to SW during CHB, the orbital period increases slightly to 14 days. The theoretically predicted masses of both components of V444 Cyg equal the observed masses if it is assumed that ~50 % of the mass lost by the primary during its RLOF also leaves the binary. Note however that this '50%' depends on the choice of the initial mass of the secondary. If we started with a 16 M_o secondary and a smaller orbital period, the masses and present period of V444 Cyg are recovered with β = 0.8.

Using equation (10.14), the period at the end of the RLOF (mass transfer) = 3.5 days: the system resembles a WNL + OB binary. Further mass loss by SW removes the remaining hydrogen rich layers. When the WR star has evolved into a WNE star (with mass 9 M_o), the period = 4.2 days.

At the end of the mass transfer, when the secondary has been entirely mixed, the O6-type component may be an overluminous star with mass M = 26 M_o and a homogeneous composition (X,Y,Z) = (0.5, 0.48, 0.02).

The future of V444 Cyg

Figures 14.5, 14.6 and 14.7 also illustrate the theoretically expected further evolution of V444 Cyg. The WR star will explode as a SN leaving a NS as remnant. When the binary is disrupted, the single OB-type runaway will become a RSG. Mass loss by SW will remove the hydrogen rich layers and the star will become a WR star with mass between 10-14 M_o, depending on whether the previous RLOF was conservative or quasi-conservative and on whether the standard accretion model or the accretion induced full mixing model was used.

In the non-disrupting case, the binary will evolve through a spiral-in phase. Since the presently observed period of the WR binary is only 4.2 days, the NS will spiral-in completely and a TZO will be formed. Further SW mass loss during the RSG phase of this TZO may remove the hydrogen rich layers: a WR_{TZ} is formed (section 12.5).

In figure 14.6 we illustrate the similarity, as far as the evolutionary history, the present evolutionary phase and the component masses are concerned, of the binary HD 163181 (section 14.1) and V444 Cyg. The former is a binary on its way to becoming a WR binary like V444 Cyg.

Can we imagine an evolutionary model in which no mass transfer occurred in the progenitor of V444 Cyg?

Let us consider this more generally, and investigate the

consequences for the WR+OB population when it is assumed that no mass transfer occurred in the progenitor systems.

We list four facts, conflicting with observations:

- if no mass transfer occurred in the progenitor systems of the observed WR+OB binaries, the mass ratio of these progenitors can be determined from the results of table 14.1. It follows that almost all of them should have had a mass ratio near to one ($q \geq 0.7$-0.8). However, the observations of

OB+OB binaries (which are considered as the progenitors of the observed WR+OB binaries) reveal the existence of at least as many systems with initial q < 0.7-0.8 as binaries with larger q-values (section 16.5),

- the observed early spectral type and luminosity class of the companion of some WR stars cannot be explained if no mass transfer (and thus no rejuvenation) did occur in the progenitor system,

- if a WR+OB progenitor becomes a WR+OB binary as a consequence of stellar wind only (the LBV scenario), starting from the observed periods of WR+OB binaries, the period of the progenitor follows immediately from equation (10.6). It can easily be checked that this scenario leads to ridiculous initial periods for a number of them.
  The second possibility to remove all the mass lost by a primary (out of the system) is an efficient use of the available orbital energy (section 10.2.2). Using equation (10.15), it follows then that the period of the progenitors of all the WR+OB binaries should have been larger than 100 days, a very uncomfortable situation in view of the observed periods of most of the observed OB+OB binaries,

- around many single WR stars, one observes shell structures with small diameter, witnesses of the previous LBV/RSG mass loss phases (section 2.8). If all the mass lost by the primary during its RLOF also leaves the binary, one could expect to see such structures around the (short period) WR+OB binaries as well. A few WR+OB binaries show evidence of shells (Marston, 1995), however their large diameter suggests that they are related to the SW mass loss and the formation of a wind blown bubble during the progenitor OB+OB phase. We conclude that

the fact that only a few (no?) short period WR+OB binaries show some evidence of shell/ring-like structures with small diameter can be considered as an indication that the secondaries in massive case A/case B_r binaries are efficient vacuum cleaners and that the RLOF process is quasi-conservative.

Obviously, the absence of evidence has to be confirmed by more observations.

For V444 Cyg in particular, the 'no mass transfer' scenario implies that

- the mass ratio of the progenitor binary q > 0.9 and the period should have been beyond 100 days,

- the secondary should be an early B-type supergiant rather than an O6-type, which is not overluminous for its mass,
- there should be a 14 M_o ring around the binary; this however has never been observed until now.

We consider the discussion above, and the consequences for V444 Cyg in particular, to be arguments in favor of significant mass transfer in most of the case A/case B_r progenitors of WR+OB binaries, i.e.

'quasi-conservative' RLOF occurred in a large number of case A/case Br progenitors of WR+OB binaries.

Many theoretical evolutionary computations of case A/case B_r close binaries have been performed in the past, assuming that 50% of the mass lost by the primary is transferred, whereas the remaining 50% leaves the system. This is however an arbitrary choice. In section 12.4, we proposed a strategy to obtain a better idea of the real value of the accretion factor β as a function of the initial parameters of the binary.

Remark

Larger convective cores during CHB of the primary (possibly as a consequence of rapid rotation of the star; however, see section 6.3) imply a larger post-RLOF mass for a given pre-RLOF mass. It can readily be checked then that

the larger the enlargement of the convective core of the CHB progenitor of the WR component of V444 Cyg, the more conservative the RLOF and the mass transfer must be in order to explain all observations of the system.

14.2.3. WR STARS WITH A LESS LUMINOUS OB TYPE COMPANION

Significant mass transfer is not expected to occur in those OB+OB binaries where

- the LBV SW mass loss was very efficient (the LBV scenario),
- the period is large enough to allow the primary to experience RSG SW mass loss prior to the RLOF (the RSG scenario),
- the binary is a case B_c type evolving through a common envelope (as discussed in section 12.1.2, a WR+OB binary remnant after a common envelope evolution in a case B_c binary should have a period of the order of days, up to a few decades).

Since mass transfer did not take place (or was significantly reduced due to the SW mass loss), the mass ratio of the binary (M_{OB}/M_{WR}) may be significantly smaller than in WR+OB binaries where mass transfer has occurred, and the OB companion will be harder to detect. The WR stars WR19, WR104 and WR119 from the catalogue of Van der Hucht et al. (1981) could be binaries with a fainter OB type companion (Williams and Van der Hucht, 1996) and might be descendants from MCBs that evolved according to one of the three scenarios mentioned above. In this case, it can be expected that shell or ring structures will be present comparable to those observed around some single WR stars.

14.2.4. WR STARS WITH A NORMAL LOW MASS COMPANION

The existence of Her X-1-like binaries, and LMXBs with a BH component, is indirect evidence that MCBs exist with extremely small mass ratio (see also section 14.6) and that their frequency is quite high.

When the primary in such a binary is massive enough and when the LBV SW mass loss has been very efficient, a WR+[normal low mass companion] binary is formed with a period of the order of days to decades.

When the primary mass is smaller than 40-50 M_o and the binary period is large enough, the binary will evolve according to the RSG scenario: the remnant is a WR+[normal low mass companion] binary with very large period (of the order of a few years).

Also a case B_c binary with initial mass ratio $q \leq 0.2$, may become a WR+[normal low mass companion] binary after a SW/common envelope phase (section 12.1.5).

Figure 14.8 illustrates the MCB evolution leading to the formation of binaries with a CHeB star (WR) + low mass companion.

If the initial mass of the CHeB (WR) star was smaller than ~40 Mo (resp. larger than 40-50 M_o), it will end its evolution as a NS accompanied by a SN explosion (resp. it will collapse into a BH and no SN occurs). If the SN explosion does not disrupt the binary, the cc+low mass star system may evolve into a LMXB, like Her X-1 (resp. a LMXB with a BH candidate). In section 14.6 dealing with the LMXBs, we will consider this type of X-ray binaries in more detail.

It is interesting to note that if the primary mass was smaller than 40 M_o and thus a SN occurred, the binary will be disrupted in most of the cases. We are then left with a single NS and a low mass runaway with peculiar space velocity exceeding 100 km/s.

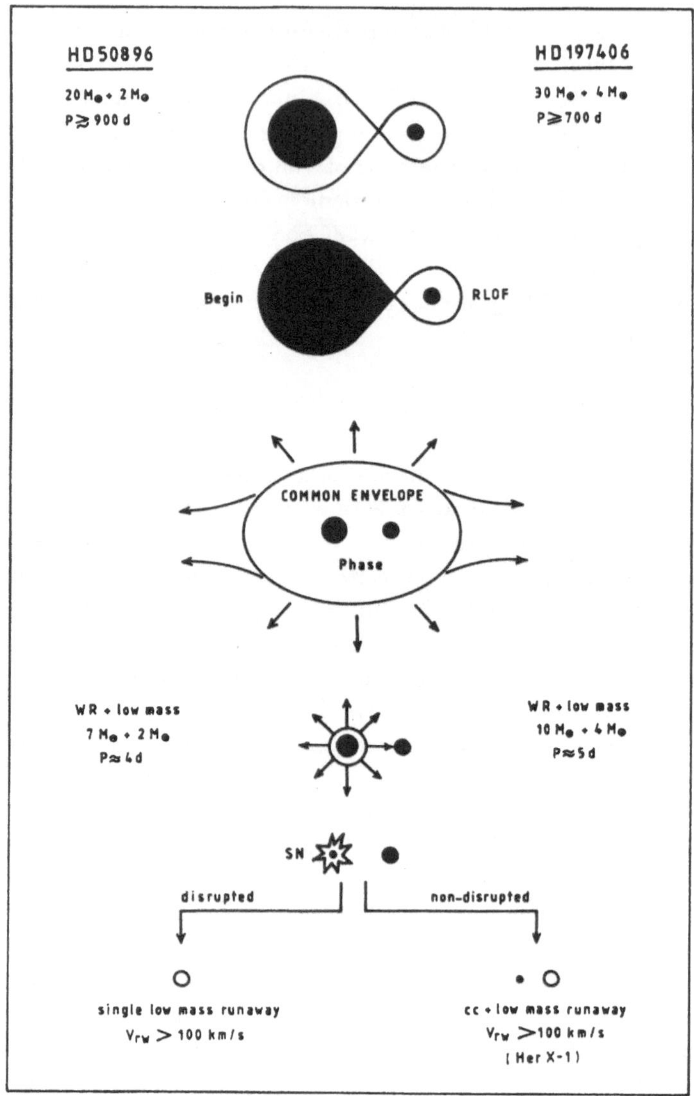

Figure 14.8: The evolutionary scenario of MCBs with small mass ratio, leading to the formation of WR stars with a low mass solar type component (possible candidates HD 50896 and HD 197406).

HD 50896 and HD 197406 are two WR candidates with a low mass companion orbiting with a period of a few days (note however that it cannot be excluded that the periodic variations in HD 50896 are due to instabilities or rotation of the star itself, St.-Louis et al. 1993a). The suggestion that the low mass companion is a compact star will be discussed in section 14.5.3. A normal low mass companion may be an alternative. Since with this scenario, a SN explosion has not occurred yet, the only way to explain the runaway status of HD 50896 and HD 197406 is by means of the cluster ejection mechanism (section 13.5). Whether or not the latter can produce a star with a runaway velocity up to 100 km/s (HD 50896) or a star ~800 pc above the galactic plane (HD 197406) remains to be demonstrated.

14.2.5. LATE O/EARLY B TYPE STARS WITH A (TOO) FAINT WR TYPE COMPONENT

Also the binary V Sagittae (section 13.2) may have had an evolutionary history similar to the one described in the previous subsection. However, particularly important here is the very low mass value of the WR like component. If confirmed, it follows that the WR phenomenon occurs in binaries over a much larger primary mass range than commonly thought.

A Case A/Case B_r MCB with primary mass ≤ 15 M_o and mass ratio $q > 0.2$ ($\beta \geq 0.5$) produces a remnant binary with a hydrogen stripped CHeB component with mass < 3 M_o and a late O-type/early B type mass gainer. Therefore, it is quite possible that a number of late O/early B type dwarfs exist which have a WR like component as the one in V Sagittae but which is too faint to be observed at normal wavelengths.

Is there another possibility to detect such systems? If the WR like component has a SW mass loss rate of a few times 10^{-6} M_o/yr, we suspect that this could be visible in the infrared. It may also be possible to detect the WR star due to the presence of helium lines, as is the case in Φ Per (section 14.1). The link with the binary Φ Per is interesting since this may indicate that we have to look among the OBe stars if we want to find faint WR components.

14.3. The formation of carbon enhanced OB type stars.

A (small) number of OB type supergiants show anomalously strong carbon lines (OBC stars, section 2.3). If this can be interpreted in terms of an increased surface carbon abundance, it is hard (impossible) to explain them by means of single star evolution. Binary evolution however offers the following two possibilities.

- Consider a binary consisting of a WC star and an OB type companion (figure 14.9). It is obvious that the amount of matter accreted by the OB

type star from the SW of the WR star is quite small. However, if the WR star is a WC type, the SW is composed mainly of helium and carbon and if the OB star is a supergiant (with a relatively large radius), although the amount of accreted matter from the SW is small, it is sufficient to enhance the surface carbon abundance significantly, up to a factor 2 to 4 (Vanbeveren, 1994).
This model is obviously still ad hoc. The OB star is losing mass by SW as well and although the $\dot{M}$ of the WC star may be significantly larger, one may wonder how matter from the WC star can reach the surface of the OB companion.

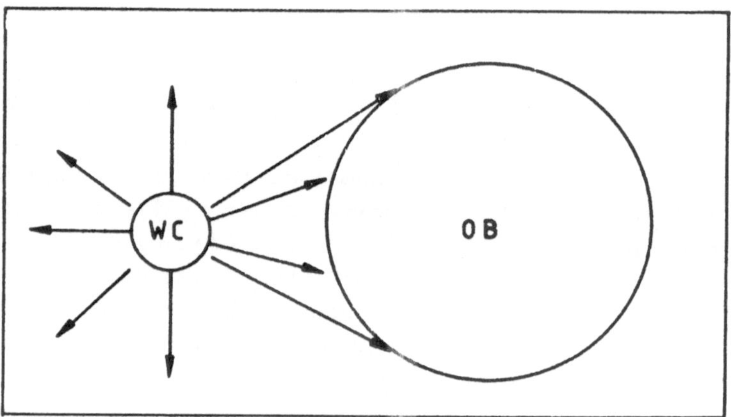

Figure 14.9: An OB type binary component accretes mass from the stellar wind of the WC type component.

- A second possibility is a binary with primary mass ≤ 15 M_o evolving according to the case BB scenario. During case BB RLOF, matter can be transferred that is enriched in carbon due to the nuclear 3α-process.

14.4. The descendants of CHeB+OB binaries

Depending on the mass, the CHeB remnant after RLOF in a MCB forms a NS (or a low mass BH, Brown et al., 1996) accompanied by a SN explosion, or the CHeB star collapses completely (without a SN explosion) and forms a massive BH. When a SN explosion occurred, the binary can be disrupted.

The further evolution of the binary (or single star) is determined by the evolution of the OB-type star. Since the OB type star was the secondary of a

MCB, it is possible that its mass and chemical composition have been modified during a previous mass transfer/accretion phase. The evolution of mass gainers was discussed in section 12.3.

When the OB type star has a compact companion, the system may become a HMXB.

14.4.1. HMXBs WITH AN OB-TYPE COMPONENT: Vela X-1, Wray 977, Cyg X-1

Reliable orbital information of Be/X ray binaries is scarce. From the spectral type of the optical components, accounting for the binary evolutionary calculations of section 12, it seems most probable that these X-ray binaries originate from case B_r systems with primary mass between 9 M_o and 20 M_o where mass transfer has played a dominant role; the rapid rotation necessary to produce a disc is due to the spin-up of the star by mass accretion. The overall evolutionary history of Be/X-ray binaries was illustrated in section 14.1 (figure 14.3).

If the SN explosion was spherically symmetric, equation (11.21) reveals that the OBe/HMXBs should have eccentricities ≤ 0.1; however the observed values are much higher (table 13.4). This can be interpreted as indirect (but strong) evidence that the SN explosion was asymmetric.

The available information derived from the masses of the OB components in standard HMXBs is more reliable. For evolutionary purposes, the three HMXBs Vela X-1, Wray 977 and Cyg X-1 are particularly interesting.

Vela X-1.

The observations

Figure 14.10 shows a high signal to noise spectrum of the B0.5 Ib component in the spectral range 4175 - 4525 Å. The spectrum is taken at orbital phase 0 (the X-ray source is behind the supergiant) in order to assure minimal X-ray contamination.

We focus on Hγ, He I 4388 and He II 4200. Using the hydrostatic NLTE model atmospheres of Herrero et al. (1992), figure 14.10 also gives the best theoretical fit which is obtained for T_{eff} = 26000 K, log g = 2.7 and $\varepsilon = N_{He}/(N_H + N_{He}) = 0.28$. Notice the larger value of T_{eff} compared to the calibration value for B0.5Ib-type stars of table 2.1.

A projected rotational velocity of 125 km.s^{-1} was derived from eleven metal lines of C, N, O, Si and Mg. Since the inclination angle of the binary is very large (>70°), v_{rot} is not much larger than its projection.

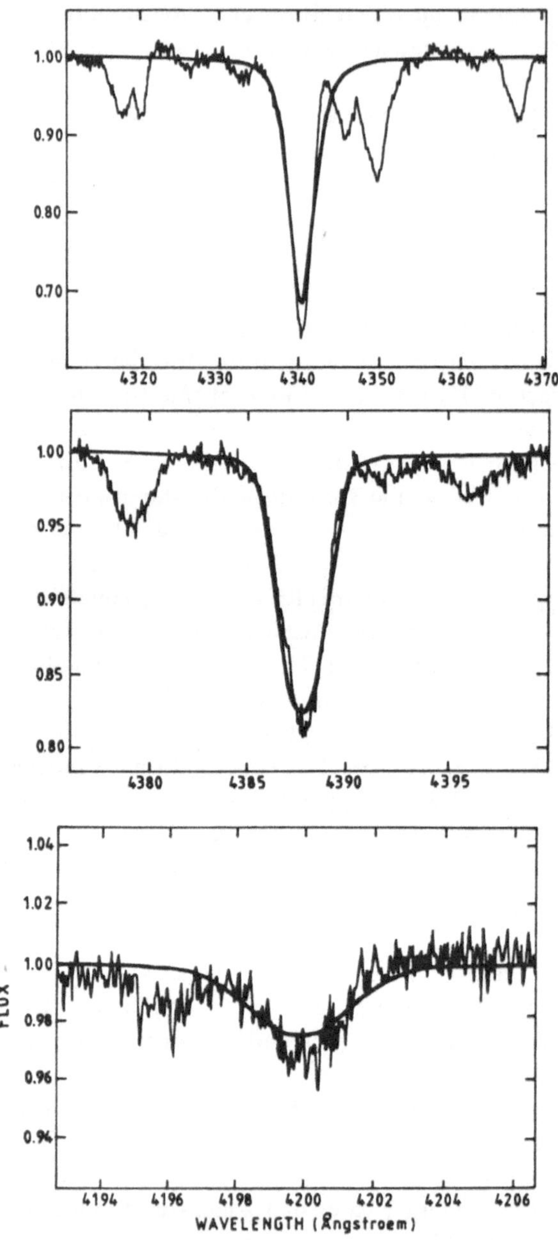

Figure 14.10: Theoretical best fits to the observed lines Hγ, He I 4388 and He II 4200 of the optical component of the HMXB Vela X-1 (from Vanbeveren et al., 1994; the spectrum was taken by M. Van Kerkwijck).

A first interesting result is the significant overabundance of He in the atmosphere of the supergiant (the solar value of $\varepsilon \approx 0.1$). If this overabundance is confirmed, we may expect that a detailed abundance analysis of the metals will also reveal a N overabundance and a CO underabundance compared to solar.

From the orbit and X-ray eclipse one obtains the mass and radius of the supergiant: $R = 28 - 35\ R_o$, $M = 21.5 - 26.5\ M_o$ (these values are 95% confidence limits, Rappaport and Joss, 1983; Joss and Rappaport, 1984).

Combining the NLTE results, the orbital and X-ray eclipse observations, it follows that $\log L/L_o = 5.5 - 5.7$ and therefore, from the relation $M_v = m_v + 5 - 5\log d$ (the distance d in parsec), d has to be in the range 1.8 - 2 kpc.

Using the annual proper motions and the radial velocity (taken from the Hipparcos Input Catalogue, Turon et al., 1992), Van Rensbergen et al. (1996) critically tested possible origins of the past path of Vela X-1 through a number of OB- and R associations in the Vela and Puppis region. It was concluded that only Vela OB1 is hit by the past path, and that Vela X-1 left the association some 2±1 million years ago. It is interesting to note that the distance of Vela OB1 is ~ 1900 pc (Blaha and Humphreys, 1989), a value close to the one obtained for Vela X-1 and thus strengthening the proposition that the progenitor of Vela X-1 was a member of the association.

The evolutionary model

Figure 14.11 shows evolutionary tracks of mass gainers, where the accretion process has been treated in the standard way (section 9.3.1). After the accretion process, the evolution of the star is continued up to the beginning of hydrogen shell burning. The models depicted in the figure have a final mass roughly equal to the observed mass of the optical star.

When the HRD position of the B0.5 supergiant is compared with these tracks, we conclude:

- evolutionary prediction of mass gainers where accretion is treated in the standard way is unable to reproduce the observed HRD position of the optical component of Vela X-1; even accounting for thermohaline mixing (section 9.3.1), the predicted H and He abundance is very close to solar and this differs significantly from the ε value derived with actual NLTE atmosphere codes.

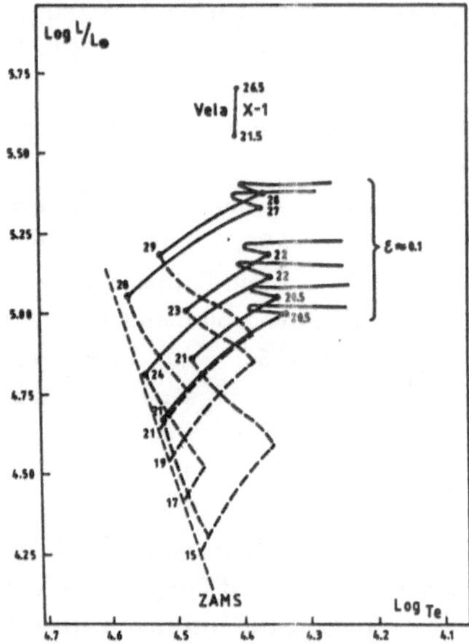

Figure 14.11: Evolutionary tracks of mass gainers, where the accretion process is treated in the standard way; the numbers along the tracks are mass values; the position of the tracks can be compared to the observed HRD position (and masses) of the optical component of the HMXB Vela X-1.

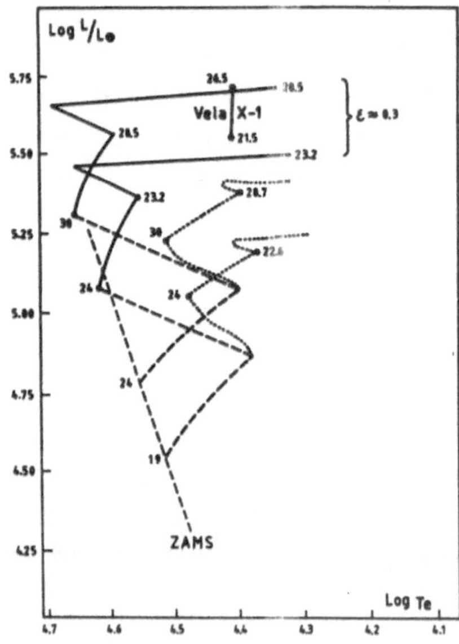

Figure 14.12: Similar as figure 14.11 but the accretion induced mixing process is used.

In figure 14.12 we illustrate the effect of accretion induced full mixing. We have chosen mass gainers where the accretion process starts at the end of CHB, corresponding to MCBs with initial mass ratio close to one. The figure also gives the predicted H to He ratio.

We conclude:

- with the accretion induced full mixing model, we can not only reproduce the observed HRD position of the B0.5 supergiant of Vela X-1, but we also explain the ε value.

Evolutionary tracks for the progenitor of the binary Vela X-1, meeting all observational properties, are given in figure 14.13 whereas figure 14.14 explains the evolutionary scenario. The model predicts then that due to the SN explosion, the X-ray binary was kicked out of the association Vela OB1 less than 3 million years ago.

The further evolution of Vela X-1 will be determined by the spiral-in process of the cc into the B supergiant. Using the model discussed in section 12.5, a TZO will be formed. When SW has been able to remove the hydrogen rich layers, the star may show up as a WR_{TZ}

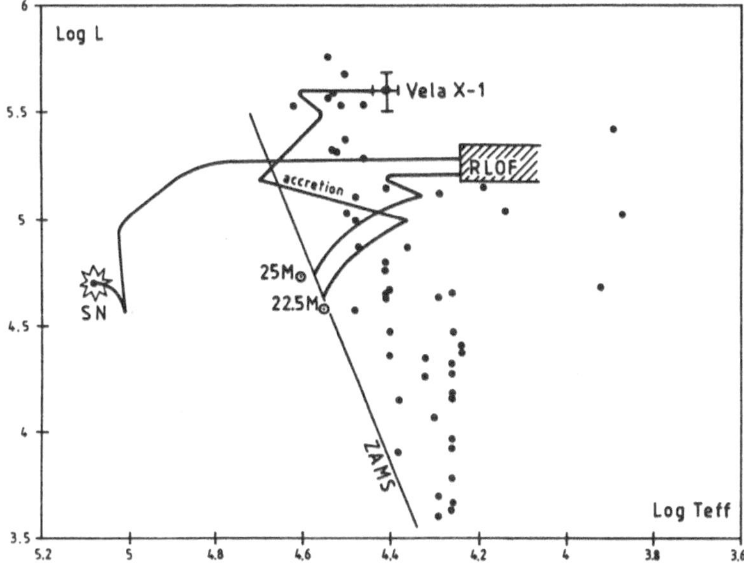

Figure 14.13: Evolutionary tracks of a MCB explaining the observed properties of the HMXB Vela X-1; we compare the HRD position of the optical component to the HRD of the members of the association Vel OB1.

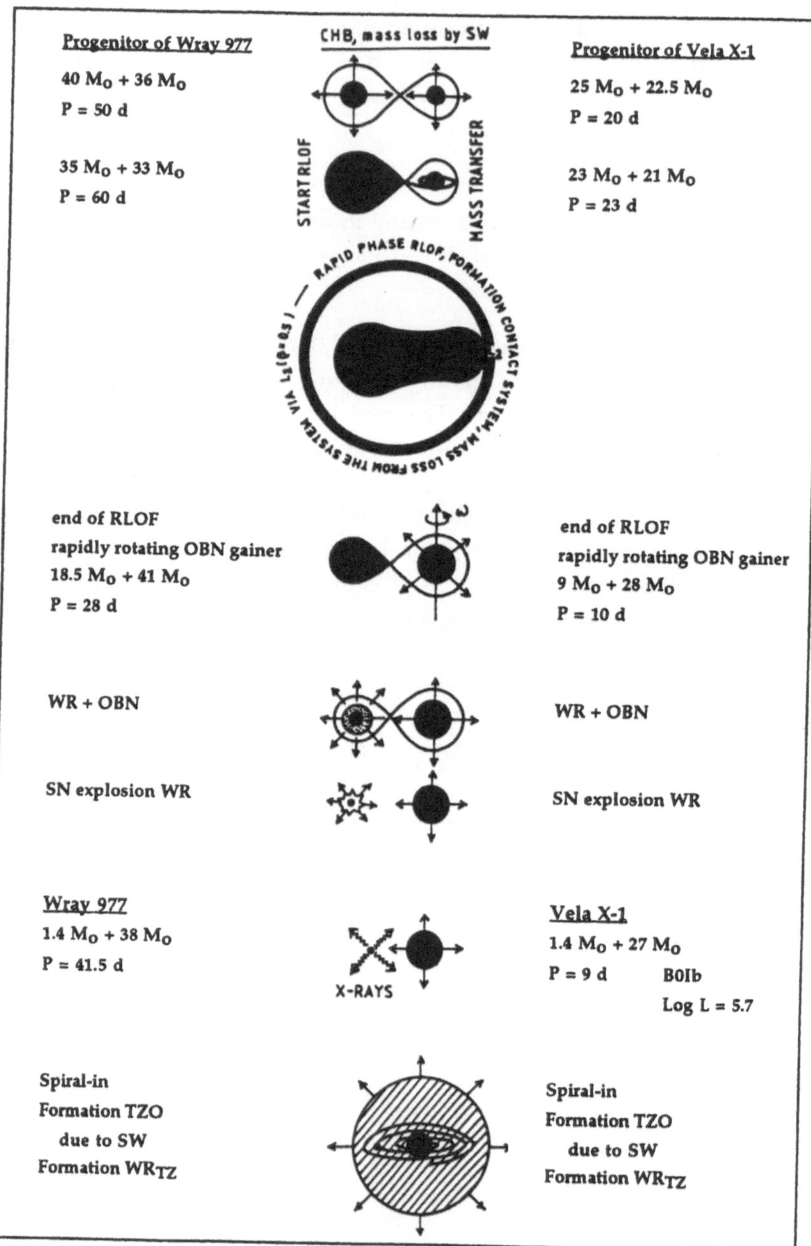

Figure 14.14: The overall evolutionary scenario of the HMXBs Wray 977 and Vela X-1.

Wray 977

The X-ray source is a pulsar, indicating that it is NS, and together with 4U1700-37, it has the highest-mass optical companion known. Sato et al. (1986) propose a mass of ~ 38 M_o. However, Kaper et al. (1995) re-investigated the system and concluded that it is probably a hypergiant with a minimum mass of ~ 48 M_o.

The evolutionary scenario is sketched in figure 14.14 as well. Accounting for the mass of the optical component, we conclude:

- if 50 % of the mass lost by the primary during its RLOF also leaves the case B_r binary, the primary of the progenitor binary must have had a mass larger than 40 M_o; if the RLOF is assumed to be conservative, this minimum mass equals 33 M_o,
- the previous conclusion means that primaries in interacting binaries with initial mass up to ~ 30 M_o (and most probably up to ~ 40 M_o) end their life as NSs (accompanied by a SN explosion),
- if the SN explosion was spherically symmetric and if the binary was circularized prior to the SN explosion, the post-SN eccentricity computed with equation (11.21) should be smaller than 0.12. This is much smaller than the observed value, suggesting that the SN explosion was not spherically symmetric.

The further evolution of Wray 977 depends on the SW mass loss of the supergiant and its effect on a possible spiral-in phase. When $\dot{M}$ is small, Wray 977 will evolve into a TZO just like Vela X-1. However if the supergiant turns into a LBV, the SW can reduce significantly the importance of the spiral-in process and a WR+cc binary will be formed.

Remark:

When the cc in the HMXB 4U1700-37 is a NS, a similar analysis with similar conclusions can be performed as for Wray 977.

Cyg X-1

A detailed analysis has been presented by Gies and Bolton (1986). Herrero et al. (1995) studied the spectrum between 4000 Å and 5000 Å of the optical component of the HMXB and compared selected hydrogen and helium lines with theoretical prediction using a NLTE, spherical model atmosphere including the effects of SW mass loss. Accounting for all uncertainties, combining both studies, the following parameters can be proposed

- the optical star is a O9.7 Iab type star with 28000 K $\leq T_{eff} \leq$ 33000 K
- Log $(L/L_o) \approx 5.4$
- the mass of the optical component M $\geq$ 17 M_o; most probably (accounting for its luminosity), the mass ~30 M_o,
- He may be overabundant (?)
- $M_{cc} \geq 7$ M_o; the most probable mass ranges between 10 M_o and 16 M_o.

Particularly interesting is the mass M_{cc} of the compact star. Its value exceeds by far the maximum mass of a stable NS and it can therefore be considered as one of the best BH candidates.

The observations of the HMXB Wray 977 seem to indicate that primaries with initial mass as high as 40 M_o end their life as a NS. This means that the BH in Cyg X-1 originated from a star with initial mass > 40 M_o and when the probable mass value quoted above is confirmed, it tells us that

some massive stars with initial mass $\geq$ 40 M_o end their life with a mass larger than 10 M_o.

We can suspect that the progenitor binary evolved according to the LBV scenario and mass transfer hardly occurred.

14.4.2. THE OB TYPE RUNAWAYS ζ Oph AND ζ Pup

Let us first remark that accounting for the asymmetry of the SN explosion, a large fraction of the OB type stars which became runaway as a consequence of the SN explosion in a binary, are expected to be really single (section 19.2.2).

The observations[8]

ζ Oph

is an O8.5V star with a very high rotational velocity ($v_e$sini = 365 km/s) and a peculiar velocity ~ 46 km/s which makes it a runaway. Its surface hydrogen abundance X_{atm} ~0.577 (Herrero et al., 1992) is significantly below the solar value (X_{atm} = 0.7). The star is located in the neighbourhood of the Sco-

[8] *The data needed to calculate annual proper motions have been taken from Röser and Bastian (1991) and Fricke et al. (1988).*

Cen OB association which consists of three subgroups: Upper Scorpius, Upper Centaurus and Lower Centaurus Crux. As demonstrated by Van Rensbergen et al. (1996), there are no single star evolutionary computations which meet the abundance constraint, the past trajectory of the star, the location of the three clusters and the assumption that the star was kicked out of one of these clusters.

ζ Pup

is (similarly to ζ Oph) a rapidly rotating O4If star ($v_e$sini = 210 km/s) with runaway velocity ≈ 65 km/s (calculated including the effect of differential galactic rotation) and X_{atm} ~0.546 (Herrero et al. 1992). Hipparcos parallax data place ζ Pup at a distance d = 429^{+120}_{-77} pc and this means that its luminosity Log $L/L_o \approx 5.8$.

There are two clusters in the direction of ζ Pup from where the star could have been ejected: Vela OB2 (marginally possible in view of the comparison between the trajectory of the star and the location of the cluster, Van Rensbergen et al., 1996) and Vela R2.

If ζ Pup originated from Vela OB2, it left the cluster some 2 million years ago. However the age of Vela OB2 is ~20 million years which is much older than the age of ζ Pup when it is considered as a normal single star.

Vela R2 has been placed at 800 pc from the Sun by Sahu (1992). If confirmed, then ζ Pup cannot have been a member of this cluster. Accepting a very large error and placing Vela R2 a factor 2 closer, the star may have been a cluster member. However, since the earliest spectral type of the cluster stars is B2V, the star formation history must have been rather peculiar. Furthermore, Vela R2 contains only few massive stars. Runaways can be produced by the cluster ejection mechanism provided that the cluster is well populated and has many binaries (Leonard and Duncan, 1988, 1990). This condition does not seem to be satisfied here.

We conclude that the statement 'ζ Pup is a normal massive single star that acquired its peculiar space velocity as a consequence of cluster ejection less than 2.10^6 years ago' is only marginally possible.

An MCB evolutionary scenario for both runaways.

Since there is no signature for the presence of a compact companion in neither of the two, we assume that the SN explosion disrupted the binary. This means that the space velocity of both runaways is related to the masses of the components of the pre-SN binary as predicted by equation (11.17).

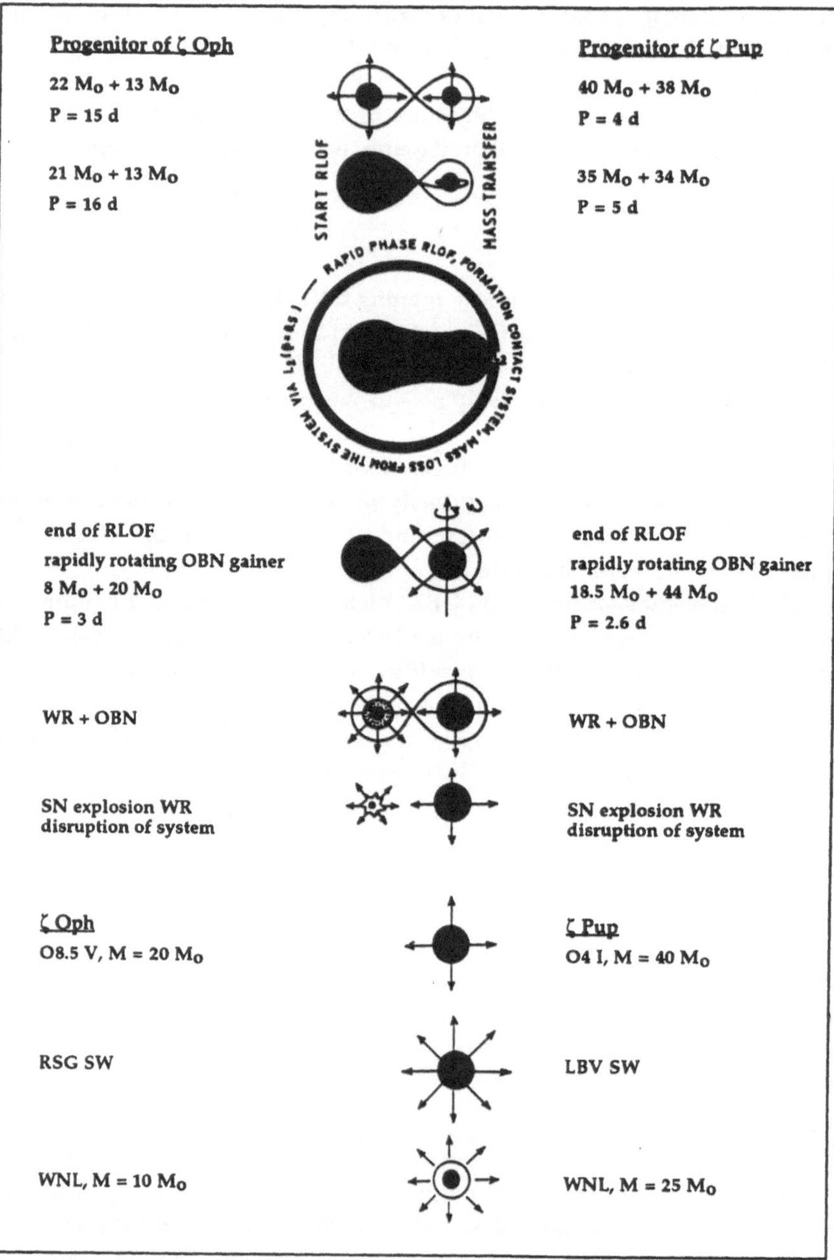

Figure 14.15: A probable evolutionary scenario of the runaways ζ Oph and ζ pup.

ζ Oph

We consider the observed value of the surface hydrogen abundance as evidence for the accretion induced mixing model (section 9.3.2). A probable evolutionary scenario is depicted in figure 14.15 whereas evolutionary tracks are drawn in figure 14.16.

ζ Oph started as the secondary of a 22 M_o + 13 M_o MCB where 50% of the matter lost by the primary during its RLOF also leaves the binary. The primary exploded after some 10 million years, but ~3 million years later the mass gainer hits the observed HRD position of ζ Oph. These results indicate that the binary originally belonged to Upp Cen Lup (a subgroup of the Sco-Cen-OB association).

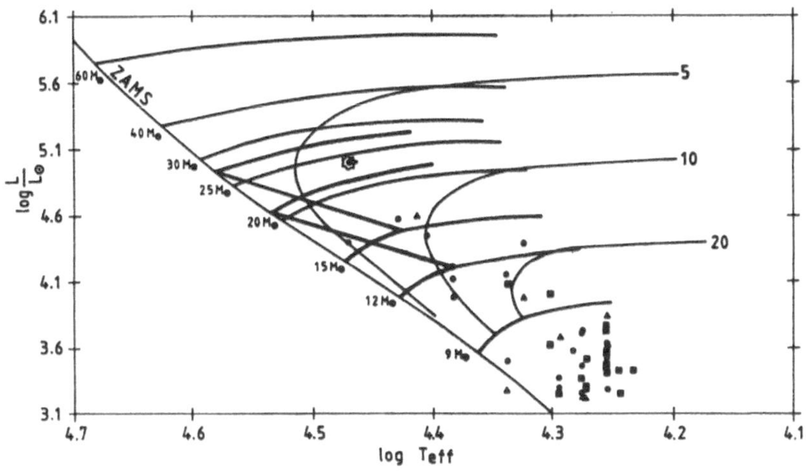

Figure 14.16: ζ Oph is a blue straggler when compared to the positions in the HRD of the other stars of the association. The stars of Upp Sco are dots; the stars of Upp Cen Lup are squares and those of Low Cen Crux triangles. Time isochrones are shown for t = 0 (ZAMS), five, ten and twenty million years. The evolutionary tracks of single stars are labeled by their initial mass on the ZAMS. The position of ζ Oph is explained by the evolution of a close binary where the accretion induced mixing process aplies and with component masses between [20+12] M_O and [25+15] M_O. The tracks followed by the secondary of these systems are traced with an extra-bold full line (from Van Rensbergen et al.,1996).

The most luminous stars of Upp Cen Lup are shown in figure 14.16 as well. This illustrates why ζ Oph can be considered as a blue straggler of the subgroup.

ζ Pup

Since the observed hydrogen abundance is very low, similarly to ζ Oph, we assume that extended mixing occurred during the accretion process. Figures 14.15 and 14.17 give a possible evolutionary history of the runaway. The star started as the secondary of a 40 M_o + 38 M_o MCB. After RLOF, the binary consists of a 18 M_o WNL star and a 44 M_o early O type dwarf. After the removal of the remaining hydrogen rich layers by SW mass loss, the binary becomes a WC+early O. We assume that the SN explosion of the WC star disrupts the system. Using the observed space velocity (= 70 km/s), it follows from equation 11.17 that the pre-SN period must have been ~4 days and this allows constraints to be placed on the initial period of the binary.

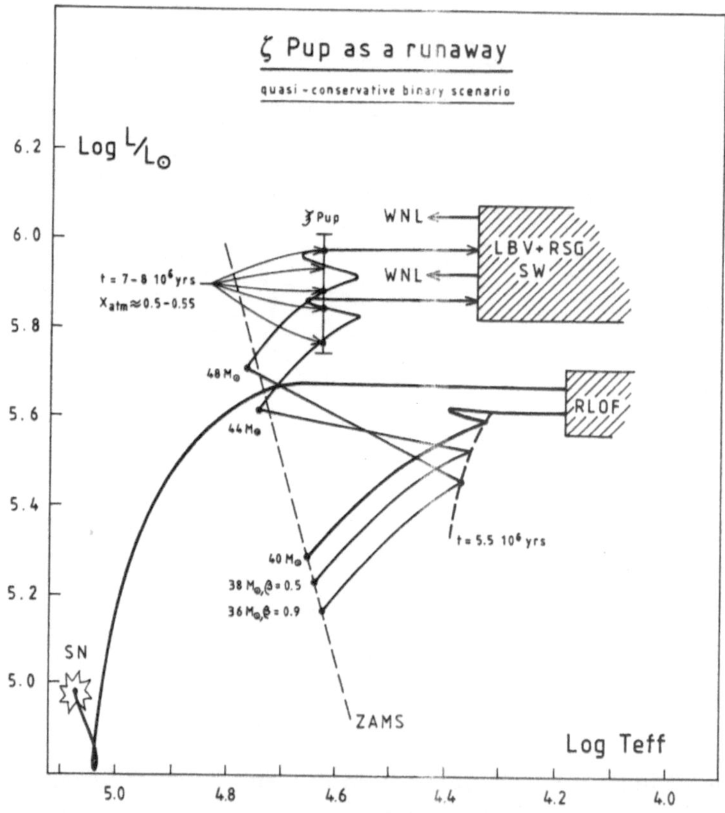

Figure 14.17: The theoretically predicted evolution in the HRD of a MCB meeting the observational contraints of the runaway ζ Pup. The bold track traces the evolution of the 40 M_o primary. We consider two possible mass gainers meeting the observed HRD position and the atmospheric chemical abundance of ζ Pup.

The mass transfer mixed the gainer thoroughly, explaining the overluminosity of ζ Pup and the high surface helium abundance. Due to this mixing process, the further evolution of the star is similar to the evolution of a single star but with initial (X,Y) abundances which differ significantly from normal i.e. (X, Y) = (0.5-0.55, 0.48-0.43). After the removal of ~15 M_o due to a LBV and/or RSG type SW, it will evolve into a 25 M_o WNL star. We expect that ζ Pup will end its life as a massive BH.

Notice the similarity with the evolution of the optical component of the HMXB Wray 977.

14.4.3. CRITICAL REMARKS

All evolutionary models of accretion stars predicting a significantly increased helium abundance at the surface, also predict a significantly increased surface N abundance (factor 5 larger than solar) and depleted C and O (factor 4 to 5 smaller than solar). The optical stars of the HMXBs and the OB runaways discussed in the two previous subsections have been extensively observed and their spectra have been studied in detail.

Why then have these stars never been classified as OBN types?

The actual NLTE model atmosphere code of Munich, where most of the ε values quoted above are based on, does not account for the effect of microturbulence. In a recent paper (McErlan et al., 1997) it was concluded that at least in early B-type supergiants microturbulence may be very important with microturbulent velocities close to the sound velocity. One of the main effects (important for stellar evolution) is that the hydrogen and helium lines in the spectra can be explained without drastically increasing the helium abundance. Although none of the stars considered above were included in the study of McErlan et al., it is interesting in order to discuss qualitatively the consequences, if in reality they would all have a normal helium abundance in their surface (and thus also normal CNO).

First, if helium is normal, we must conclude that accretion was not able to mix the whole star. It is straightforward to understand that, except for quantitative differences, the overall model for the runaways ζ Pup and ζ Oph is very similar as the one proposed in section 14.4.2. Since both stars probably are post-SN runaways, both stars are rapid rotators and ζ Oph is a blue straggler of Upp Cen Lup, we conclude that binary mass transfer occurred causing rejuvenation and spin-up but no extended mixing.

Among the evolutionary scenarios of the HMXBs discussed in section 14.4.1, only the model for Vela X-1 depends explicitly on the value of ε. The position in the HRD of the B0.5Ib supergiant hardly depends on the NLTE analysis, so that the overluminosty remains a fact independent from whether or not He is overabundant in the atmosphere. The question then is:

if the surface helium abundance in the optical star of the HMXB Vela X-1 is normal, what causes its overluminosity (with respect to the orbital mass)?

At present, we have found no evolutionary model (with or without rotation) meeting all observations of Vela X-1 but where the optical star has a normal He abundance at the surface. We are therefore inclined to conclude:

if the overluminosity of the optical component of the HMXB Vela X-1 is confirmed, it can be explained by evolution only if its surface helium abundance is significantly larger than normal; we thus expect the atmosphere to be rich in N, depleted in C and O.

14.5. The descendants of OB+cc binaries: the CHeB (WR)+cc candidates HD 50896, HD 197406 and Cyg X-3

If the OB+cc binary survives the spiral-in phase, the remnant will be a CHeB+cc binary where the CHeB component may show up as a WR star.

As was discussed in section 14.4.1, the HMXBs Wray 977 and Cyg X-1 may evolve into a CHeB (WR)+cc system with a period of a few days/decades.

The evolution of the other known HMXBs will be governed mainly by the spiral-in process of the NS (section 10.2.1.2). As outlined in section 12.5, only when the period of the OB+cc binary is larger than ~ 100 days, complete spiral-in of the NS can be avoided. The value of the pre-spiral-in period depends on the efficiency of the process which converts orbital energy into escape energy of the outer layers of the OB type star; the limiting period quoted above holds for maximum efficiency [the parameter α in equation (10.16) equals one] but it can be considerably larger if the efficiency is smaller. Since all HMXBs have much lower periods, it follows that the formation of a CHeB+cc binary from this class is doubtful.

There are three WR+cc candidates: Cyg X-3, HD 50896 and HD 197406. The first one has a very high X-ray luminosity whereas the latter two do not emit X-rays at all. In the next subsections we will demonstrate that these X-ray observations constrain the possible model of the three candidates.

14.5.1. THE FORMATION OF X-RAYS IN WR+cc BINARIES

The overall formation process of hard X-rays in X-ray binaries was discussed in section 13.3.3. In the case of WR+cc candidates, it seems logical to start from the assumption that the cc accretes mass from the SW of the WR companion.

When the conditions for X-ray emission are fulfilled (i.e. either the cc is a BH, or the cc is a NS with a spin period larger than or equal to the equilibrium period defined in section 13.3.3), the observed X-ray flux will be only a fraction of the emitted flux since the radiation crosses the wind from the location of the X-rays emitting compact component r_c to the observer.

In section 14.5.2 we show how one can calculate the absorption coefficient k(E) at any energy E. The energy dependent optical depth of the location at which the X-ray radiation is emitted for any isothermal wind model for which ρ(r) is known, follows from

$$\tau(E) = k(E) \int_{r_c}^{\text{observer}} \rho(r)dr \tag{14.1}$$

The density in the wind of the mass losing star follows from

$$\rho(r) = \frac{\dot{M}}{4\pi r^2 v_{wind}(r)} \tag{14.2}$$

where the wind velocity as a function of r is computed by adopting the usual semi-empirical formalism

$$v_{wind}(r) = v_{wind}(A) + \left[v_\infty - v_{wind}(A)\right]\left(1 - \frac{A}{r}\right)^\beta \tag{14.3}$$

v_∞ denotes the terminal wind velocity; typical values for β range between 0.7 and 1.5 (Hamann and Koesterke, 1996).

The emitted energy I(E) is now attenuated into the observed one

$$I(E)_{obs} = I(E).e^{-\tau(E)} \tag{14.4}$$

whereas the observed X-ray luminosity in a particular pass band $[E_0\text{-}E_1]$ can be determined from

$$L_{X,obs} = \int_{E_0}^{E_1} I(E)_{Obs}\, dE \tag{14.5}$$

14.5.2. THE X-RAY OPACITY IN THE WIND IN HMXBs

The X-ray opacity in a wind of a massive star has been computed in detail by Vanbeveren et al. (1982) using an admixture of *moderately ionized species* in the wind. This applies for a wind where the temperature is not higher than 10^5 K.

Stevens and Willis (1988) remade these calculations for the absorption in the wind of the WR-star HD 50896. They argued that the ionization should be calculated in detail as outlined below in this section. However, such calculations have never been done in the wind of a WR star.

In the present work, we have repeated the Vanbeveren at al. (1982) computations using updated atomic data and we will demonstrate that calculations where the detailed ionization balance is not followed in detail, give yet very reasonable results as far as the opacity of X-rays in the > 2 KeV region is concerned.

The X-ray opacity is determined by

- the chemical composition of the wind.
- the ionization balance in the wind.
- the X-ray absorption coefficients of the particles in the wind.

The chemical composition.

The He content Y of a wind depends upon the hydrogen content X which can vary between hydrogen exhaustion (X=0) and solar abundance (X=0.7). For C,N and O one assumes either the solar composition either the CNO equilibrium values. The sum of the mass fractions of these five elements covers more than 99% of the total mass. For the other elements which can contribute to the opacity we take the cosmic abundances.

The ionization balance.

The degree of ionization in the stellar wind is determined assuming statistical equilibrium. This ionization equilibrium is only achieved if the highest velocity of the ions is smaller than the velocity of the slowest ionization or recombination rate. One assumes that this requirement is valid in the wind.

The number density (n_e) of free electrons determines the importance of Thomson-scattering and depends critically upon the type of ionization balance which is achieved. Its non-LTE values can be obtained as follows.

Let n_{ij} be the number density (cm^{-3}) of the j-th ionization stage of the i-th element. The number of transitions from (i,j) into the continuum κ=(i,j+1) and vice-versa per sec and per cm^3 are then evaluated from:

1. the number of photo-ionizations:

$$n_{ij}R_{ij\kappa} = n_{ij}\frac{4\pi}{h}\int_{\nu_{ij}}^{\infty} \alpha_{ij\kappa}(\nu)\frac{J_\nu}{\nu}d\nu \qquad (14.6)$$

i.e. the number of photoionizations $n_{ij}R_{ij\kappa}$ (sec^{-1} cm^{-3}) depends on the radiation density J_ν (ergs cm^{-2} sec^{-1} hz^{-1}) and the photo-ionization cross-sections $\alpha_{ij\kappa}(\nu)$ (cm^2),

2. the number of collisional ionizations:

$$n_{ij}C_{ij\kappa} = n_{ij}n_e\int_0^{\infty}\sigma(v)f(v)v dv \qquad (14.7)$$

i.e. the number of collision ionizations depends on the cross- section σ (cm^2) for producing the transition (ij ---> κ) and the electron temperature. Both ionization processes include auto-ionization so that the continuum κ includes the doubly excited states of the species (i,j).

3. the number of radiative recombinations is calculated as:

$$n_{i,j+1}n_e\alpha_{\kappa ij}(T) \qquad (14.8)$$

i.e. the number of radiative recombinations depends on the electron temperature through the radiative recombination coefficient $\alpha_{\kappa ij}$. This recombination process also includes the doubly excited states of the species (i,j) into its continuum so that dielectric recombination is taken into account.

Three body recombination at a rate $n_{i,j+1}n_e^2C_{\kappa ij}(T)$ can be disregarded because the electron number density in a typical stellar wind is still too low for this process to occur.

The ionization equilibrium is then achieved for each atomic species i by balancing exactly the reaction rates (14.6), (14.7) with (14.8). Every transition has to be summed over all subshells (i,j). The quantity ν_{ij} is the frequency corresponding to the ionization edge of this subshell.

The number density of free electrons determines the importance of Thomson-scattering.

The absorption coefficient.

The extinction coefficient per centimeter at location r and frequency ν, $\kappa_\nu(r) = k_\nu(r)\rho(r)$ is the sum of all relevant bound-free true absorption coefficients $\kappa_{bf}(r,\nu)$ and of electron scattering $\kappa_e(r) = \sigma_e.n_e(r) = k_e(r).\rho(r)$. The cross section for electron scattering is independent of wavelength and given by

$$\sigma_e = 0.665 \text{ barn} = 0.665 \times 10^{-24} \text{ cm}^2 \qquad (14.9)$$

For a fully ionized gas also k_e is constant and equals:

$$0.19(1+X)\frac{\text{cm}^2}{\text{g}} \qquad (14.10)$$

The absorption coefficient $\kappa_{bf}(r,n)$ characterizes the contribution to the extinction of bound-free photo absorption from a bound state b into the continuum f. For one bound state b this local quantity is calculated with the corresponding cross section $\alpha_{bf}(n)$ through:

$$\kappa_{bf}(r,\nu) = \alpha_{bf}(\nu).n_b(r) = k_{bf}(r,\nu).\rho(r) \qquad (14.11)$$

where $n_b(r)$ is the local number density of radiation absorbing bound state b.

In order to obtain the total bound-free extinction at frequency ν integrated over all constituents in the wind, one has to add quantities as defined in (14.11) over all bound states which at a given location contribute to an absorption at frequency ν. Summed over all atomic and ionic species one thus obtains:

$$\kappa_{bf}(r,\nu) = k_{bf}(r,\nu)\rho(r) = \sum_b n_b(r)\alpha_{bf}(\nu) = [\sum_{Z,b} \frac{X_Z}{Z.m_H}.\frac{N_b}{N_Z}.\alpha_{bf}(\nu)]\rho(r) \qquad (14.12)$$

X_Z is the abundance by mass of the species with atomic mass Z and m_H is the proton mass; N_b/N_Z is the fraction of atoms (or ions) of species Z which are in the appropriate bound state b. The abundance X_Z of an element can be taken as constant throughout the wind, whereas the fraction N_b/N_Z is a local quantity and determined by the ionization and excitation needed to obtain the bound state b. The most simplifying assumption of an isothermal wind can lead to a value of N_b/N_Z which is constant throughout the wind. In that case the term $k_{bf}(r,\nu) = k_{bf}(\nu)$ is independent from r. Moreover, the decay times of the excited bound states are much shorter than the mean life time of the ground level against photo-ionization, so that to a very good approximation all bound-free absorbtions of photons come from the ground states of the atoms and ions in the wind.

Below the ionization threshold the bound-free photo absorption cross sections $\alpha_{bf}(\nu)$ vanish. The $\alpha_{bf}(\nu)$ for any atom or ion above the ionization threshold can be approximated with Kramers' law:

$$\alpha_{bf}(n',\nu) = 2.815 \times 10^{29}\, N \frac{g_{II}(n',\nu)}{n'^5 \nu^3} Z_{eff}^4 \text{ cm}^2 \qquad (14.13)$$

In this expression N equals the number of equivalent electrons in the bound shell, n' is the principal quantum number of these electrons and the energy of the electrons in that shell is given by $E_{n'} = -\frac{Z_{eff}^2}{n'^2}$ Ryd

The deviation of the Gaunt factors $g_{II}(n',\nu)$ from unity is usually disregarded.

Bethe and Salpeter (1957) give an exact formula for the bound-free absorption cross section per absorbing K- and L-electron. For absorption through one K-shell electron they find:

$$\alpha_{bf}(n',\nu) = \frac{128}{3}\frac{\pi e^2}{m c}\left(\frac{\nu_1}{\nu}\right)^4 \frac{1}{\nu_1}\frac{e^{-4q'\operatorname{arccot} q'}}{1-e^{-2\pi q'}} \text{ cm}^2 \text{ with } q' = \sqrt{\frac{E_1}{E-E_1}} \qquad (14.14)$$

ν_1 and E_1 are the frequency and energy at the ionization threshold.

The absorption through any species can be calculated summing up expressions like (14.14). Such calculation considers each electron as moving in a central field.

Recent calculations of bound-free coefficients have been published in which every possible absorbing electron was considered and for which the wave function of each electron was calculated selfconsistently.

Reilman and Manson (1979) calculate photo-absorption cross sections for atoms and ions with $Z \leq 30$ in the energy range between 5 eV and 5 keV whereas analytical fits are presented by Verner and Yakovlev (1995). The latter reproduce the tables of Reilman and Manson accurately and yield the cross sections above 5 keV which will be needed in the subsequent calculations. The simple parameterization of Verner and Yakovlev also reproduces the exact analytical solution (14.14) for the ions of the iso-electronic sequence of hydrogen.

Apart from H, He, C, N and O the total amount of 30 chemical species considered here have been taken with their solar photospheric abundaces (Anders and Grevesse, 1989) and constitute more than 99,9 % of the total mass.

The absorption in the keV- region is largely dominated by the absorption from the K-shell. The absorption through all the other shells has, however, been included in our calculation. For atoms with "many" electrons the absorption behaviour of the inner electrons is not much influenced by the exact number of outer electrons. Only when the atom is left with only one electron and the K-shell absorbs as in the hydrogenic case, following expression (14.14), does the cross section reduce significantly and lose approximately 50% of its absorbing power. Obviously, when completely stripped, the atom loses all of it. We therefore conclude:

> *the value of the absorption in the keV region ($k_{bf}(\nu)$) depends only marginally on the ionization equilibrium actually attained. Significant changes occur only when dominant chemical species turn into hydrogenic ions or are completely stripped off.*

Table 14.1 shows the contribution of the different species to the absorption coefficient in the region [2-20] keV. Since we will apply the theory mainly to WR+cc candidates, we have taken a hydrogen exhausted wind with $X = 0$ and $Y = 0.98$ but any other mixture with $X \neq 0$ and $X+Y = 0.98$ can be calculated in a similar way. In figure 14.18 we compare the absorption coefficient for $X = 0$ and 0.7; C, N and O have been taken with their equilibrium abundances (section 3.3.1) and all the atoms or ions are in their ground state. The table illustrates clearly the statement above that $k_{bf}(\nu)$ does not depend very much on the degree of ionization. The absorption coefficient is only significantly altered when one of the major contributors to the absorption turns into an ion of the hydrogen isoelectronic sequence or has lost all its electrons, in which case its contribution to the absorption vanishes abruptly. Apart from noticable differences at the lower edge of the [2-20]keV band, between admixtures of different ionizations, the absorption coefficient starts to melt dramatically away from the moment that C is completely ionized.

Since the ionization in the WR wind will only be moderately high

a reliable estimate of the hard X-ray spectrum of a HMXB produced by accretion of stellar wind mass by a compact star can be calculated without knowing the detailed ionization balance in the wind.

TABLE 14.1: The absorption coefficient k_{bf} for different ionization of the contributors (referred to in the last column) with X +Y = 0.98, equilibrium abundances for C,N,O and cosmic abundances for the other elements. The value of k_e must be added to k_{bf} in order to obtain the total absorption coefficient.

Energy (keV): 2	3	4	5	6	7	8	10	12	14	16	18	20	
22.146	7.227	3.179	1.681	0.986	0.626	0.915	0.504	0.301	0.194	0.132	0.094	0.069	(neutral; a)
17.830	6.083	2.737	1.471	0.872	0.558	0.872	0.484	0.290	0.188	0.128	0.091	0.067	(neutral; b)
19.601	6.561	2.922	1.557	0.917	0.585	0.882	0.489	0.293	0.189	0.129	0.092	0.068	(1 x ion.; a)
16.580	5.769	2.618	1.413	0.840	0.538	0.853	0.475	0.286	0.185	0.126	0.090	0.066	(1 x ion.; b)
15.480	5.480	2.506	1.360	0.811	0.521	0.835	0.469	0.282	0.183	0.125	0.089	0.066	(2 x ionized)
15.572	5.512	2.519	1.366	0.807	0.523	0.840	0.470	0.283	0.184	0.125	0.089	0.066	(3 x ionized)
15.449	5.480	2.505	1.359	0.802	0.520	0.839	0.470	0.283	0.183	0.125	0.089	0.066	(4 x ionized)
15.029	5.375	2.466	1.339	0.793	0.515	0.836	0.468	0.282	0.183	0.125	0.089	0.066	(5 x ionized)
10.774	4.578	2.122	1.161	0.690	0.450	0.777	0.436	0.263	0.170	0.116	0.083	0.061	(6 x ionized)
7.368	3.558	1.695	0.946	0.567	0.374	0.743	0.421	0.256	0.167	0.114	0.082	0.060	(7 x ionized)
7.100	3.474	1.661	0.929	0.558	0.369	0.739	0.420	0.255	0.166	0.114	0.081	0.060	(8 x ionized)

(a) All species [Z≤30] without hydrogen (X = 0)
(b) All species [Z≤30] with hydrogen (X=0.7)
k_e has to be added as follows:
X = 0 yields k_e =0 (neutral) ; k_e=0.095 (1x ionized); k_e= 0.19 (all other cases)
X = 0,7 yields k_e =0 (neutral) ; k_e =0.296 (1x ionized); k_e = 0.324 (all other cases)

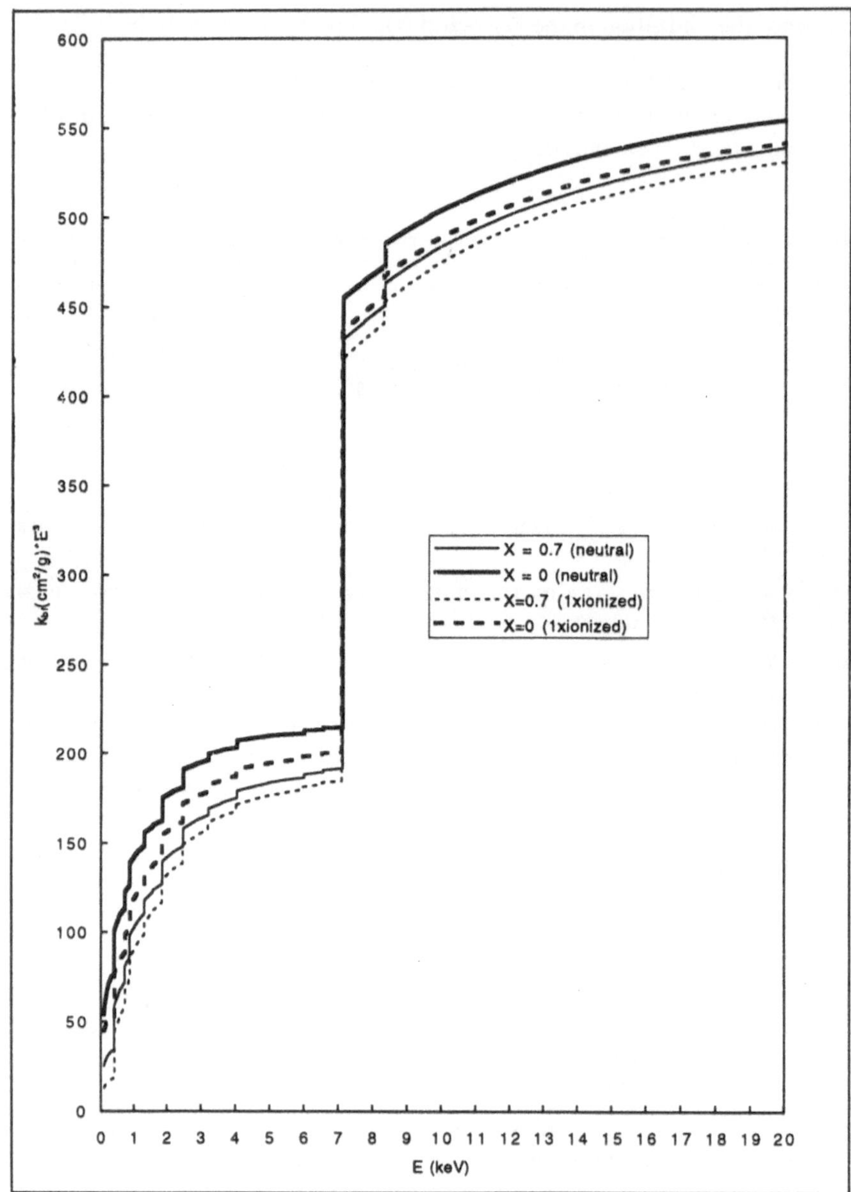

Figure 14.18: Comparison of k_{bf} (in cm^2/g) between a hydrogen exhausted neutral gas and a gas in which hydrogen has its solar abundance. The dashed lines correspond to the case where every atom is ionized once. The gas contains atoms with their solar photospheric abundances, except for CNO which have been taken with their equilibrium abundances. Observe many absorption edges out of which the Fe-edges near 7.1 keV are predominant. The vertical axis values have been multiplied by E^3 to blow up the scale at higher energies.

14.5.3. THE EXPECTED X-RAYS IN WR+cc BINARIES

HD 50896 and HD 197406

Obviously if HD 50896 and HD 197406 are WR+cc binaries, no hard X-rays are expected when the cc is a too fast spinning NS ($P_s < P_{eq}$, section 13.3.3).

Suppose now that the cc is a BH or a NS with $P_s \geq P_{eq}$. The period of variability of HD 50896 (resp. HD 197406) equals 3.8 days (resp. 4.3 days) (section 13.3.3). Combining the results of sections 14.5.1 and 14.5.2, it is straightforward to compute the expected X-ray luminosity (in different pass-bands), for different values of all parameters in the model, for WR+cc binaries with orbital period between 3 and 5 days. Accounting for the distance of HD 50896 (the Hipparcos distance ~0.4-1 kpc, Van der Hucht et al., 1997) and HD 197406 (Moffat and Seggewiss, 1980) and the sensitivity of present X-ray detectors, both stars should have been observed as relatively bright hard X-ray sources, but they are not. The latter conclusion was also drawn by Vanbeveren et al. (1982) although older opacity data tables were used. We therefore conclude

- if HD 50896 and HD 197406 are WR+[low mass component] binaries, the observed lack of hard X-ray radiation can only be explained when the low mass companion is a normal solar type star or when the low mass component is a NS rotating fast enough to prohibit mass accretion.

Cyg X-3

We will try to give an answer to the following question:

accounting for all observed features of the HMXB Cyg X-3, is it possible to find a model where the optical component is a 'normal' Population I WR star with a spherically symmetric stellar wind?

a. General

Apart from the variation as a function of orbital phase, the X-ray flux varies in time (see e.g. the review of Bonnet-Bidaud and Chardin, 1988). Using a distance of ~ 10 kpc (Dickey, 1983) its 2-20 keV X-ray luminosity varies between ~10^{37} erg/s (low state) and ~2.10^{38} erg/s (high state). The observed high and low state spectrum are shown in figure 14.19. Here, we will focus on the high state.

The absence of a X-ray eclipse in the quasi-sinusoidal variation of the X-ray emission -with Period = orbital period of ≈ 0.2 days- indicates a rather low inclination (i = 20-60°) although Van Kerkwijk (1993) argues a value i = 74°.

As the compact star revolves around the WR-star within the WR-wind the X-ray emission $L_{X,obs}$ will vary in a sinusoidal way with the orbital period.

We choose the phase Φ so that $L_{X,obs}(\Phi=0)$ is at minimum intensity and $L_{X,obs}(\Phi=\pi)$ is at maximum intensity. It is clear that $L_{X,obs}(\Phi=\pi/2)$ = $L_{X,obs}(\Phi=3\pi/2)$ will show the mean intensity which for a spherical symmetric WR-wind equals the constant value which $L_{X,obs}$ would have at zero inclination.

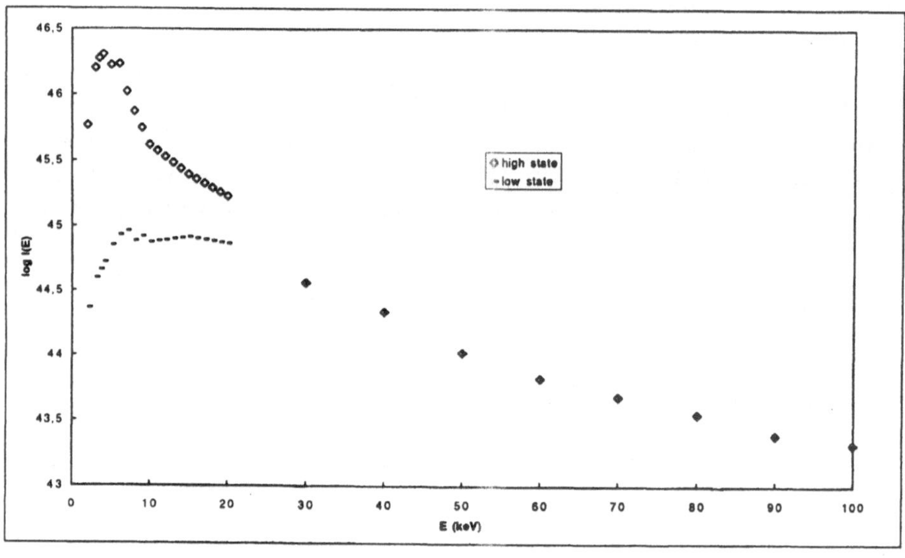

Figure 14.19: The observed typical high and low state spectrum of the HMXB Cyg X-3 between [2-100] keV. The high state is scaled so that L_X[2-20] keV = 2.10^{38} erg/s at a distance of 10 kpc (from Bonnet-Bidaud and Chardin, 1988).

Schmutz et al. (1996) derived a velocity amplitude K = 480 ± 20 km/s. The mass function f(m) =2.3 M_o. With a realistic WR mass value, it follows that the mass of the cc is significantly above the mass value of all known NSs, and it can therefore be considered as a BH candidate. The former result is still quite uncertain so that we have chosen two models: one where we ignore the mass function and simply assume that the compact star is a NS with mass equal to

1.4 M_o, and a second model where we account properly for the mass function and where the cc is assumed to be a BH candidate.

Van Kerkwijk et al. (1992) measured the variation of the orbital period $\dot{P} = 1.6 \cdot 10^{-6}$ yr^{-1}. If the optical star is losing mass by a spherically symmetric stellar wind, it follows from equation 10.6 that there is a relation between the mass loss rate and $\dot{P}$, which in the case of Cyg X-3 corresponds to

$$\dot{M} = 0.8 \cdot 10^{-6}(M_1 + M_2) \tag{14.15}$$

the masses in M_o, $\dot{M}$ in M_o/yr.

b. <u>The intrinsic X-ray luminosity of Cyg X-3</u>

With the accretion model of section 14.5.1, it is straightforward to compute L_{acc} for different values of the parameters of the model. We use bremsstrahlung [equation (13.18)] to describe the intrinsic X-ray spectrum since this explains the observations for energies E > 20 keV best.

The temperature in expression (13.18) is chosen such that the observed energy dependence of the X-rays is explained. The observations show that the photon flux distribution at high energies $\div E^{-3}$ (Bonnet-Bidaud and Chardin, 1988) so that $I(E)_{obs} \div E^{-2}$. Since the intrinsic X-ray I(E) radiation has to follow approximately the same power law, it can readily be checked that a bremsstrahlung spectrum with a temperature around kT = 4 keV will describe the observations best and will be used in the subsequent calculations.

With this value of kT, using definition (13.14) for L_X and (13.18) for I(E) one obtains L_X(2-20 keV) = $0.6L_{acc}$

c. <u>The theoretically predicted X-ray luminosity of Cygnus X-3</u>.

A WR-star is surrounded by a WR-wind having a density structure given by the expressions (14.2) together with (14.3).

The observed energy distribution $I(E)_{obs}$ at $\Phi = \pi/2$ and $\Phi = 3\pi/2$ of the X-ray luminosity is calculated with (14.4) and the X-ray luminosity at mean epoch (between minimum and maximum) with (14.5).

The optical depth $\tau(E)$ at the location of the cc can now be calculated using definition (14.1) together with table 14.1 for any ionization admixture.

Since we have shown that k(E) is constant over a wide range of degree of ionization we have taken as a representative case the values of k(E) from table 14.1 for an admixture in which all chemical species are ionized twice.

We have computed the expected value of $L_{X,obs}$(2-20 keV) as a function of the parameters in the model. In a first attempt, we have chosen different values for the stellar wind mass loss rate independently of the constraint given by equation (14.15). Furthermore, although the masses of most of the WR stars in binaries are larger than 5 M_o, accounting for the mass value of the WR component in the binary V Sagittae (section 13.2) we also allow for models where the WR mass is smaller than 5 M_o. We thus looked for solutions in the following parameter space:

- model 1: the mass of the compact star equals 1.4 M_o
  model 2: the mass of the compact star satisfies the mass function of the system,
- the mass of the WR star ranges between 1 M_o and 20 M_o,
- $\dot{M}$ of the WR star ranges between 5.10^{-7} M_o/yr and 5.10^{-5} M_o/yr,
- the efficiency factor ξ in equation (13.10) between 0.1 and 1,
- v_{wind}(A) between 500 km/s and 2000 km/s,
- $v_\infty > v_{wind}$(A) between 1000 km/s and 3000 km/s,
- the parameter β in the velocity law ranges between 0.7 and 1.5.

Comparison with the observed value of 2.10^{38} erg/s already excludes a large number of parameter values. For the models that are left, we then calculate the expected X-ray spectrum which would be observed after absorption through the WR-wind and compare it with the observed spectrum of figure 14.19. The results are shown in figure 14.20 for two typical models with $L_{X,obs}$(2-20 keV) = 2.10^{38} erg/s. The figure illustrates why again a large number of parameter sets are unlikely.

From the models that remain after the comparison, we conclude that

- the observed X-ray luminosity and distribution of X-rays over energies in Cyg X-3 can be explained only if the compact star is a BH with a mass that satifies the mass function proposed by Schmutz et al. (1996) and if the SW mass loss rate of the CHeB component < 10^{-6} M_o/yr.
  The best correspondence is achieved when the WR star has a mass equal to 2 M_o (analogous to the WR star in the binary V Sagittae) and the compact companion has a mass ≈ 8 M_o.

However, this best model (as well as all other models with a small stellar wind mass loss rate that explain more or less the observed X-ray luminosity and distribution of X-rays over energies) does not meet the constraint given by equation (14.15), so that we are forced to admit that *it is not possible to find a model that meets all observed properties of the HMXB Cyg X-3 where the optical component is a 'normal' Population I WR star with a spherically symmetric stellar wind.*

The latter corresponds to the conclusion of Mitra (1996) who used a simplified absorption mechanism.

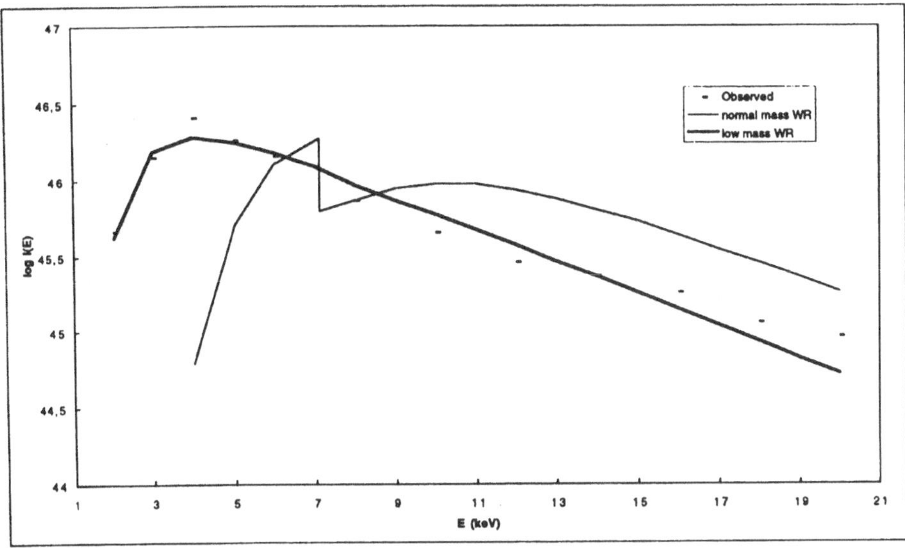

Figure 14.20: The observed [2-20] keV spectrum of the HMXB Cyg X-3 compared to the theoretically predicted one for two models: the full bold line is considered as a good fit and holds for the parameters [M_{BH} = 8 M_o, M_{WR} = 2 M_o, $\dot{M}_{WR}$ = 5.10^{-7} M_o/yr, ξ = 0.35, $v_w(A)$ = 1000 km/s, v_∞ = 1500 km/s]; the thin line is representive for a bad fit and corresponds to [M_{BH} = 14.6 M_o, M_{WR} = 10 M_o, $\dot{M}_{WR}$ = 10^{-5} M_o/yr, ξ = 1, $v_w(A)$ = 1000 km/s, v_∞ = 2000 km/s].

An alternative model for Cyg X-3 is the following: the optical component is a $\leq 2\ M_o$ WR-like star similar to the one in the binary V Sagittae. If the star is at the end of its CHeB it will expand, reach its Roche lobe for a second time (case BB, section 12.2.2) and start mass transfer. The very strong X-rays may be capable of driving part of the mass lost by the CHeB component out of the binary. If it is assumed that Cyg X-3 is in this phase, equation (14.15) is obviously no longer valid and the constraint against a small stellar wind mass loss disappears.

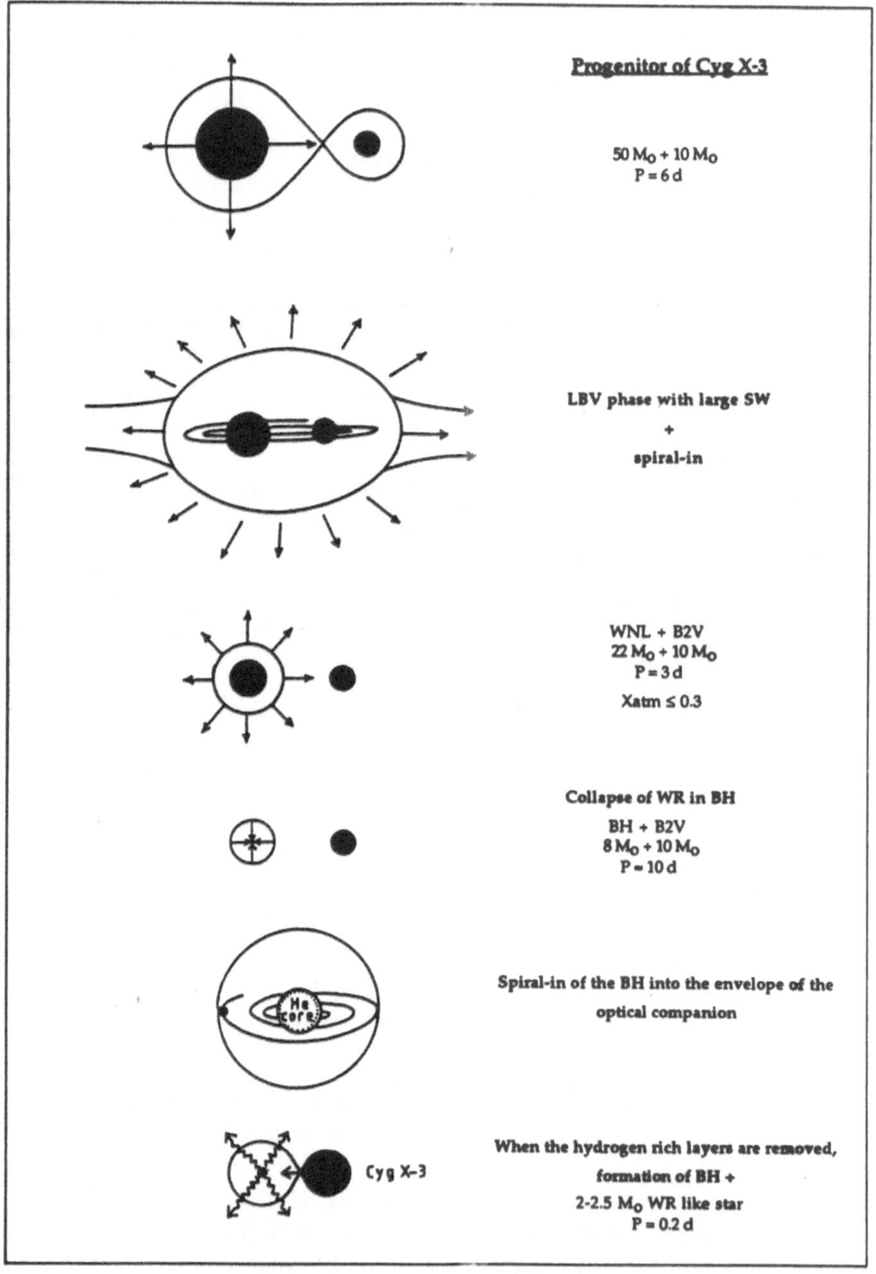

Figure 14.21: An evolutionary model for the MCB producing a system like Cyg X-3.

An MCB evolutionary model

Cyg X-3 is a CHeB+cc candidate where the CHeB component is a WR-like star with mass $\leq 2\ M_o$ similar to the one in the binary V Sagittae (section 13.2). Using the mass function of the system advanced by Schmutz et al. (1996), it follows that the mass of the cc $\geq 5\ M_o$ and is thus very likely a BH candidate. A probable evolutionary scenario is sketched in figure 14.21. We start with a binary with a secondary with mass ~10 M_o and a 50 M_o primary. The primary first loses a lot of mass during its LBV phase, but since its initial mass is not too different from 40 M_o, we assume that despite the LBV SW mass loss, a spiral-in phase occurs but with a largely reduced efficiency of course. At the end of its evolution, the primary collapses into a 8 M_o BH (thus no SN). When the 10 M_o secondary reaches its Roche lobe, the further evolution is governed by the common envelope/spiral-in process. When most of the hydrogen rich layers of the secondary are removed, we are left with a 2-2.5 CHeB star which may show up (in the IR) as a WR like star.

At the end of CHeB (during He shell burning), the WR like star expands and reaches its Roche lobe. The resulting mass transfer makes a strong X-ray source like Cyg X-3.

Together with RLOF mass transfer, the WR like star may lose mass by a spherically symmetric SW, but the $\dot{M}$ must be smaller than $10^{-6}\ M_o/yr$ to assure a small X-ray absorption efficiency.

14.5.4. THE FORMATION OF WR STARS WITH A cc IN THEIR CENTER

From the discussion in the previous subsections, it should be clear that the formation of a CHeB+cc binary (thus of a WR+cc) is very rare and that most of the OB+cc binaries will merge to form Thorne-Zytkow objects (TZO) (section 12.5). If, while being a RSG, a TZO loses its hydrogen rich layers, the remnant may resemble a WR star, although it has a compact star in its center. At least part of these stars may be runaway stars.

14.6. The evolution of LMXBs: evidence for MCBs with very low mass ratio.

Her X-1

This LMXB consists of a X-ray pulsar (a NS with a 95% probability mass of $1.4\pm0.3\ M_o$, Van Kerkwijk et al., 1995) and an A-type companion with a mass $\approx$ 2 M_o. Her X-1 is located at a height z = 3 kpc out of the galactic plane. Since a 2 M_o star has an age around 800 million years, starting from inside the galactic disc, the z-component of its velocity at the moment of explosion must have been larger than 120 km/s in order to reach its present height against the gravitational pull of the Galaxy.

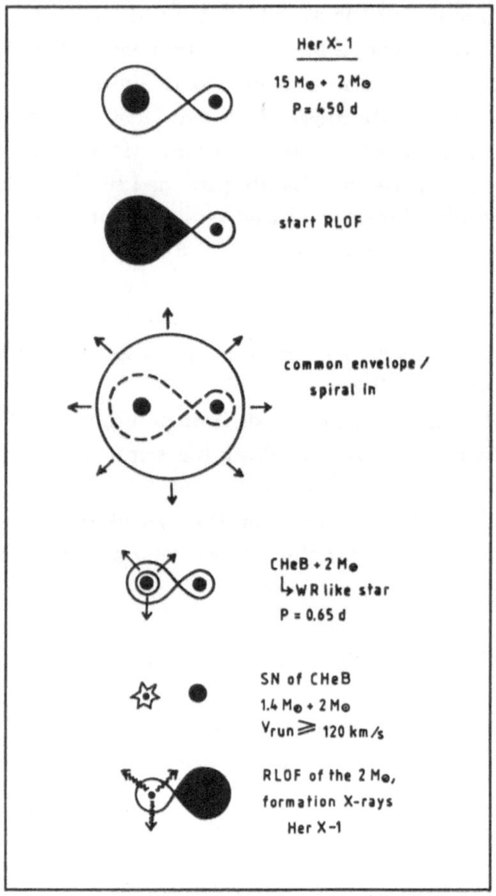

Figure 14.22: The MCB evolutionary model explaining the observed properties of the LMXB Her X-1.

The only binary model that explains the observed properties of the LMXB Her X-1 is illustrated in figure 14.22. It is similar to the original suggestion made by Sutantyo (1975) (see also Verbunt et al., 1990). The initial system consists of a massive star with mass around 15 M_o and a 2 M_o companion, with a period larger than 1 year. The system evolves through a common envelope phase (section 12.1.2) and at the end of it, it consists of a ~ 4 M_o CHeB star and a 2 M_o A-type star with a period ~1 day. The CHeB star explodes leaving behind a NS with mass ~ 1.4 M_o in an eccentric orbit around the A-type star. The system will acquire a peculiar space velocity of the order of 100 km/s. As the A-type companion evolves and reaches its own Roche lobe, the initial stage of the mass transfer transforms the NS into a X-ray source.

LMXBs with a BH component

Observed properties of three LMXBs with a BH candidate are given in section 13.3.3. They are located within 3 kpc from the Sun. Assuming a mass ~0.7 M_o for the optical stars in the three LMXBs of table 13.8, it follows that the mass of the X-ray source is larger than 3.4 M_o, which is considerably above the maximum mass for NSs.

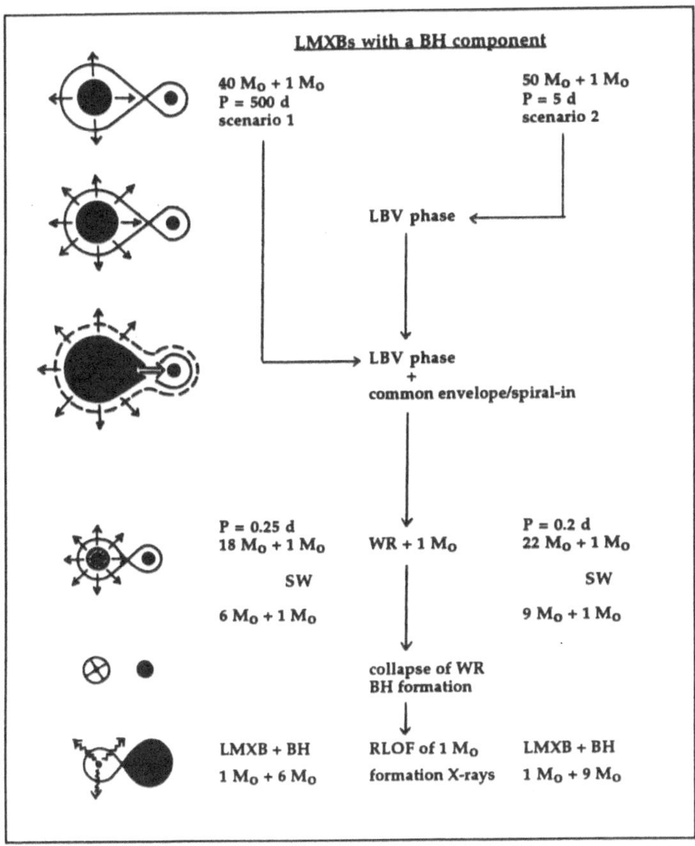

Figure 14.23: Two possible evolutionary scenarios for the LMXBs with a BH component.

Accounting for the probable evolutionary scenario of the HMXB Wray 977 (section 14.4.1), the minimum mass for BH formation in a MCB component ≥ 40 M_o. Two possible scenarios for these LMXBs are shown in figure 14.23. The first one with the 40 M_o primary is comparable to the suggestion made by De Kool et al. (1987). The system has a low mass secondary and a period of ~500 days. Due to the extreme mass ratio, the binary evolves through a common envelope phase. The latter phase stops when the primary has lost most of its hydrogen rich layers. Using the evolutionary computations of section 12, it then has a mass ≥ 18 M_o and resembles a WNL star whereas the binary period ~0.25 days. The star starts losing mass by SW at a rate which is typical for a WR star. Just prior to the collapse of the FeNi-core, the remaining mass of the star is ~ 5 M_o. If the collapse results in the formation of a BH, no SN occurs and thus the system remains bound.

In the second scenario the initial primary mass equals 50 M_o and the initial period P = 5 days. The primary evolves through a LBV phase but we assume that the SW is not large enough to avoid the common envelope phase completely. The total mass lost during the latter is very much reduced due to the previous SW phase, and although we started with a binary with a period of 5 days, the period at the end of the common envelope/LBV phase ~ 0.4 days. From thereon, the evolution is similar to the first scenario.

It is important to realize that

- if stars with initial mass > 40 M_o lose a significant part of their hydrogen rich layers by a LBV type SW and end their life as a BH, most of the binaries with a > 40 M_o primary and a low mass secondary will avoid merging (which would occur if the binary went through a normal common envelope/spiral-in phase), and most of them will evolve into a LMXB with a BH component.

Overall conclusion concerning the LMXBs

Accounting for the fact that most of the binaries with small mass ratio, progenitors of Her X-1 like systems, will merge during spiral-in, or will be disrupted during the SN explosion, the existence of only one which has survived the spiral-in and the SN can be considered as an indication that a large number of MCBs with small mass ratio exist.

Also the LMXBs with a BH candidate suggests the existence of very massive close binaries with very small mass ratios. We will discuss the mass ratios of MCBs in more detail in section 16.5.1.

14.7. SN-type of a MCB component: SN 1987A

We expect the following SN types resulting from MCB evolution:

- the final collapse of the FeNi-core of a primary with initial mass > 40 M_o is not accompanied by a SN explosion but leads to the formation of a BH,
- all primaries with initial mass ≤ 40-50 M_o of MCBs that avoid merging produce type I SN explosions. If SW and/or a second RLOF phase has been able to bring He burning products to the surface, the explosion will probably result into a type Ic,
- due to accretion and the effect of semi-convection, a mass gainer of a binary with mass ratio close to one, remains a blue star throughout its entire life, producing a SN event like SN 1987A; how close to one depends on the model in order to treat semi-convection,
- when merging of a secondary with a primary is comparable to an accretion process, mergers can also be expected to produce SN 1987A like events,
- when the SN explosion disrupts the binary (as will be shown in section 16 and 19, most of the MCBs are disrupted), and if the initial mass ratio of the binary was not too close to one, the SN type of the single (runaway) OB-type mass gainer corresponds to the SN types predicted by single star evolution,
- if the SN does not disrupt the binary, and if during the ensuing spiral-in phase merging is avoided, the remnant CHeB star explodes as a type I,
- the final fate of TZO rests on speculation; possibly the star-matter around the cc is removed by SW and the cc reappears again. They do not contribute to the SN population.

Binary scenarios for the progenitor of SN 1987A

The fine tuning required to reproduce all observations of SN 1987A makes single star scenarios rather doubtfull. Two binary evolutionary models:

- the progenitor was a former 10-16 M_o secondary in an early case B_r MCB with mass ratio close to unity evolving in a quasi-conservative way. How close to unity depends on the way semi-convection is treated (section 12.3). When the SH criterion is used, the mass ratio must be large enough that the RLOF of the primary starts if also the secondary is a post-CHB, hydrogen shell burning star. The evolution is illustrated in figure 14.24. The initial system is a 12 M_o + 11.98 M_o close binary, with LMC initial abundances, evolving in a conservative way (De Loore and Vanbeveren, 1992; Vanbeveren, 1994). Accretion starts when the secondary is a hydrogen shell burning star. After accretion, the star remains blue throughout its entire life and explodes as a blue supergiant. It is imperative for this scenario that the initial mass ratio of the binary has to be very close to one and therefore the frequency of occurrence should be quite small

When semi-convection is treated as a diffusion process with small efficiency, the mass ratio restriction becomes much less severe. In figure 14.25, we show the evolution of a binary with mass ratio q = 0.9 and with a 12 M_o primary. Semi-convection is treated with the diffusion approximation with small efficiency [α = 0.01 in equation (3.23)]. As can be noticed, also in this case the mass gainer remains blue throughout its entire life,

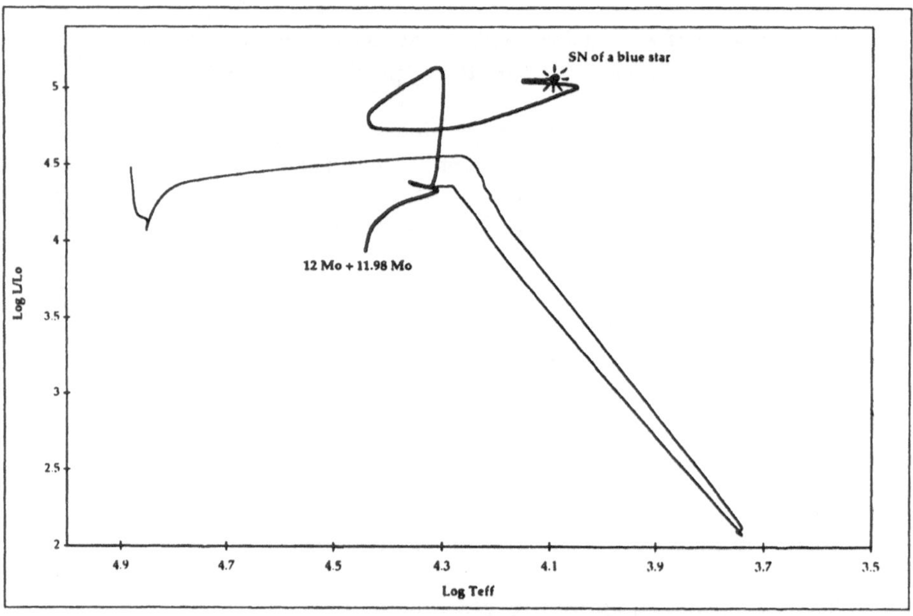

Figure 14.24: The conservative evolution of a 12+11.98 M_o close binary where accretion is treated in the standard way and semi-convection by using the SH criterion; the mass gainer explodes as a blue star.

- the progenitor is a merged binary (Podsiadlowski et al., 1990). Merging may occur when the binary has a small mass ratio or when the binary evolves in a highly non-conservative way, accompanied by the loss of substantial orbital angular momentum. When the secondary is swallowed by the primary, a situation similar to the one sketched above may occur, i.e. if merging is similar to accretion of a mass equal to the mass of the secondary, a merger may remain in the blue part of the HR diagram.

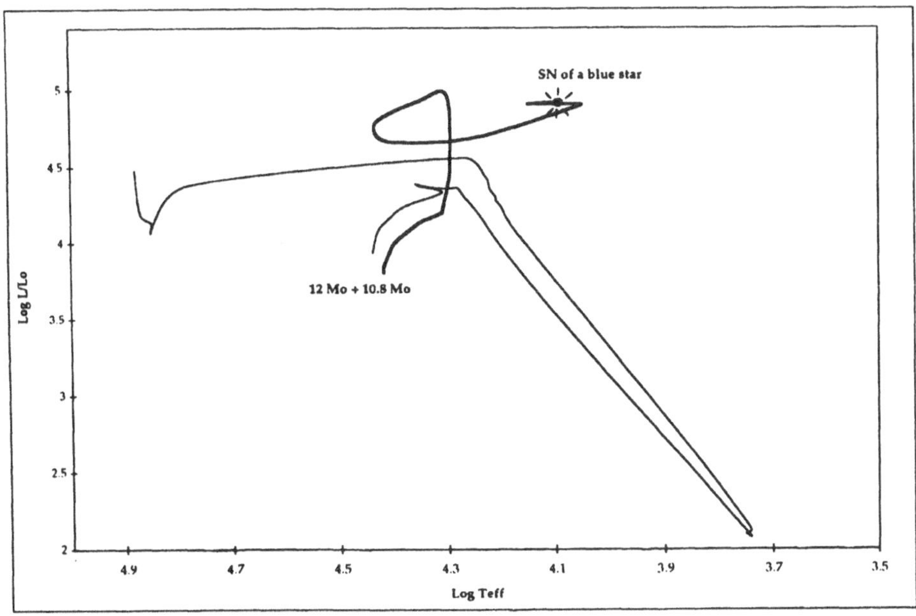

Figure 14.25: The conservative evolution of a 12+10.8 M_0 close binary where accretion is treated in the standard way and semi-convection by using the diffusion approximation with small efficiency. The mass gainer explodes as a blue star.

14.8. Summary: the overall MCB evolutionary model

Accounting for the evolutionary computations of MCBs discussed in section 12, and the conclusions resulting from a comparison between these computations and observations, we give in figures 14.26a,b our preferred overall MCB evolutionary model.

if $M_1 \geq$ 40-50 M_o

if $q \leq 0.2$

case A	**larger period binaries**
spiral-in	***LBV-scenario***
mergers possibly populating the BHG	WR +OB dwarf or low mass (solar type) companion ($P \geq 10$ days)

if $q > 0.2$

case A	**larger period binaries**
(quasi)-conservative RLOF	***LBV-scenario***
WR +OB	WR + OB ($P \geq 10$ days)

if $M_1 <$ 40-50 M_o

if $q \leq 0.2$

case A, early case B_r	**larger period binaries**
spiral-in	***spiral-in and/or RSG stellar wind***
mergers	sdO/WD/WR like star/WR + low mass (solar type) companion
possibly populating the BHG	depending on the importance of the RSG stellar wind mass loss, either small period or period of the order of a year

if $q > 0.2$

case A, case B_r	**larger period binaries**
(quasi)-conservative-RLOF	***common-envelope and/or RSG stellar wind***
WR + OB	sdO/WD/WR like star/WR + OB
(1 day $\leq P \leq$ 120 days)	depending on the importance of the RSG stellar wind mass loss, either period of the order of 1-100 days or period of the order of a year

Figure 14.26a: The overall MCB evolutionary model prior to the final collapse of the primary.

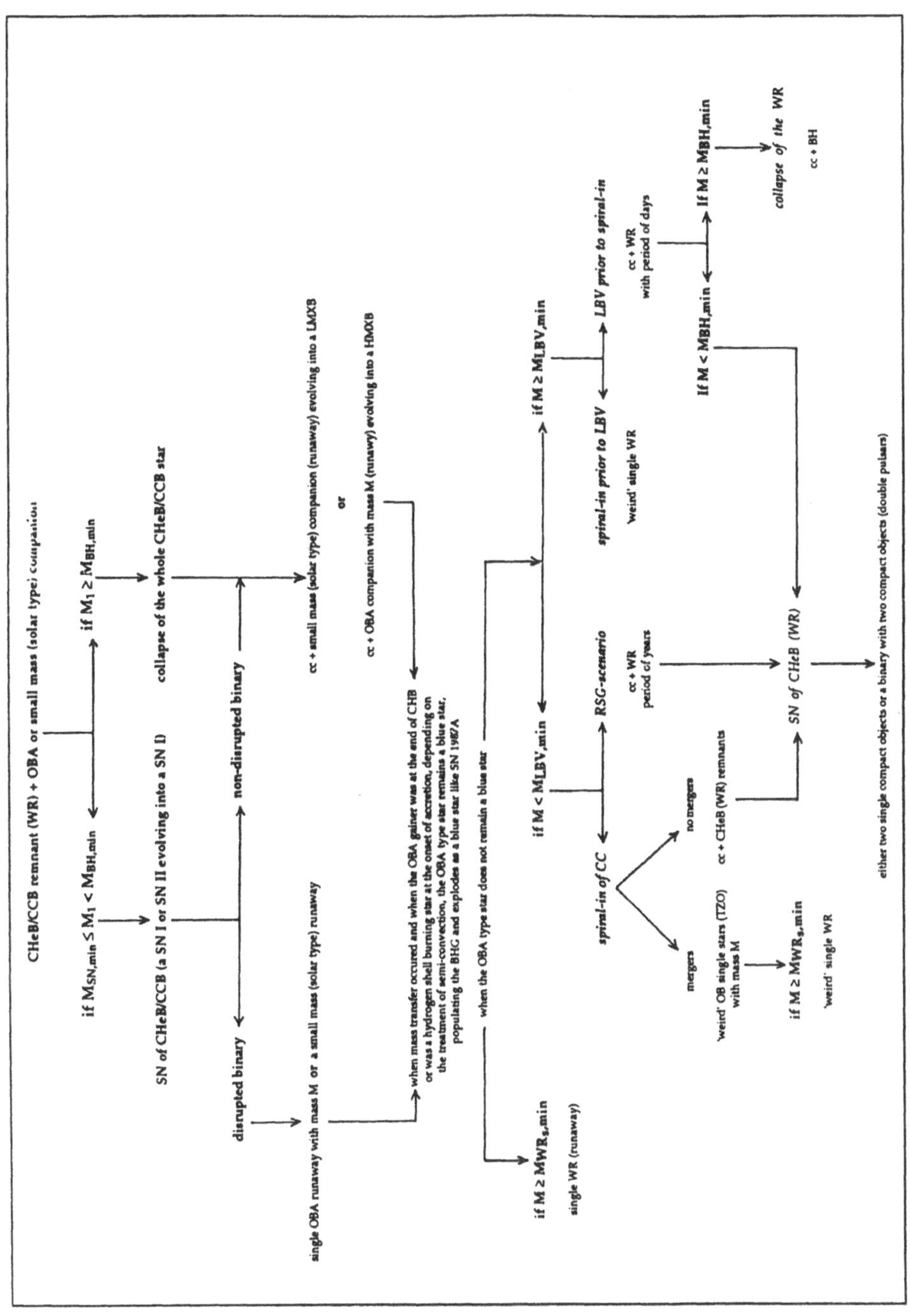

Figure 14.26b: The overall MCB evolutionary model after the final collapse of the primary.

The following abbreviations are used:

$M_{WR_b,min}$ = the minimum initial mass a primary must have to evolve into a normal WR binary component; if we trust the observed luminosities of WR stars, the minimum mass of a WR star ~5 M_o corresponding to an initial primary mass ~18 M_o. Note, however, that the masses of the WR components of the observed WR+OB binaries seem to indicate that they all originated from primaries with initial mass ≥ 30-35 M_o,

$M_{WR_s,min}$ = the minimum initial mass a single star must have to evolve into a normal WR star via RSG SW mass loss; for the Galaxy this is ~15-20 M_o,

$M_{LBV,min}$ = the minimum initial mass a star must have so that it will lose a significant part of its hydrogen rich layers by a LBV type SW reducing the mass lost by a possible RLOF; for the Galaxy, its value ≥ 40-50 M_o,

$M_{SN,min}$ = the minimum mass a primary of a CB must have in order to end its life as a NS accompanied by a SN explosion; when convective core overshooting during CHB is small, $M_{SN,min}$ ≈ 8-10 M_o; rapid rotation during CHB may lower this limit,

$M_{BH,min}$ = the minimum mass a primary of a CB must have in order to end its life as a BH; when convective core overshooting during CHB is small, $M_{BH,min}$ ≈ 40-50 M_o; rapid rotation during CHB may lower this limit.

A 'low mass WR' star is a faint WR-like star with mass < 5 M_o as observed in V Sagittae or in Cyg X-3.

We use the term 'quasi-conservative RLOF' for case A and case B_r MCBs which means that the RLOF is accompanied by mass transfer and possibly also by mass loss from the system, expressed in terms of the parameter β [defined in relation (10.7)]. The few observations of WR+OB binaries seem to indicate that some mass has to leave a case B_r binary, but certainly not all of it. How much mass will leave the binary and how this depends on the binary parameters such as primary mass, orbital period and mass ratio, is unclear at present.

MASSIVE CLOSE BINARIES

From detailed modelling of observed binaries which are representative for the class they belong to, we arrive at the following general evolutionary conclusions:

- the orbital parameters of the sdO6+Be binary Φ Per cannot be explained if overshooting during CHB is large and/or if the RLOF is non-conservative,
- the lack of circumstellar rings/shells with small diameter around the majority of the WN+OB binaries (V444 Cyg considered as prototype) indicates that the RLOF in the progenitors was quasi-conservative,
- all the WR components of the observed WR+OB binaries originate from O type stars with initial mass larger than 30-35 M_o if convective core overshooting during CHB is small,
- the mass of the BH component of the HMXB Cyg X-1 indicates that stars with initial mass larger than 40 M_o end their life with a mass > 10 M_o; this restricts the possible stellar wind mass loss models during the CHeB (WR) phase of a massive star, single or binary component,
- the NLTE analysis of the optical component of the HMXB Vela X-1 indicates that large scale mixing has been very important for the evolution of the progenitor, probably due to mass transfer and the resulting spinning-up,
- the orbital masses of the HMXB Wray 977 compared to evolutionary models let us conclude that binary components with initial mass up to 40 M_o form NSs accompanied by a SN explosion,
- the existence of LMXBs like Her X-1 and those with a BH candidate indicate that a significant fraction of all massive stars may have a low mass/intermediate mass companion; most of these binaries merge and therefore this lower mass companion largely affects the evolution of the massive star,
- when semi-convection is a slow mixing/diffusion process, a very probable scenario for the progenitor of SN 1987A is provided by binary evolution, i.e. a binary with primary mass ~10-15 M_o and with mass ratio $q \geq 0.9$ evolving in a quasi-conservative way.

CHAPTER IV

MASSIVE STAR POPULATION NUMBER SYNTHESIS

15. General

Assuming a formation rate of single stars as a function of mass, a formation rate of binaries as a function of primary mass, mass ratio and orbital period, using a model for the evolution of single stars and of the evolution of binaries, it is a straightforward procedure to estimate the number of stars with different subtypes provided that these subtypes are clearly defined and can be linked unambiguously to a stellar evolutionary phase.

The theory of star formation does not allow the determination of the binary frequency, the mass ratio and period distribution of interacting binaries, the mass distribution function of single stars and of primaries of binaries. We thus have to rely on observations.

The evolution of a MCB after the SN explosion of the primary depends on whether the binary was disrupted or not, i.e. on the asymmetry of the SN explosion, expressed in terms of a kick velocity the compact star gets during the SN explosion. The runaway velocity of single NSs (pulsars) may reflect the kick velocity that they receive at birth due to the SN explosion.

Of course we want to compare the theoretical results to observations, and if possible observations in regions of different metallicities. Particularly important are the number ratios B/R of blue to red supergiants; the ratio WR/O of WR stars to O-type stars; the ratio WC/WN of WC type stars to WN type; the WR + OB binary frequency; the q and P distribution of WR + OB binaries; the number of O-type runaways; the number of HMXBs and LMXBs with a MCB evolutionary history; the number of WR+cc binaries; the formation rate of binary pulsars; and the number distribution and formation rate of the different types of SN explosions.

16. Observations used as input in a population model

A population model needs as input the star formation rate of single stars and close binaries, the binary frequency, the mass ratio and period distribution of close binaries, and the runaway velocity of single pulsars (= the kick velocity a compact star gets due to the asymmetry of the SN explosion).

16.1. The initial mass function (IMF)

The initial mass function is the distribution function which describes the probability that a star with a mass in the range [M; M+dM] is formed per year. In order to have an idea what an IMF for stars with initial mass larger than M_{min} looks like, one considers a region where it can be assumed that star formation was continuous and constant over a period larger than or equal to the lifetime of the smallest mass M_{min}. One counts the number of stars within a mass interval [M; M+dM]. This number is divided by the (average) evolutionary lifetime of the mass interval.

An IMF for the solar neighbourhood ($\leq$ 3 kpc from the Sun) has been proposed by Humphreys and McElroy (1984) for stars with initial mass larger than 15 M_o, i.e.

$$\mathrm{IMF} \propto (M / M_o)^{-2.5\pm0.2} \tag{16.1}$$

which is not significantly different from the overall IMF proposed by Scalo (1986). Furthermore, accounting for incompleteness effects, the authors conclude that the IMFs of the LMC and of the SMC are similar to the galactic one.

Two remarks are appropriate:

- as discussed in section 5.1, a massive star may remain embedded in its parent cloud for 1-2 million years. Therefore dividing the number of massive stars in a mass interval by the whole average lifetime may not be correct. As a consequence, the real massive IMF may be less steep,
- depending on the binary frequency, on the mass ratio distribution of MCBs and on the evolution of a binary during RLOF, a significant fraction of a population of massive stars may look like single stars, but are rejuvenated mass gainers of close binaries. The IMF obtained by direct number counts may therefore lose quite some significance.

16.2. The IMF of starburst regions

It was shown in Vanbeveren (1982) that the IMF of stellar aggregates could be considerably less steep than the IMF of a galaxy as a whole. This can be understood as follows.

Suppose that most of the stars form in aggregates (Reddish, 1978) with total mass M_A and that the mass distribution of aggregates (IMF_A) satisfies

$$IMF_A \propto M_A^{-\alpha} \tag{16.2}$$

whereas the mass distribution of massive stars within an aggregate follows the law

$$IMF \propto M^{-\beta} \tag{16.3}$$

Due to the finite value of M_A, and due to the fact that the sum of the masses of the stars which are formed in the aggregate cannot exceed M_A, a maximum mass M_{max} exists for the stars in the aggregate which obviously depends on M_A. It is easy to show that if $\beta > 1$, then

$$M_A \propto M_{max}^{\beta-1} \tag{16.4}$$

Since a star with mass M can only be formed in an aggregate with $M_{max} \geq M$, it follows that the overall initial mass function IMF_T for a galaxy composed of stellar aggregates, is given by

$$IMF_T \propto M^{-b} \int_{M_{A^*}}^{\infty} IMF_A(M_A)dM_A = M^{-\beta} M_{A^*}^{1-\alpha} \tag{16.5}$$

with $M_{A^*} = M_A(M=M_{max})$ and $\alpha > 1$.

Combining equations (16.4) and (16.5), it follows then that

$$IMF_T \propto M^{\alpha(1-\beta)-1} \tag{16.6}$$

Reddish (1978) proposes $\alpha \approx 1.5$-2. This means that in order to arrive at $IMF_T \propto M^{-2.5}$ [as in relation (16.1)], the initial mass function of the (massive) stars within an aggregate should obey the relation $IMF \propto M^{-2}$ ($\alpha = 1.5$) or $\propto M^{-1.75}$ ($\alpha = 2$) which is considerably less steep than the overall IMF_T.

16.3. The IMF of massive single stars and of primaries of MCBs

The observations of massive stars (primarily of O-type stars) are too poor to make a meaningful comparison between the IMF of real single stars and the IMF of primaries of MCBs. For simplicity, one generally assumes that both IMFs are similar. As outlined in Vanbeveren (1982), the necessary condition for the similarity is that the binary frequency and the mass ratio distribution are independent of the total mass of the binary.

The existence of a maximum mass M_{max} of a star in a stellar aggregate, given the total mass of the aggregate (section 16.2), could affect the binary frequency and binary distribution in an aggregate. If the maximum mass is also the maximum mass of the sum of the masses of both components in a binary, it is clear that the frequency of binaries in the mass range above $M_{max}/2$ will decrease. Even more, as an illustration, consider an aggregate where the maximum single star mass = 60 M_o. It is thus possible that a 40 M_o + 20 M_o can form but a 40 M_o + 30 M_o will not form. This means that the existence of a maximum mass in an aggregate will also affect the mass ratio distribution of the binaries with primary masses larger than $M_{max}/2$.

16.4. The overall MCB frequency in the solar neighbourhood

First remark that the binary frequency has not to be confused with the frequency of stars in binaries. The latter obviously accounts for the secondaries, i.e. the frequency of stars (of a certain subtype) in binaries is larger then the binary frequency. Furthermore, we will demonstrate in section 19 that the observed binary frequency can differ significantly from the binary frequency at birth of a population of stars.

Garmany et al. (1980) studied the binary frequency among all known O type stars brighter than m_v = 7 and north of -50° (a total of 67 O type single stars or primaries of binaries). Deleting the O type HMXBs (section 13.3) from the sample, one concludes that 33% of O stars are the primary of a massive close binary (±13% accounting for small number statistics).

In order to get an idea of the binary frequency in the B0-B3 spectral range we proceed as follows: if N = the number of stars, d = the distance of the star (relative to the observer), the real shape of the space distribution of the stars in the solar neighbourhood will be achieved between a disc-shaped distribution and a spherically shaped distribution.

Assuming a spherically homogeneous space distribution

$$\frac{dN}{dr} \propto d^2 \tag{16.7}$$

The apparent magnitude m_v of the star depends on its distance d

$$m_v = M_v - 5 + 5\log d(pc) \tag{16.8}$$

Hence, for constant M_v, the number of stars has to increase as function of m_v according to

$$\frac{dN}{dm_v} \propto 10^{\frac{3}{5}m_v} \tag{16.9}$$

If the stars are distributed homogeneously in a disc rather than in a sphere, it follows that

$$\frac{dN}{dr} \propto d \tag{16.10}$$

and thus

$$\frac{dN}{dm_v} \propto 10^{\frac{2}{5}m_v} \tag{16.11}$$

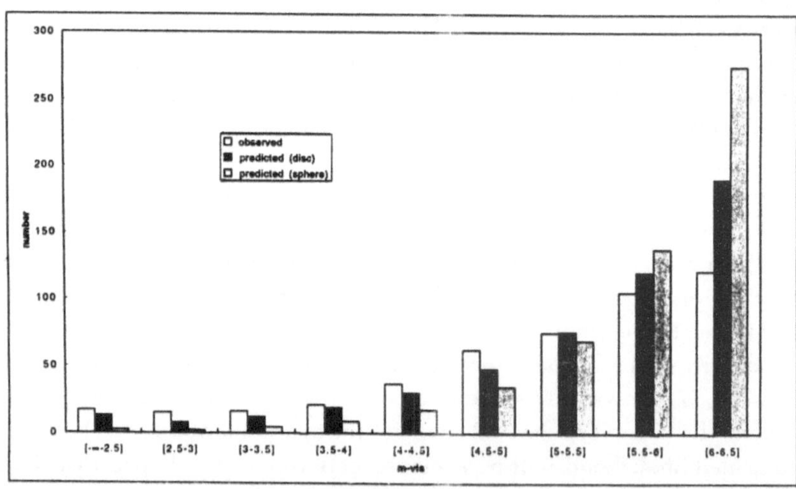

Figure 16.1 The observed number of early B type stars as a function of m_V is compared to a disc model and a spherical model of the space distribution of the stars. The spherical model has to be rejected whereas the disc model is acceptable.

The 'Bright Star Catalogue' = BSC (Hoffleit and Warren, 1991) contains 511 B0-B3 [III-IV-V] stars. The observed number distribution as a function of m_v has been compared to the predictions (16.9) and (16.11). Figure 16.1 shows the comparison between models and observations assuming completeness up to $m_v = 6.5$.

We conclude:

- the dependence of the number of B0-B3 stars on m_v clearly suggests the disc-like distribution of the stars.
- a χ^2 analysis reveals further that the number of B0-B3 stars in the BSC is complete up to $m_v = 6$ (a 95% confidence hypothesis),

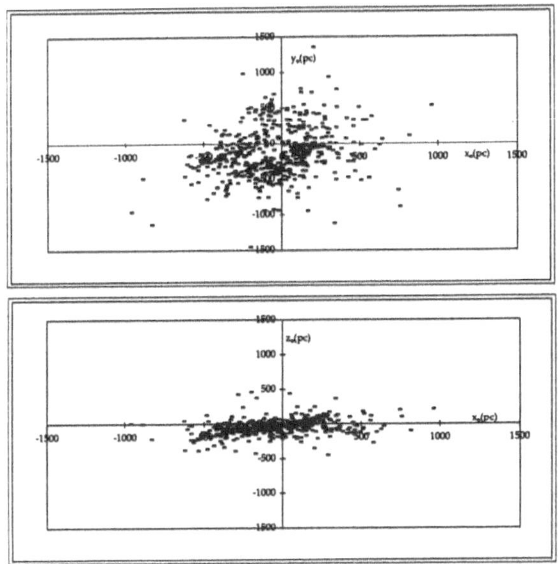

Figure 16.2: The early B stars of the BSC are distributed in a disc around the sun. This can be seen from the distribution in the galactic (x_0y_0)-plane viewed from above vs. edge on (x_0z_0)-plane.

The validity of the latter statement can easily be verified since the coordinates of the 511 B0-B3 stars are given in the BSC. If one determines the distance from the observed m_v-values and the M_v-values as given by the calibration by Humphreys and McElroy (1984) (table 2.1) one obtains the position of each star in the Galaxy. The apparent distribution of the stars in the galactic (xy)-plane and the distribution perpendicular to the galactic plane [e.g. the xz-plane] confirm the fact that unevolved early type B stars are localized in a narrow disc-shaped region, ±200 pc from the galactic plane.

The real space distribution of the stars -after the stars' distances have been corrected for interstellar reddening- is very similar and is shown in figure 16.2.

There are 59 B0-B3 binaries in the DAO8 (Batten et al., 1989) brighter than m_v = 6. The BSC (which is complete up to m_v = 6) contains 348 B0-B3 stars, i.e. ~17 % of the B0-B3 stars in the BSC are primaries of close binaries listed in the DAO8.

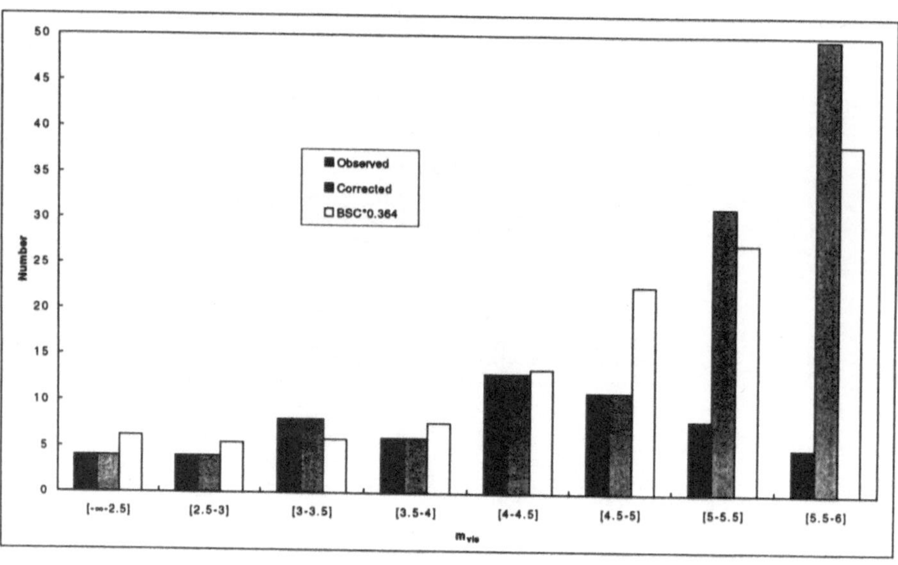

Figure 16.3 The number of observed early B primaries in the DAO8 is incomplete above m_V=5. Correcting for incompleteness enhances the fraction of binaries up to ~ 36%. The number of early B primaries being ~ 36% of the number of early B stars in the BSC is also shown.

However, the DAO8 catalogue appears to be very incomplete. This is illustrated in figure 16.3 where we give the number of B0-B3 primaries in the DAO8 catalogue as a function of m_v and where one observes that the distribution deviates significantly from that of figure 16.1, at least for systems with $m_v \geq 5$. Furthermore, a significant number of binaries, listed in the BSC are not in the DAO8. As an illustration, counting all B0-B3 stars in the BSC which are classified as binaries in this catalogue, again restricting to $m_v = 6$, it follows that the binary frequency ~ 66 % (using $m_v = 6.5$ as limit one obtains 61 %; considering all stars, one arrives at a ~ 60 % binary frequency).

Of course the BSC catalogue also contains the wide binaries which may not interact and we are obviously interested in the interacting binary frequency. In order to obtain a more realistic number we proceed as follows.

In figure 16.3 we have also drawn the expected (N-m_v) distribution when we assume that the DAO8 is complete for $m_v < 5$ and we correct the number of stars in the m_v-bin [5-6] assuming a binary fraction independent of m_v and a disc distribution for all early Bs. From this, we conclude that there should be 127 spectroscopic binaries (most of them interacting) with $m_v \leq 6$ and with properties which are similar to the properties of the 59 binaries which are contained in the DAO8. This means that among the B0-B3 sample, the real interacting binary frequency is more like 127/348, i.e. ~ 36 %.

This number could be biased, since a primary is brightened by the presence of its companion. If one assumes a mean brightening with $0^m.25$ (a brightening with $0^m.75$ would be achieved when all secondaries and primaries have the same magnitude), one should rather compare the number of binaries brighter than $m_v = 5.75$ to the number of single stars brighter than $m_v = 6$. Extrapolated from the DAO8 catalogue one obtains for a complete sample 104 binaries brighter than $m_v = 5.75$, whereas the complete sample from the 'Bright Star Catalogue' contains jointly 120 singles brighter than $m_v = 6$ and 201 binaries brighter than $m_v = 5.75$.

With these assumptions, and using the same method as described above, we conclude that ~32 % of the B0-B3 stars should be the primary of an interacting close binary. It is interesting to notice that this is remarkably similar to the binary frequency for the O-type stars.

Let us remark that the observations discussed above give us a hint about the MCB frequency in the solar neighbourhood. However, whether or not this frequency is universal is a matter of faith.

16.5. The observed mass ratio and period distribution of pre-RLOF MCBs in the solar neighbourhood

16.5.1. THE OBSERVED MASS RATIO DISTRIBUTION

16.5.1.1. Method of determination

We distinguish double lined spectroscopic binaries (SB2s) from single lined spectroscopic binaries (SB1s).

Whereas the mass ratios for SB2s are known directly from observation, the statistical distribution of the mass ratio for SB1s can be determined from the knowledge of the mass function f(m)

$$f(m) = \frac{M_2^3 \sin^3 i}{(M_1 + M_2)^2} \tag{16.12}$$

and the assumption of random distribution for the angle i between the line of sight and the axis perpendicular to the orbital pane.

We use the calibration given in table 5.1 to estimate the mass of a primary when its spectral type and luminosity class is known. When the luminosity class is not known, we adopt class V.

From this mass a possible value of the mass ratio can be calculated from relation (16.12) for every possible value of the inclination angle i.

For a random distribution of the angle i the fraction (df) of stars with i-values in the range [i - i+di] is given by:

$$df = \sin i \,.\, di \tag{16.13}$$

It is clear that the fraction of stars between two limiting values of i is given by:

$$f[\, a - b \,] = \cos a - \cos b \tag{16.14}$$

so that e.g. f[0°-90°] = 1; f[0°-30°] = 0.134; f[30°-60°] = 0.366; f[60°-90°] = 0.5

Following Abt & Levy (1985) one can sample the possible inclination angles with an inclination angle of 15° with a probability of 13.4%, an angle of 45° with a probability of 36.6% and an angle of 75° with a probability of 50%. Three mass ratios are therefore known with their probability.

Here, however we have performed a more refined evaluation of relation (16.12) by determining for each star a range [q_{min} - 1] with the inclination angle in the range [90° - i_{min}]. For each bin [q_1 - q_2] one obtains a range [i_1 - i_2] which occurs with a probability given by relation (16.14): $\cos i_2 - \cos i_1$.

Figure 16.4 shows that the latter method (quoted henceforth as "our method") gives q-distributions which differ from those as obtained with the sampling method of Abt & Levy, in that their method overestimates artificially the fraction of binaries in the range q = [0.8 - 1] and underestimates in the bins [0.4 - 0.6] and [0.6 - 0.8].

We will further use "our method" to determine the mass ratio distributions of binaries with non evolved O-type and B-type primary

The obtained values of q depend of course on the assumed value of M_1 which can differ from the real value. However, starting from a given value of the mass function f(m), a large uncertainty on M_1 will not introduce a large change for the obtained statistical distribution of the mass ratio q. Especially for massive primaries the uncertainties on M_1 will not alter our conclusions significantly.

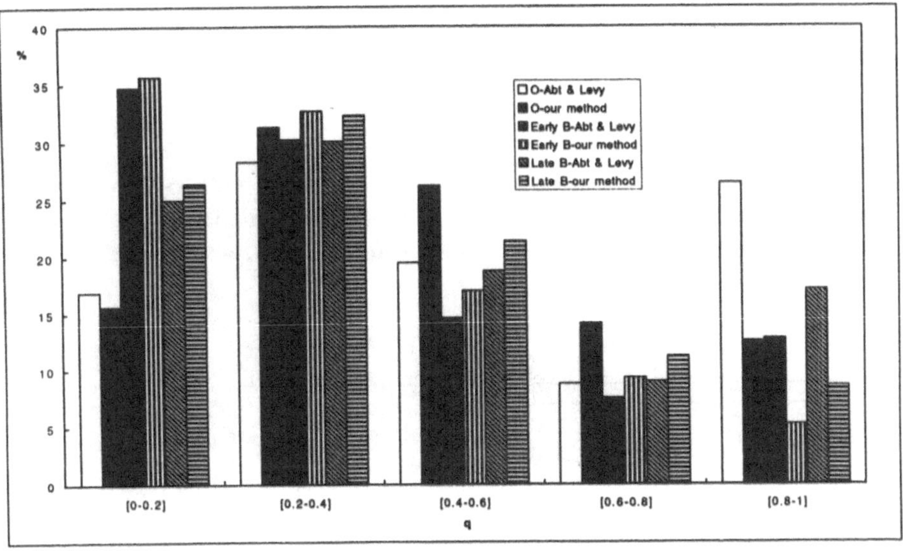

Figure 16.4: Distribution of mass ratios for 14 SB1s with an O-type primary, 74 SB1s with an early B-type primary and 80 SB1s with a late B-type primary, showing that the Abt & Levy method overestimates the number of binaries with large mass ratios.

16.5.1.2. <u>O-type primaries</u>

We have taken the 16 SB2s and 14 SB1s of table 13.1a with a non evolved O-type primary. Low q- values are evidently under-represented among the SB2s since these simultaneously observable stars cannot differ too much in mass. Low mass ratios are, at the other hand, better represented by the SB1s, where it is not astonishing that the non observable secondary has a much lower mass than the primary. Figure 16.5. includes the mass ratio distribution of 30 stars with a non-evolved O-type primary.

We conclude:

- the observed mass ratio distribution of O-type binaries indicates that systems with q in the range [0 - 0.2] are rare,
- the observed mass ratio distribution of O-type binaries reveals that systems with two components of similar mass (q in the range [0.8 - 1]) are remarkably frequent.

Although we used an updated set of observations, our conclusions agree very well with the conclusions made by Garmany et al. (1980).

16.5.1.3. B-type primaries

We have taken the 39 SB2s and 74 SB1s with a non evolved early B-type primary (listed in tables 13.1b & 13.1c) with 40 SB2s and 80 SB1s with a non evolved late B-type primary. Figure 16.5 includes the mass ratio distribution of these 233 stars

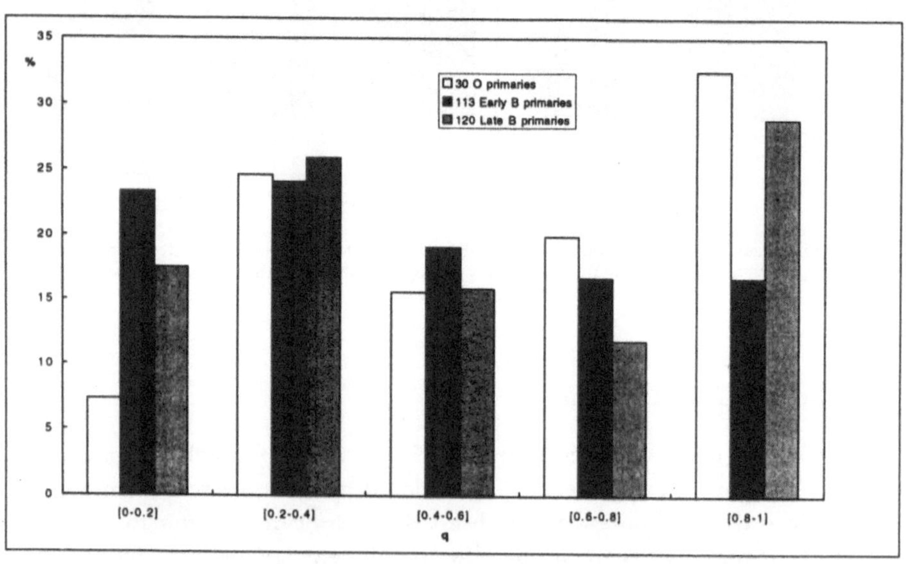

Figure 16.5: Distribution of mass ratios for 30 SBs with an O-type primary, 113 SBs with an early B-type primary and for 120 SBs with a late B-type primary

We conclude:

- the mass ratio distribution of the entire set of early B-type SBs [SB1s + SB2s] shows that every possible mass ratio occurs with an almost constant probability; the same is true for the late B type binaries.

16.5.2. THE OBSERVED PERIOD DISTRIBUTION

The observed period distribution of the unevolved O type binaries and of the early B type (B0-B3) and late B type binaries from the DAO8 catalogue are given in figure 16.6.

Comparing the periods of the unevolved MCBs in our sample with the minimum period a binary must have in order to avoid RLOF, we conclude that most of them will interact.

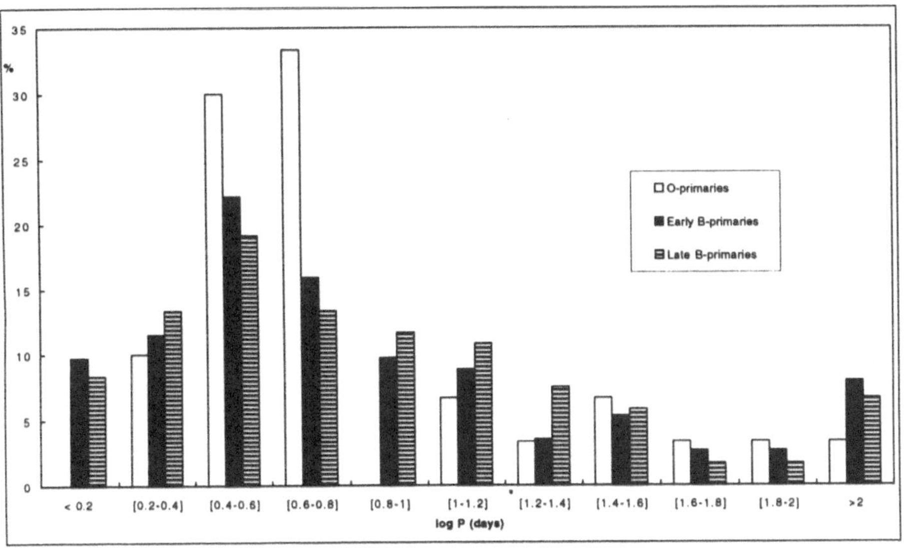

Figure 16.6 Observed periods of binaries with an O type primary, an early B type primary and a late type B primary

16.5.3. THE IMPLICATIONS OF SELECTION EFFECTS

Short period binaries and/or binaries with large mass ratio are easier to detect than binaries with longer periods and/or binaries with small mass ratio. This obviously biases the observed q and P distributions and it underestimates the binary frequency.

A detailed study of the implications of these selection effects on the q distribution of all spectroscopic binaries in the DAO8 catalogue has been presented by Hogeveen (1991, 1992) (see also Halbwachs, 1987). He concludes that the true overall q distribution for all spectroscopic binaries can be described by the relation:

$$\Phi(q)\begin{cases} \propto q^{-2} & \text{if } 0.3 \le q \le 1 \\ = C & \text{if } q < 0.3 \end{cases} \tag{16.15}$$

Remark that, accounting for the small survival probability during the SN explosion of a MCB with small mass ratio and for the spiral-in process after the SN explosion leading to mergers in most of the cases, the existence of X-ray binaries like Her X-1 and the X-ray binaries with a low or intermediate mass optical star and a BH component, can be considered as indirect evidence for the existence of a significant number of MCBs with small mass ratio.

According to Popova et al. (1982) (see also Vereshchagin et al., 1987, 1988), the semi-major axis A of the relative orbit of (all) binary systems, is distributed according to:

$$\Pi(A)dA \propto \frac{dA}{A} \tag{16.16}$$

and thus the period (P) distribution according to:

$$\Pi(P)dP \propto \frac{dP}{P} \tag{16.17}$$

extending up to periods of 10 years.

It is not meaningful to make such a detailed statistical study for the O type and the early B type binaries separately, since their number is too small. However, if the foregoing distributions also apply there, this does have a number of curious results.

Figure 16.6 shows that MCBs with observed period larger that ~100 days are very scarce. If however the period distribution of OB type binaries follows equation (16.17) as well, we have to conclude that we are missing at least ~30% of the interacting MCBs, i.e. the interacting MCB binary frequency > 45%.

Suppose now that we detected all massive systems with mass ratio between 0.6 and 1. If equation (16.15) applies for the mass ratio distribution, we expect about 2.5 times more stars in the $0.3 \leq q \leq 0.6$ range compared to the number of stars with $q \geq 0.6$. From figures 16.4 and 16.5 we conclude that again we are missing at least 30-40% of the binaries, i.e. the frequency now reaches a value > 60% and we yet have not counted the possible number of binaries with small mass ratio ($q < 0.3$).

When a similar calculation is performed but with a flat mass ratio distribution, apart from the binaries with very small mass ratio, the observed numbers in the two mass ratio bins ($0.3 \leq q \leq 0.6$ and $q \geq 0.6$.) are approximately correct, so one only has to account for the period distribution and for the systems with $q \leq 0.3$.

Accounting for the foregoing discussion we conclude

more than 30% (possibly up to 60%) of all massive stars are born as the primary of an interacting binary with mass ratio $q > 0.2$

and, in addition

a non-negligible fraction of all massive stars may have a low/intermediate mass companion.

16.6. The runaway velocities of single pulsars

A 1% asymmetry in the neutrino momentum flux during the SN explosion of a massive star, is sufficient to give a NS a kick velocity ~400 km/s (Shklovskii, 1969; Sutantyo, 1978; Bailes, 1989). Unfortunately, the physics of SN explosions do not yet allow reasonable guesses to be made of the possible asymmetry. We thus have to rely on observations. It is possible to derive the space velocity of a large number of pulsars. This space velocity is then linked to the kick velocity a compact remnant gets due to an asymmetric SN explosion.

Estimates of pulsar transverse velocities v_t (km/s) are based on the measurement of the proper motion μ ("/year) and the distance d (pc)

$$v_t = 4.74\mu d \tag{16.18}$$

Using a new distance scale for pulsars determined by Taylor and Cordes (1993), Lyne and Lorimer (1994) estimated transverse velocities for 99 sources. The whole sample has a mean $<v_t> = 300 \pm 30$ km/s. Starting from the v_t histogram we can apply a similar method as outlined in section 2.4 for the rotational velocities, in order to obtain the 3-dimensional runaway velocity (v_p) distribution (we thus assumed that the observational errors follow a Gaussian distribution with $\sigma = 50$ km/s). This is shown in figure 16.7. Observe that the average value ~ 400 km/s. If the runaway velocitiy distribution of pulsars reflect the kick velocity (v_{kick}) distribution a NS receives at birth, the average v_{kick} may be a factor of 3 larger than a value proposed by Gun and Ostriker (1970). As will be discussed in section 19.2.2, this means that a large fraction of all pre-SN binaries will be disrupted during the SN explosion.

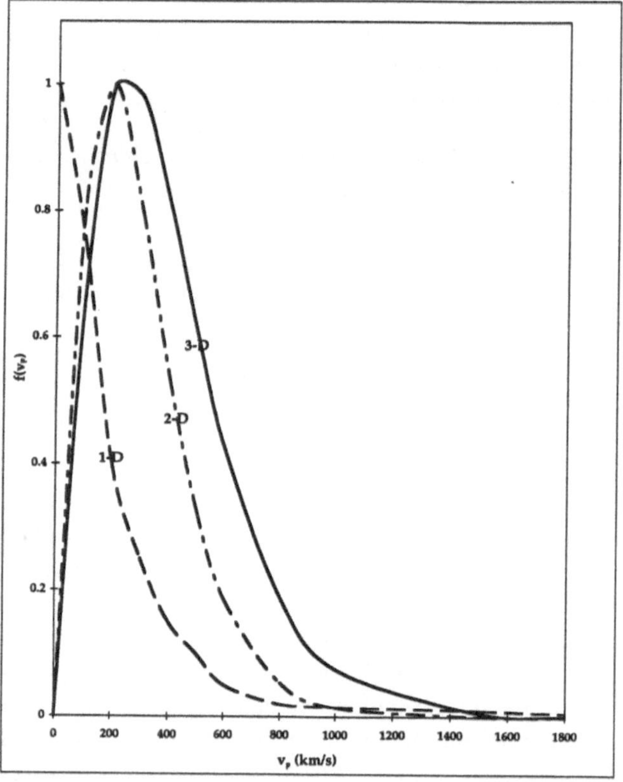

Figure 16.7: The transverse velocity v_t histogram and the estimated 3-dim (and 1-dim) runaway velocity v_p distribution for 99 pulsars; the figures are normalized in such a way that the maximum value equals 1.

The values discussed above may still be somewhat underestimated (Cordes, 1986) due to the following selection effect

- after a few million years, a fast moving pulsar will have a significantly greater distance to the galactic plane than at birth. This larger distance reduces the likelihood of detection, i.e. pulsars with large velocity will be underrepresented.

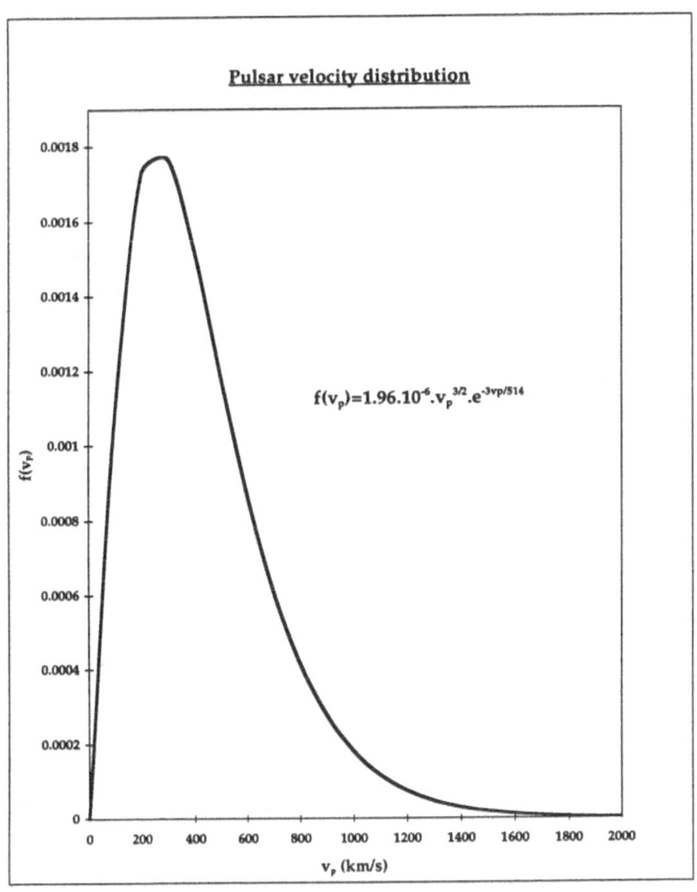

Figure 16.8: The velocity distribution function of pulsars with characteristic age smaller than 3 million years.

As discussed by Bhattacharya and Srinivasan (1995), from the observations of 600 known radio pulsars, there seems to be no evidence for significant magnetic field decay in a pulsar, i.e. the assumption 'B_s is constant in time' is not unreasonable. Using equation (13.20), it follows that the characteristic age τ of a pulsar (= the characteristic spin down age) is given by

$$\tau = \frac{P}{2\dot{P}} \tag{16.19}$$

with P the pulse period and $\dot{P}$ its variation.

Using this age determination, it was noted by Lyne and Lorimer (1994) that the consequences of the selection effect discussed above are minimal when only pulsars younger than ~3 million years are considered. Their sample then contains 29 pulsars meeting the latter restriction. The average transverse velocity of this subsample equals 345 ± 70 km/s. In figure 16.8 we show the 3-D pulsar velocity (v_p) distribution using the same method as for the whole sample (fig. 16.7). It has a mean $<v_p>$ = 450 ± 90 km/s which is only 13 % above the mean value when the whole pulsar sample is considered. The distribution can very well be described by

$$f(v_p) = 1.96 \cdot 10^{-6} v_p^{3/2} e^{-3v_p/514} \tag{16.20}$$

Remarks:

1. Small proper motions are hard to measure and errors may be substantial. These errors may not at all be Gaussian (as assumed above). The effect of such errors has been discussed in detail by Hartman (1997) and it was concluded that a distribution with an average around 300 km/s (rather than 450 km/s) cannot be excluded. However, a detailed discussion on the effect of measurement and statistical errors and selection effects has been presented by Lorimer et al. (1997). They confirm the large (450-500 km/s) average runaway velocity of pulsars. Current proper motion measurements are enclosed in several VLBI programs and will give more conclusive answers.

2. A significant fraction of single pulsars may be former binary components where the SN explosion disrupted the system. The runaway velocity of these pulsars is a vector sum of the kick velocity due to the asymmetrical SN explosion and of the former orbital velocity. In section 19 we will give an answer to the following question:

given an initial distribution of v_{kick} velocities, how does the effect of binaries affect the final (observed) runaway distribution of single pulsars?

17. The observations which need to be reproduced by a massive star population model

17.1. The B/R, WR/O and WC/WN number ratios

The definition of the WR/O and WC/WN number ratio is obvious. The B/R ratio is the number ratio of the OBA supergiants and the RSGs brighter than $M_{bol} = -7^m.5$ (e.g. Langer and Maeder, 1995).

We make a distinction between regions where star formation was constant in time and starburst regions.

17.1.1. CONTINUOUS STAR FORMATION REGIONS

Constant star formation means constant within the time scale of the lowest mass considered. When massive stars are concerned the lowest mass is 9±1 M_o and thus the time scale is ≈ 30 million years.

Using the catalogues of Humphreys and McElroy (1984), Van der Hucht et al. (1981, 1988), Breysacher (1981) complemented by Lortet (1991), Azzopardi and Breysacher (1979), table 17.1 summarizes the metallicity dependence of the number ratios WR/O, WC/WN and B/R. Note that for the galactic WR+OB binaries with period P ≤ 112 days, the ratio WN/WC ≈ 2 (section 13.2) which differs considerably from the overall value.

TABLE 17.1: The observed WR/O, WC/WN and B/R number ratios in different stellar environments (different values of the metallicity Z); SN = Solar neighbourhood within 3 kpc from the Sun.

Z	WR/O	WC/WN	B/R
0.002(SMC)	0.017	0.13	4
0.008(LMC)	0.04	0.26	10
0.013(outer MW)	0.03	0.5	14
0.02(SN)	0.1	1	28
0.03(inner MW)	0.2	1.2	48

17.1.2. STARBURST REGIONS

A starburst region is a region where in a recent past, for a short period, a sudden and important increase of the star formation rate has occurred. In the case of a massive star starburst it can easily be understood that the number ratios discussed in the previous subsection may be very different from those in regions with a constant star formation rate. Moreover, the precise values depend on the time elapsed since the beginning of the burst. A galaxy where such an event has occurred is called a 'starburst galaxy'.

In distant galaxies the O-type stars and the WR stars cannot be resolved as individual stars. However certain features in the integrated spectra can be attributed to the presence of WR stars. A method to derive a WR/O number ratio from these spectra has been derived by Kunth and Sargent (1981) and applied to the dwarf galaxy Tol 3. The idea is the following:

- One observes the spatially integrated flux of the nebular H_β line (possibly corrected for extinction); from simple recombination theory one obtains the total number of Lyman continuum photons (N Lyc) being emitted. The N Lyc is a direct measure of the number of exciting stars, i.e. the product of the number of O-type stars and the N Lyc at each spectral subtype. It can thus be expected that there is a relation between H_β and the distribution of O-type stars. Obviously, in order to have realistic results, one needs the correct IMF of the starburst and the distribution of subtypes, which can only be obtained by using evolutionary models,

- when one observes a broad emission feature at He II 4686 Å, this is attributed to the presence of (large?) numbers of WR stars (WN). A correct number can only be deduced if the WN distribution is known and, as for the number of O-type stars, evolutionary models have to be used.

Kunth and Sargent did not account for evolutionary effects and simply adopted an average IMF of stars during their CHB. They conclude that WR/O $\approx$ 0.3-1 for Tol3, far above typical values for regions of constant star formation.

Since this pioneering work, 40-50 galaxies were detected showing a broad He II emission feature at 4686 Å. From long-slit optical spectra, Vacca and Conti (1992) estimated the WR/O number ratio in 14 starburst galaxies using the same method as the one used by Kunth and Sargent for Tol 3.

Evolutionary effects were introduced by a number of research groups (see e.g. Robert et al., 1993; Leitherer and Heckman, 1995; Leitherer et al., 1995; and references therein) and the term 'population synthesis models' was introduced. However, in all studies only the evolution of single stars is considered. In section 18 we will describe how binaries have to be included, while in section 19 we will illustrate to what extent binaries are important for population synthesis.

Starburst galaxies may contain several distinct starburst regions. Since most known starburst galaxies are very distant (NGC 1569 is probably the closest one known and has a distance estimate of 2.2±0.6 Mpc, Israel, 1988) spectra may contain information of several starburst regions with different ages as well as the stellar information of the whole galaxy. Features which are attributed to the presence of very young stars may be visible together with WR and/or RSG features but it is hard to tell whether or not they come from one single starburst (as an illustration, see the discussion in Gonzales-Delgado et al., 1997, and in De Marchi et al., 1997).

The interpretation of the observations of starburst galaxies also depend critically on the time scale of starformation in starbursts, i.e. on whether or not stars in a starburst form coevally. Since the individual stars in starburst regions cannot be resolved at present (again due to the large distance), direct evidence is lacking.

Although the number of stars in starburst regions of starburst galaxies is significantly larger than the number of stars in clusters and associations in the Milky Way or in the MCs, the study of the latter may provide important hints.

The stellar content of massive star clusters and associations in our Galaxy has been studied by Humphreys and McElroy (1984). Two OB associations in the LMC (i.e. LH117 and LH118) and one in the SMC (NGC 346) have been discussed in detail by Massey et al. (1989a,b). Parker et al. (1992) presented results for the LMC associations LH 9 and 10.

In figure 17.1 we show the HR diagram of the most massive stars of Vela OB1 in the Galaxy and of the four LMC aggregates. In section 14.4.1 we argued that the HMXB Vela X-1 originates from Vela OB1. We therefore show the HRD position of the optical star of the binary as well. When the HRD position of the massive stars in these aggregates is compared with evolutionary tracks and time-isochrones, one has to conclude that

massive stars in an aggregate in the Galaxy and in the MCs form over a timescale of several million years, i.e. coeval star formation does not apply.

Particularly interesting is that in all of them there seems to exist a class of stars that are younger than the aggregate turn off (defined as there where the observed star sequence tends to bend towards the red). The authors suggest that this may be an indication that star formation was bimodal, the highest mass stars being formed later. We will show however that this phenomenon is a natural consequence of the effect of binaries on the evolution of starburst regions.

In table 17.2 we present a list of galactic stellar aggregates containing WR stars (from Lundstrom and Stenholm, 1984; however based on the Hipparcos distance determination of the WR+OB binary γ^2Vel, we excluded Vel OB2). The number of WR stars is given as well as the number of RSG members. We conclude

only ~27% of the galactic massive stellar aggregates containing WR stars also contain RSGs.

TABLE 17.2: Galactic stellar aggregates containing WR stars; the number of WR and RSG members are given.

Stellar aggregate	#WR	#RSG
Pis 20	1	0
Ser OB2	1	0
Ser OB1	2	1
Sco OB1	2	0
Sgr OB1	3	1
Cen OB1	1	0
Tr 27	1	1
Pis 24	1	0
Tr 16	1	0
Coll 228	1	0
Cyg OB3	3	0
Car OB1	2	0
Cyg OB1	5	4
Cyg OB2	3	0
Vul OB2	1	0
Be 87	1	0
NGC 3603	1	0
Cep OB1	4	2
Cas OB1	1	0
Pup OB2	1	0
NGC 2439	1	0

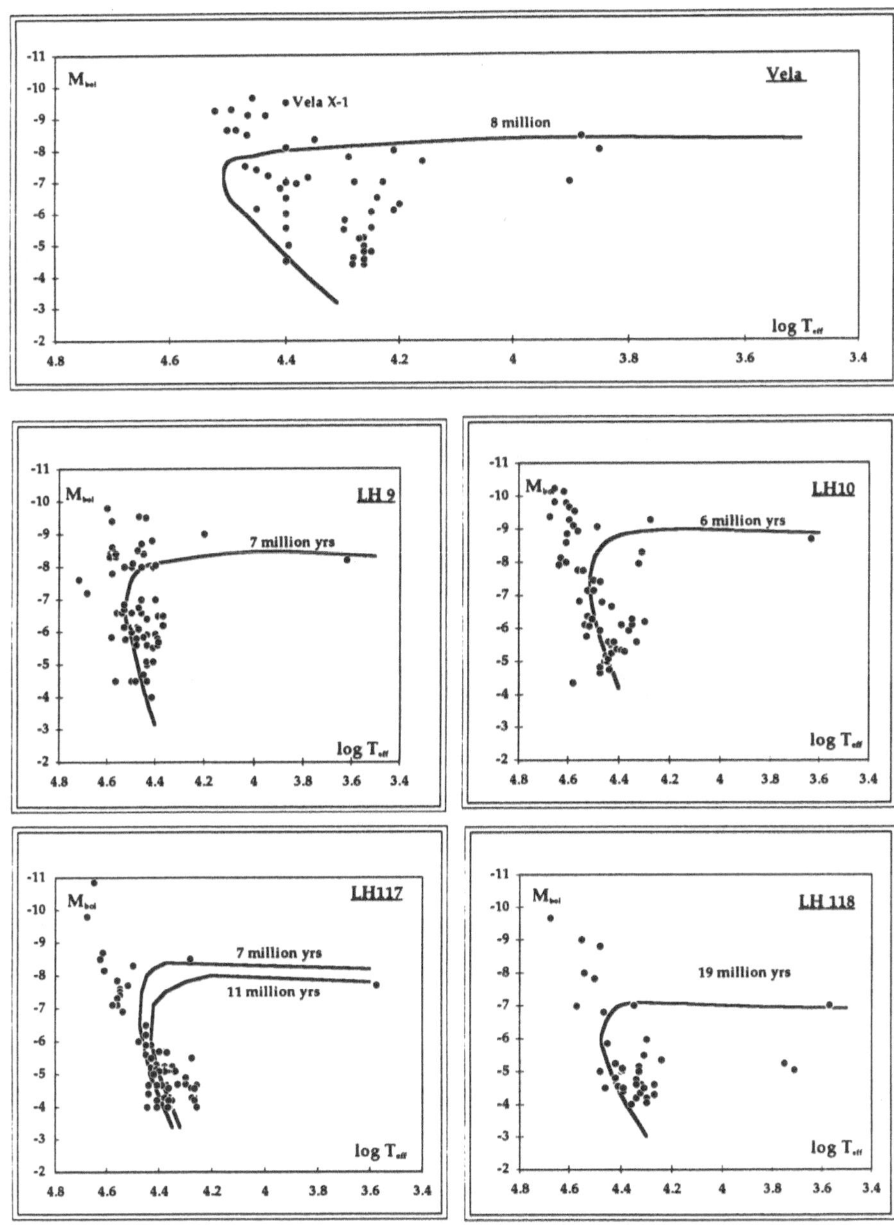

Figure 17.1: The HR diagram of the most massive stars in the stellar aggregates Vela OB1 in the Galaxy, LH 9, 10, 117 and 118 in the LMC.

17.2. WR+OB binaries

17.2.1. THE WR+OB FREQUENCY

The observed WR+OB binary frequency within 2.5 kpc from the sun ~42% (Moffat, 1995). This may increase up to 48% if all large period WR+OB candidates are confirmed. A warning is appropriate here: the distance of WR stars which do not belong to a stellar aggregate (a so called field WR star constituting about 50 % of all WR stars) is based on a calibration where the WR subclass is linked to an absolute visual magnitude (table 2.2). The crudeness of this calibration obviously introduces an uncertainty in the distances, thus on the decision whether a WR star is located within 2.5 kpc from the sun and thus on the above quoted binary frequencies.

The WR+OB frequencies given above should be considered as lower limits. Some WR stars that have been considered as single until now may hide an OB type companion which is fainter than in other WR+OB binaries where the OB type star is by far the brightest star. A number of candidates were listed by Williams and Van der Hucht (1996). Furthermore, as discussed in section 14.2.4 and 14.6 (see also section 16.5.3), a significant number of MCBs may have very small mass ratio so that there may be WR stars with normal intermediate mass/small mass companions.

Orbital periods are known for only few WR+OB binaries in the Magellanic Clouds however about 50% of the WR star spectra show evidence of absorption lines. One could conclude that the WR+OB binary frequency in the LMC is ≤ 50%.

In the SMC almost all WNE and WC stars are binary members.

Let us finally remark that none of the 9 observed WN8/9 stars in the Galaxy and in the LMC (there are no WN8/9 stars in the SMC) is a binary member with an OB type companion. These stars seem to have large runaway velocities, they do not belong to stellar aggregates and they show a very high degree of variability (Cherepashchuk and Moffat, 1994).

17.2.2. THE q AND P DISTRIBUTION OF WR+OB

Using the results of table 13.3, figures 17.2a and 17.2b show the observed q and P distribution of the galactic WR+OB binaries. Due to the small number of systems, one has to account for significant uncertainties, however, notice the apparent period gap between ~ 100 days and ~1000 days.

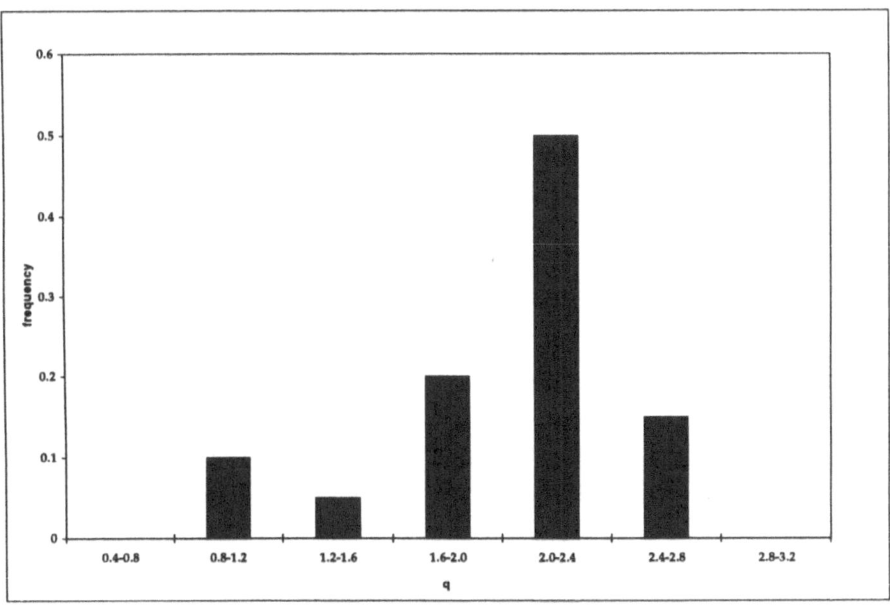

Figure 17.2a: The observed mass ratio distribution of WR+OB binaries.

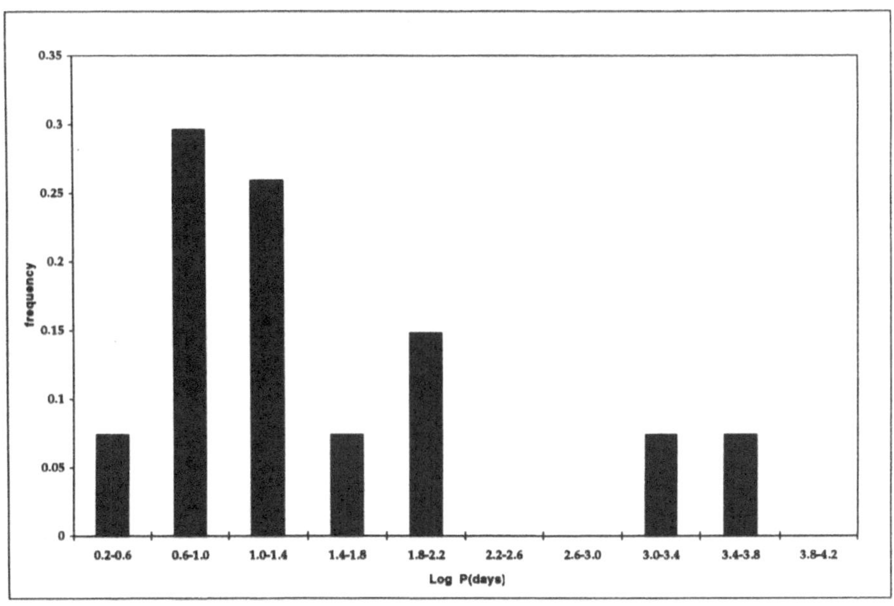

Figure 17.2b: The observed period distribution of WR+OB binaries

17.3. O-type runaways

An O-type star is classified as a runaway when its peculiar space velocity v_{pec} is larger than 30 km/s (section 13.5). Since v_{pec} is the combination of the radial component and of the transverse component, its value is often not known. Therefore the real percentage of observed runaways among OB type stars is still highly uncertain. The careful study of Gies (1987) promotes a percentage not larger than 10%.

17.4. The number of WR stars with a compact companion

As discussed in section 14.5.3, HD 50896 and HD 197406 are two WR+cc candidates for which the observed lack of hard X-rays can be explained if the cc is a rapidly spinning NS (according to St. Louis et al. 1993a, the periodic variation in HD 50896 can equally well be attributed to rotation of the WR star itself). The mass losing component in Cyg X-3 may be a WR like star, but it is unlikely that it is a 'normal' one, comparable to other WR stars (section 14.5.3). We thus have to conclude that the number of WR+cc binaries is very small, i.e. ≤ 5% of the total sample.

17.5. The radio pulsar population

The radio pulsar population consists of solitary pulsars originating from massive single stars, binary pulsars, but also of solitary pulsars with a binary history.

17.5.1. SOLITARY RADIO PULSARS WITH A BINARY HISTORY

Radio pulsars with a binary history are defined as NS remnants after a SN explosion in a non-disrupted binary, where the pulse period has been alterred by mass transfer from a normal component. We distinguish the following three classes of solitary radio pulsars with a binary history:

- when the binary is not disrupted during the first SN explosion, an OB+pulsar system remains. If the OB type star loses mass by stellar wind, accretion by the pulsar causes the emission of hard X-rays: a HMXB is formed. The observations of HMXBs have been summarized in section 13.3. When the binary survives the spiral-in phase, a CHeB+cc is formed and when the CHeB component (possibly a WR star) explodes, either a binary

pulsar emerges either the system is disrupted, adding to the pulsar population two single pulsars, where one is a radio pulsar with a binary history.
Due to the fact that the CHeB lifetime of a massive star $\leq 10^6$ yrs, one can expect that major changes of the pulse period of the cc did not occur, i.e. the solitary pulsars may have pulse periods similar to those observed in HMXBs.
In section 19.2.6 we will show that the 'Population Number Synthesis' model with a realistic binary frequency predicts that $\leq 5\%$ of the total pulsar population are pulsars that originally belonged to HMXBs that became single as a consequence of a second SN explosion in a MCB,

- if, during the spiral-in phase of the OB+NS binary, merging cannot be avoided, a TZO is formed. Due to an efficient SW mass loss phase, all star-matter will be blown away and the NS reappears,
- the single millisecond pulsars. They are probably the descendants from the LMXBs where the low mass component has been evaporated (section 13.4.2).

17.5.2. THE FORMATION RATE OF DOUBLE NEUTRON STAR SYSTEMS IN THE GALAXY

If a MCB survives the two SN explosions, a double NS system (NS+NS) can be formed. There are (presently) three such binaries known in our Galaxy (section 13.4). Using the estimated lifetime of a binary pulsar, Van den Heuvel (1993) proposes a formation rate of the order 10^{-5} yr^{-1} of galactic binary pulsars with the properties of the 3 binaries on which the rate is based.

The total binary pulsar formation rate could be larger than the value given above. This can be illustrated as follows:

The orbital period P of a binary consisting of two relativistic objects, will decrease due to the emission of gravity waves as predicted by Einstein's general relativity. The period variation depends on the period as (Shapiro and Teulkolsky, 1983):

$$\dot{P} \propto -\frac{1}{P^{\frac{5}{3}}} \tag{17.1}$$

The three observed binary pulsars have an orbital period of between 0.42 and 12 days. For such periods the time scale before coalescence due to gravity-wave-emission is longer than the expected lifetime of a pulsar [= the characteristic spin down age τ, given by equation (16.19)]. Therefore Van den Heuvel used the latter lifetime for the three pulsars in order to derive an average formation rate. Suppose however, that a considerable fraction of the binary pulsars have a much smaller orbital period and thus a much larger period variation [as can be seen from relation (17.1)]. Such binaries may coalesce much faster due to gravity-wave-emission, therefore decreasing their detection probability.

Accounting for this effect will then obviously increase the 'observed' binary pulsar formation rate. Using the observed eccentricity of PSR 1519-16 as an average eccentricity for the NS+NS sample (this has to be confirmed by detailed computations of course) and the expressions for the evolution of the semi-major axis and eccentricity of Wagoner (1975), the coalescense will occur within the lifetime of the pulsar when its period is less than 2 hours.

Anticipating, the Brussels PNS model follows the period evolution of the binary sample in detail. In this way we are able to predict the period distribution of the NS+NS binaries (see section 19). From this distribution it can be expected that no more than ~5-10% of the NS+NS systems will coalesce within the pulsar lifetime.

17.6. The SN rate

It was first pointed out by Tammann (1970, 1974) that the SN rate in galaxies is proportional to the galaxy blue B luminosity L_B (see also Cappellaro et al., 1993). SN rates are therefore expressed in so called SN units, i.e. 1 SNu = 1 SN $(100yr)^{-1}(10^{10}\ L_{B_o})^{-1}$ with $L_{B_o} = L_B/L_o$.

Probably the most thorough discussion on SN rates in different types of galaxies has been presented by Cappellaro et al. (1993) using the Third Reference Catalogue of Bright Galaxies (de Vaucouleurs et al., 1991). After correcting for selection effects on the SN discovery, they propose the rates summarized in table 17.3.

Muller et al., (1992) predict a SN number ratio type Ib+c/type II = 1.39. This is remarkably different from the results of table 17.3, which predicts an overall number ratio type Ib(+c)/type II ~0.25. One may think of two reasons for this difference:

a. SNIb/c have been clearly separated from SN Ia only after 1985 and most of the SNs on which the statistics of table 17.3 is based date from before 1985,

b. the study of Muller et al. is based on a CCD search; CCDs have a very high red sensitivity and since SN type $I_{b/c}$'s are significantly redder at maximum light than type IIs or Ias, the number of type Ib/cs could have been overestimated by Muller et al.

TABLE 17.3: SN rate per unit blue luminosity in different types of galaxies (from Cappellaro et al., 1993).

galaxy		SNe			SNu			
type	number	Ia	Ib/c	II	Ia	Ib/c	II	All
E	263	3			0.11			0.11±0.06
S0	437	4			0.15			0.15±0.08
S0a, Sa	274	5.2	0.8	1.0	0.3	0.15	0.19	0.64±0.24
Sab, Sb	432	5.0	1.2	3.8	0.12	0.12	0.36	0.60±0.19
Sbc	202	2.8	0.4	4.8	0.22	0.10	1.19	1.51±0.53
Sc	173	5.9	2.7	6.4	0.50	0.54	1.45	2.49±0.64
Scd, Sd	351	5.3	0.3	6.4	0.48	0.09	1.87	2.43±0.70
Sdm-Im	304	1.5	1.1	1.4	0.2	0.3	0.4	0.90±0.45

It may be important to remark that the SN ratios are based on SN events detected in different galaxies, i.e. the observed number ratios are average values holding for galaxies which may have different metallicities. As discussed in sections 4 and 12, the evolution of massive stars depends on the initial Z. Even more, suppose that the binary frequency is not universal but changes from galaxy to galaxy (may be a function of metallicity). In this case, an average ratio cannot be used in order to compare with theoretical prediction holding for one particular galaxy. This will be illustrated in more detail in section 19.4.

18. The population number synthesis (PNS) model

Population number synthesis (PNS) models for massive stars with a realistic frequency of MCBs have been constructed by different groups:

- Dewey and Cordes (1987),
- Meurs and Van den Heuvel, 1989; Pols et al., 1991; Pols and Marinus, 1994,
- Tutukov, Yungelson and Iben, 1992; Iben, Tutukov and Yungelson, 1995 a,b,

- Lipunov, Postnov and Prokhorov, 1996; Jorgensen, Lipunov, Panchenko, Postnov, Prokhorov, 1997;
- Dalton and Sarazin, 1995a, b,
- Portugies-Zwart, 1995, 1996; Portegies-Zwart and Verbunt, 1996,
- the Brussels PNS model, Vanbeveren, 1988; Vanbeveren, 1995; Vanbeveren et al., 1997, Van Bever and Vanbeveren, 1997; De Donder et al., 1997, Van Bever and Vanbeveren, 1998, De Donder and Vanbeveren, 1998.

In the present work we will primarily discuss the Brussels model. However, we strongly advise an interested reader to study the other papers, since they contain much pioneering work.

18.1 The overall evolutionary model of single stars and MCBs

In this section we summarize our preferred evolutionary model of single stars and of MCBs, which is used in the PNS code.

18.1.1. SINGLE STARS

Evolutionary computations of single stars with small convective core overshooting are used (section 4).

Figures 18.1, 18.2 and 18.3 show the lifetime of the O phase, the yellow and red supergiant phase, the WR phase and the WC phase (WN = WR - WC) as a function of initial mass for the Galaxy and for the Magellanic Clouds.

A CHeB star that has not yet lost most of its hydrogen rich layers, will be observed either as an A-type supergiant (YSG), either as a RSG. The time spend as YSG or as RSG depends on the treatement of convection and semi-convection in the envelope of the star but the sum (t_{YRSG}) is largely independent from the convection criteria.

Figure 18.2 also illustrates that t_{YRSG} depends critically on the SW mass loss rate during the RSG phase.

In section 12.2, we discussed the conditions for a hydrogen deficient CHeB star to be observed as a WR star. We give the WR lifetimes when it is assumed that

- a hydrogen deficient CHeB star will always be observed as a WR star, i.e.
$$t_{WR} = t_{CHeB}(X_{atm} \leq 0.3),$$

- a hydrogen deficient CHeB star will be observed as a WR star when its mass is larger than 5 M_o, i.e.

$$t_{WR} = t_{CHeB}(X_{atm} \leq 0.3, M \geq 5 M_o), \qquad (18.1)$$

- a hydrogen deficient CHeB star will be observed as a WR star when its mass is larger than 8 M_o, i.e.

$$t_{WR} = t_{CHeB}(X_{atm} \leq 0.3, M \geq 8 M_o).$$

The WR time scales for massive single stars in a galaxy with Z=0.002 (which can be applied in the SMC, the LMC being intermediate between Galaxy and the SMC) have been computed assuming that the RSG mass loss rates depend on metallicity according to equation (2.12), and the WR mass loss rates are metallicity independent; however for the LBV phase, we consider two possibilities:

- independent of the metallicity, the LBV mass loss is assumed to be large enough to prohibit redwards evolution of stars with initial mass larger than ~40 M_o

  or

- the LBV mass loss depends on the metallicity as predited by the radiation driven wind theory [equation (2.12)].

Two interesting conclusions:

- the WR lifetimes of stars with initial mass smaller than ~40 M_o decrease with decreasing metallicity and becomes almost zero in a galaxy with Z = 0.002,

- the WR lifetime of stars with initial mass larger than 40 M_o depends critically on whether or not the SW during the LBV phase depends on metallicity.

Theoretical estimates of the SN rates depend on the minimum mass a single star must have in order to collapse into a BH. Evolutionary computations with a metallicity Z = 0.02 predict that BH formation may occur once the initial mass of the star is larger than 20-30 M_o (section 4.1.2). Since the stellar wind mass loss rate of a massive star is smaller when the metallicity is smaller, the minimum mass of BH formation may be smaller than in our Galaxy.

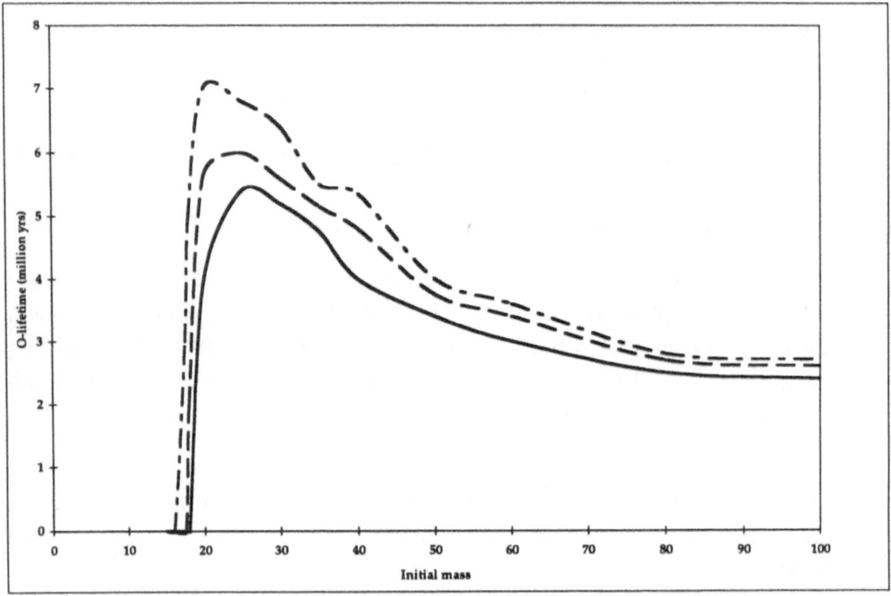

Figure 18.1: The O-lifetime as a function of initial mass for (X,Z) = (0.7,0.02) (full line), (X,Z) = (0.74,0.008) (dashed line) and (X,Z) = (0.76,0.002)(dashed-dotted line).

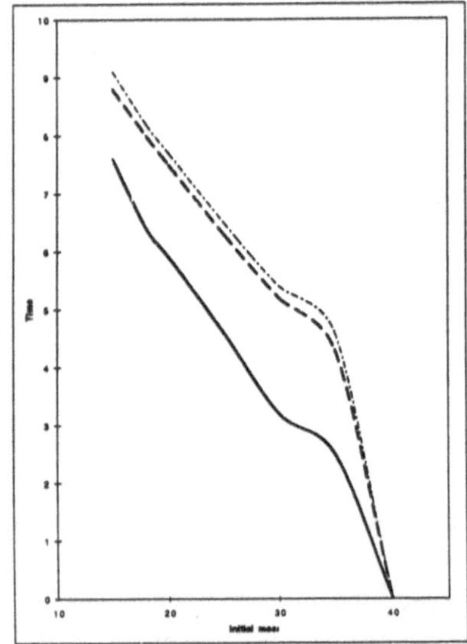

Figure 18.2: The yellow and red supergiant time scale (t_{YRSG} in 10^5 yrs) as a function of the ZAMS mass for the Galaxy (full line), for the Galaxy but the RSG mass loss has been lowered by a factor 2 (dashed line), for the SMC (dashed-dotted line).

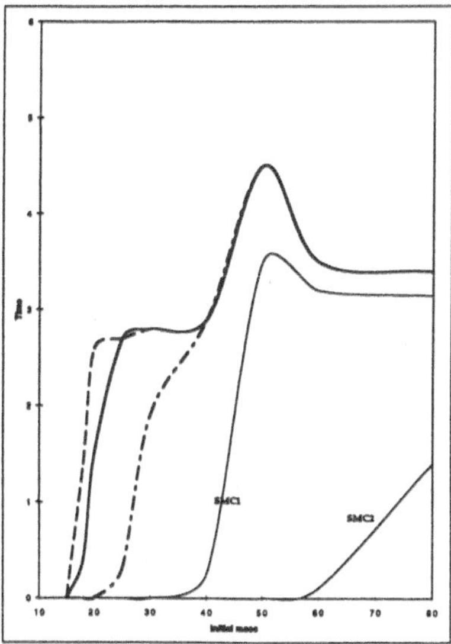

Figure 18.3a: The WR lifetime (in 10^5 yrs) of single stars as a function of initial mass for (X,Z) = (0.7, 0.02) (thick lines) and for (X,Z) = (0.76, 0.002) (thin lines). Dashed line (resp. full line and dashed-dotted line) correspond to $t_{WR} = t_{CHeB}$ (resp. $t_{WR} = t_{CHeB}(M \geq 5\ M_o)$ and $t_{WR} = t_{CHeB}(M \geq 8\ M_o)$) as explained in the text. The curves SMC1 (resp. SMC2) correspond to the case where the LBV stellar wind mass loss is assumed to be independent from metallicity (resp. depend on metallicity as predicted by the radiatively driven wind theory).

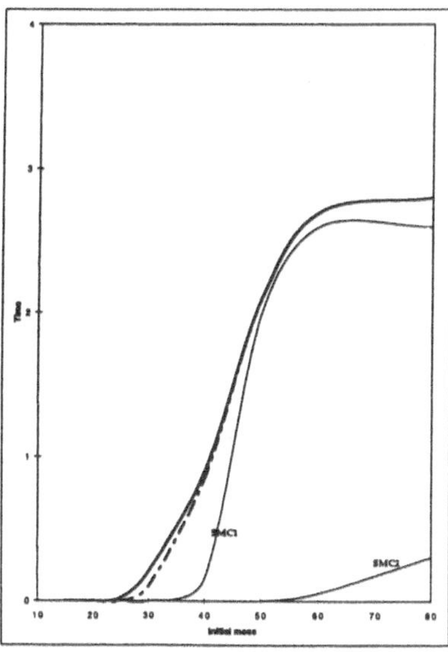

Figure 18.3b: Similar to fig. 18.3a but for the WC lifetime. Note that the case $t_{WR} = t_{CHeB}$ gives the same timescale as the case $t_{WR} = t_{CHeB}(M \geq 5\ M_o)$.

18.1.2. MCBs

The overall evolutionary model of MCBs used in the PNS code is described in section 12 and after comparison with observations of individual binaries, the model is summarized in section 14.8.

We account for

- the LBV stellar wind once the mass of the primary is larger than 40 M_o.

  To investigate the metallicity dependence, we consider the two possibilities as for single stars (previous subsection). Remind that when the LBV SW is too small to prevent a rapid expansion, the star does not escape the RLOF.

- common envelope/spiral-in evolution when $q \leq 0.2$.

  Most of these binaries merge. The mergers are treated as stars that have accreted an amount of mass equal to the mass of the low mass component, and their further evolution is the evolution of a single star with mass equal to the sum of the two masses,

- common envelope/spiral-in description or the RSG scenario for case B_c/case C binaries,

- quasi-conservative evolution of case A/case B_r binaries with $q > 0.2$

  The mass lost from the system during the RLOF of the primary (with initial mass $\leq 40\ M_o$) is parametrized; the mass leaves the binary through L_2 and therefore the period variation is determined by equations (10.14) and (10.15),

- the reaction of the secondary due to mass transfer.

  Accretion is treated either with the standard accretion model, or with the accretion induced mixing model (section 9.3 and section 12.3). The latter is used when accretion occurs through the formation of a Keplerian disc, discussed in section 9.1.

  In order to determine the number of OB type mass gainers, we performed a detailed set of MCB evolutionary computations in which the evolution of the gainer after the mass transfer process was followed till the end of its own CHB phase. The lifetimes as well as the computations are available upon request.

- the effect of SW during CHeB, discussed in section 12.2.

Primaries in VMCBs, case A, case B_r or case B_c MCBs, that have experienced either LBV/RSG SW mass loss, or a RLOF/common envelope/spiral-in phase, are hydrogen poor stars at the beginning of CHeB.
When we combine the evolutionary computations during CHeB of section 12.2, the mass relations (12.2) and the evolutionary computations during CHB, figures 18.4 and 18.5 give the WR and WC lifetime as function of initial primary mass. Correponding to the single star lifetimes of figures 18.2 and 18.3, we also give the lifetimes when the relations (18.1) apply.
Notice that

- the WR time scales of WR binary components depend only marginally on the initial metallicity, contrary to the single star timescales,
- the WR time scales of WR binary components with initial mass smaller than 40-50 M_o depend critically on the adopted minimum mass that a hydrogen deficient CHeB remnant after RLOF must have in order to be observed as a WR star.

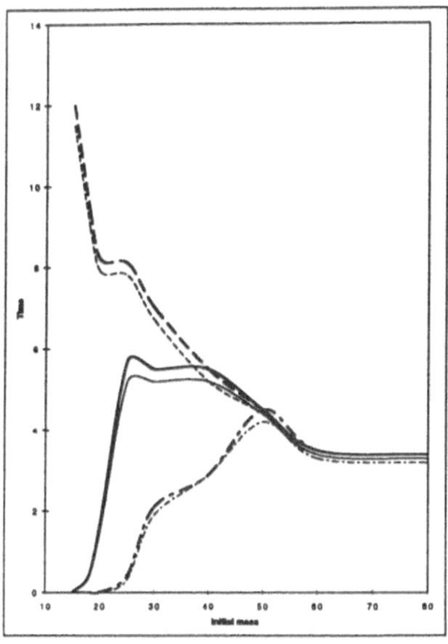

Figure 18.4: The same as to figure 18.2 but for primaries of interacting binaries.

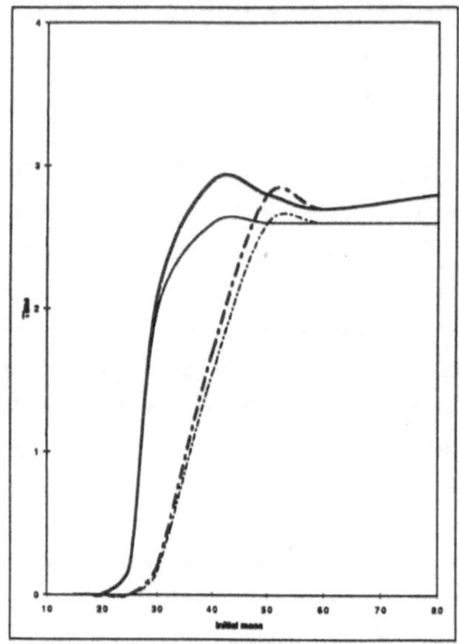

Figure 18.5: The same as to figure 18.3 but for primaries of interacting binaries.

- the formation of BHs.

  Corresponding to theoretical evolution of MCBs and the observations of the standard HMXB Wray 977, we assume that stars with initial mass larger than 40-50 M_o collapse completely to form a BH. As a SN does not occur, the binary will not be disrupted and an OB+BH binary is formed.
  Since MCB evolution is largely metallicity independent, the minimum initial mass of a binary component that is needed to form a BH is largely Z-independent as well,

- the effect of the SN explosion on the binary when the final collapse results in the formation of a NS.

  If as a consequence of the asymmetry the binary is disrupted (section 11), one is left with a single pulsar (NS) and an OB type mass gainer which has a peculiar space velocity. The further evolution of the OB star is that of a single star with a chemical structure which was modified by a previous accretion process. If the SN explosion does not disrupt the binary, we are left with an OB type mass gainer with a peculiar velocity and a NS companion,

- the evolution of OB+cc binaries.

The cc in an OB+cc binary can thus be a BH or a NS. Our PNS model accounts for the following three evolutionary paths:

- the mass of the OB type star is larger than 40-50 M_o

  When the period of the OB+cc binary is larger than a few days, the OB type star becomes a LBV and starts losing mass by SW at a very high rate. The star escapes the RLOF or the RLOF mass loss is very much reduced. After the removal of the hydrogen rich layers of the OB star, the binary will evolve into a WR+cc binary with a period of the order of days or decades,

- the mass of the OB type star is smaller than 40-50 M_o but the period is large enough that the OB star reaches the RSG phase before it engulfs the cc.

  In this case we account properly for the effect of SW during the RSG phase of the optical star, which means that we allow for the formation of hydrogen poor CHeB stars with a cc orbiting with a very large period,

- the OB+cc binary does not belong to one of the two classes above.

  Here we use the spiral-in prescription as discussed in section 12.5.

18.2. Method of computation

Given the mass of a single star, depending on the IMF and on the adopted evolutionary scenario, we compute in a straightforward way its contribution to the number of O-type stars, of WR stars and of RSGs.

Given a close binary with initial primary mass M_1, the initial mass ratio q and period P, depending on the IMF for primaries of close binaries, the mass ratio and period distribution, we determine its contribution to the number of O-type binaries. Due to the SW mass loss during CHB the period variation is given by equation 10.6. Depending on its evolution during the RLOF for which we follow the evolution of the binary period using the formalism of section 10.2, we determine the contribution to the number of mergers, to the number of WR+OB binaries, and to the number of CHeB+OB binaries. After RLOF, the variation of the binary period depends on the SW mass loss during CHeB. When the primary explodes, given an adopted value of v_{kick} and an adopted direction of the kick, we determine what the binary looks like after the SN explosion. When the SN explosion disrupts the binary, the further evolution of the OB-type single star is determined by the adopted single star scenario. If the binary is not disrupted, we use the spiral-in prescription in order to estimate the further evolution. We assume that the direction of the kicks during the SN is isotropic whereas the v_{kick} distribution is given by the distribution of pulsar velocities. In this way we determine the contribution to the number of OB-type runaways (with and without a cc), to the number of single WR but with a binary history, to the number of CHeB+cc binaries and thus when the conditions are fulfilled, to the number of WR+cc, to the number of TZOs. Finally, when the CHeB star in a CHeB+cc binary explodes, using the same assumptions as during the first SN explosion, we determine the contribution to the number of binaries with two compact stars, i.e. the number of NS+NS (binary pulsars), BH+NS or the number of BH+BH.

The SN types produced by single stars and by MCBs were discussed in sections 5.1.5 and 14.7. The PNS model determines the rates in a straightforward way.

18.3. The parameters of the PNS model

- The IMF of single stars and the IMF of primaries of binaries. For reasons of simplicity, we assume that both are the same. Although deviations from this assumption were commented in section 16.3, this assumption will hardly influence the conclusions of the population synthesis.

- The MCB frequency at birth, the mass ratio and period distribution.

  The period distribution of massive close binaries is assumed to satisfy equation (16.17). For the mass ratio distribution, we either adopt a flat distribution, either the distribution predicted by equation (16.15).

TABLES 18.1: The frequency at birth case B_r binaries with $q \geq 0.2$ and of binaries with period $P \leq 100$ days and $q \geq 0.2$ in a population of binaries with periods up to 10 years; we separately consider binaries with initial primary masses between 8 M_o and 40 M_o and between 17 M_o and 40 M_o.

q-distribution	8-100 M_o		17-100 M_o	
	case B_r	P≤100d	case B_r	P≤100d
flat	0.72	0.45	0.75	0.5
Hogeveen	0.6	0.38	0.63	0.38

Using an IMF with α = 2.7 and the results of figure 8.2, table 18.1 gives the frequency at birth of case B_r binaries with $q \geq 0.2$ and of binaries with period $P \leq 100$ days and $q \geq 0.2$ in a population of binaries with periods up to 10 years. These rates are given relative to the total massive binary frequency at birth f (binaries with periods up to 10 years).
Remark that since most of the observed MCBs have a period smaller than ~100 days and a mass ratio larger than 0.2 (due to selection), and since the observed MCB frequency ~33% (section 16.4), it follows from table 18.1 that the total binary frequency at birth may be very large (>70%).

- the effect of mass loss from the system during the RLOF [the parameter β, defined by relation (10.7)] of the primary in a case B_r MCB is studied as given in figure 18.6. The value of β in a binary with period P (corresponding to a case B_r) and $q \geq q_{min}$ is assumed to be constant = $\beta_{max} \leq$ 1. For $q \leq 0.2$, $\beta = 0$ obviously. Between q_{min} and $q = 0.2$, a linear variation of β is adopted. The final results do not very much depend on this variation but they may depend on q_{min}.
It is probable that β_{max} decreases as the period of the case B_r binary increases, approaching zero when the period is so large that the RLOF starts when the outer layers of the primary are convective (case B_c). Moreover, it may also be a function of the primary mass. The effect of this parameter has to be discussed in detail and one hopes that despite the uncertainty, at least some conclusions do not critically depend on it.

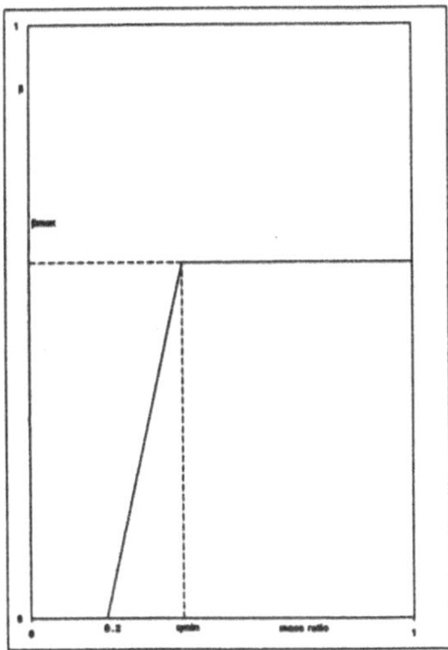

Figure 18.6: The variation of the parameter β as a function of initial binary mass ratio.

When the RLOF of the primary in a case B_r binary is non-conservative, we consider the loss of mass through the second Lagrangian point during a contact phase as the most probable mode. Therefore, the period variation can be estimated with equations (10.14) and (10.15). Since the formalism predicts a period decrease during the RLOF, a situation cannot be excluded that massive case B_r binaries merge. We assume that the evolution of these mergers is similar to the evolution of a single star with mass equal to the sum of the masses of both components,

- the parameter that determines the evolution of case B_c/C binaries is the parameter α in relation (10.16) in combination with formula (10.18). Anticipating, case B_c/C binaries do not significantly contribute to the overall population of massive stars and therefore the parameter α only marginally affects the results,

- when a SN explosion happens, we first assume that the effect of the asymmetry (v_{kick}) can be described by the distribution (16.20). As remarked in section 16.6, significant errors in proper motion determinations cannot be excluded. The calculations have therefore been repeated with a distribution function

$$f(v_p) = 2.7 \cdot 10^{-5} v_p^{3/2} e^{-v_p/60} \tag{18.2}$$

predicting an average $v_{kick} = 150$ km/s.

- if it is assumed that no SN explosion occurs when the star ends its life as a BH, it is obvious that the minimum initial mass for BH formation ($M_{BH,min}$) critically determines the heavy metal enrichment of a galaxy and the number of SN type I and type II. Moreover, a binary with primary mass larger than this minimum mass will obviously not be disrupted, will thus not produce runaways but, depending on the details of the spiral-in process, may produce WR+BH binaries.
For single stars (resp. binary components) $M_{BH,min} \geq$ 20-30 M_o (resp. 40-50 M_o),

- when a SN explosion occurred and did not disrupt the binary, depending on the mass of the OB-type star, the further evolution is governed by the spiral-in process which depends on the parameter α in formula (10.16) in combination with relation (10.19). Unlike for case B_c/C systems, the value of α here determines critically the PNS results of post-SN systems.

When as a consequence of the spiral-in process a TZO is formed, the further evolution and the effect on PNS can be studied by considering two limiting possibilities:

- model TZO_1: although the binary will merge, it is assumed that sufficient mass was removed during the spiral-in so that the post-spiral-in star is a hydrogen deficient CHeB star with a cc in its center (and when the CHeB star is massive enough, a WR star with a cc in its center: WR_{TZ} stars). The remaining WR time scale is assumed to be equal to the WR time scale of a normal post-RLOF WR star with the same mass,
- model TZO_2: if the binary merges, we take the pre-spiral-in mass of the OB-type star and we assume that its further evolution is similar to the evolution of a normal single star, although it has a cc in its center. In this case the number of WR_{TZ} stars depends on the adopted single star evolutionary scenario and in particular on the RSG SW mass loss of single stars,

- the PNS model is able to predict star and SN-type numbers in regions where star formation was continuous, at least during a time equal to the lifetime of the smallest mass in our sample. However it is also possible to follow the evolution of starburst regions where the duration of the starburst is an input parameter. In this case, the stars are allowed to form at random, satisfying the adopted initial mass, binary frequency, binary mass ratio and period distribution.

19. PNS model calculations

19.1. The expected evolution of the observed galactic WR+OB sample.

We first assume that all the WR components of the 16 massive, short period WR+OB binaries of table 14.1 collapse to form a NS with a mass = 1.4 M_o accompanied by a SN explosion. From the evolutionary computations during CHeB of post-RLOF primary remnants (section 12.2), it is straightforward to estimate the mass of the WR star at the moment of core collapse.

Using the formalism discussed in section 11 with the distribution of SN kick velocities given by equation (16.20), we calculate the probability that the system is disrupted assuming that the direction of the kick during the SN explosion is random. When the system is not disrupted, we determine the post-SN period distribution of the OB+NS system. For each OB+NS system we then calculate the further evolution, first during the CHB of the OB type star with SW mass loss, and then either through a LBV phase, a RSG phase or a spiral-in phase. In this way we obtain the probability that a CHeB (WR)+NS is formed.

TABLE 19.1: For the observed WR+OB binaries, accounting for the evolutionary computations of CHeB primary remnants after the RLOF in a MCB, we compute the probability that the binary remains bound after a possible SN explosion of the WR star assuming that a NS with mass = 1.4 Mo remains; we then follow the bound systems through the spiral-in phase and this allows us to calculate the probability that a WR+NS system can be formed. We then assume that the WR components collapse to form a BH with mass = the mass of the WR star at the end of CHeB (i.e. no SN occurs and thus obviously all systems remain bound and the period = the period prior to the collapse). These systems are then followed through the spiral-in phase. When merging can be avoided (resp. cannot be avoided), a WR+BH binary is formed with probability = 1 (resp. 0).

WR nr	HD (name)	P(days)	Sp. Type	Prob. survival	Prob. WR+NS	Prob. WR+BH
21	90657	8.3	WN4+O4-6	0.3	0.04	0
31	94546	4.8	WN4+O8V	0.38	0	0
47	E311884	6.3	WN6+O5V	0.2	0	0
97	E320102	11.6	WN3-4+O5-7	0.39	0.05	0
127	186943	8.6	WN4+O9/B0	0.4	0.06	0
133	190918	111.8	WN4+O9Ib	0.1	0.1	1
139	193576	4.2	WN5+O6	0.4	0.02	0
151	CX Cep	2.1	WN5+O8	0.4	0	0
153	211853	6.7	WN6+O6	0.3	0.02	0
155	214419	1.6	WN7+O	0.4	0.02	0
9	63099	14.7	WC5+O7	0.4	0.05	0
11	68273	78.5	WC8+O	0.2	0.1	1
30	94305	18.8	WC6+O6/8	0.3	0.06	0
42	97152	7.9	WC7+O7V	0.3	0.04	0
79	152270	8.9	WC7+O5-8	0.3	0.04	0
113	168206	28.7	WC8+O8-9	0.2	0.0	0

The results are summarized in table 19.1. We conclude:

- when a kick velocity distribution is used with $<v_{kick}>$ = 450 km/s [equation (16.20)] a large fraction (≥ 70 %) of the observed WR+OB binaries is expected to be disrupted during the SN explosion of the WR star,

- the expected number of normal WR+NS systems resulting from the observed WR+OB sample is extremely small; most of them have periods of a few hours; periods larger than a day are very rare,
- almost all OB type stars (with or without a cc) after the SN explosion have space velocities larger than 30 km/s, i.e. they will all be classified as runaways. Note however that only a minority will have a NS companion.

From the mass estimates of the 18 WR+OB binaries listed in table 14.1, it cannot be excluded that the majority of the WR components collapse to form a BH and no SN occurs. When we assume that the mass of the WR star at the end of CHeB equals the mass of the BH, we also determined the possibility that afterwards the binary survives the spiral-in phase, and thus the possibility that a WR+BH binary will be formed. The computations show that only the binaries HD 190918 and HD 68273 may evolve into a WR+BH binary. The others merge.

19.2. The predicted distribution of stars in regions of continuous star formation with a realistic binary formation frequency

We assume a total binary frequency at birth f, a period and mass ratio distribution, and an IMF for single stars and primaries of close binaries. As constraint we use the fact that the observed MCB [$q>0.2$, $P\leq100d$] frequency in the Solar neighbourhood $\approx$ 0.3 - 0.4.

19.2.1. THE LARGE NUMBER OF STARS OBSERVED IN THE BLUE HERTZSPRUNG GAP (BHG)

As discussed in section 5.1.6, it is hard to explain the large number of stars in the BHG region of the HR diagram by means of single star evolution. Here we offer two binary possibilities:

a. Binary mergers.

All but a few percent of the interacting MCBs with $q \leq 0.2$ merge. If the RLOF in case B_r binaries with $q > 0.2$ is non-conservative, mass can (only?) leave the binary through L_2 and this implies an orbital period reduction [relation (10.14)]. It follows that in addition to the number of mergers resulting from $q \leq 0.2$, also part of the case B_r binaries with $q \geq 0.2$ may merge as well. The majority of them will merge while the massive component is a hydrogen shell burning star. If the merging process resembles accretion (section 12.3), it can be expected that some of the mergers (which look like single stars) will remain in the blue part of the HR diagram and thus at least part of the stars which are observed in the BHG could be binary mergers. These stars may show the products of CNO burning at their surface.

b. Accretion stars.

When the mass ratio of a case B_r binary is large ($q > 0.7$-0.9), when semi-convection is a slow mixing process and when the RLOF is quasi-conservative, it can be expected that a non-negligible fraction of the mass gainers will further evolve as discussed in section 12.3, i.e. they will remain blue stars during their entire CHeB phase populating the BHG. These stars may explode as blue supergiants producing an event like SN 1987A.

19.2.2. THE DISTRIBUTION OF O-TYPE STARS PREDICTED BY THE PNS MODEL

A realistic O-type star population consists of

- O_s = number of O-type single stars, formed as single stars
- O_b = number of O-type primaries in a binary with an OB or a low mass component. We separately consider the number of O-type primaries in a binary with $q > 0.2$ and orbital period $P \leq 100$ days. Since it is this number that may be comparable to the presently observed number, we use $O_{b,o}$
- O_{+CHeB} = number of O-type mass gainers with a CHeB (WR) companion
- O_{sb} = the number of single post-SN O-type mass gainers where the SN disrupted the binary; we separately consider those with a space velocity ≥ 30 km/s (the runaways)
- O_{cc} = the number of post-SN O-type mass gainers with a compact companion (NS or BH); also here we separately consider those with a NS companion and with a space velocity ≥ 30 km/s.

The predicted frequency of each subclass is given in table 19.2 for the Galaxy. The results for the MCs are very similar.

TABLE 19.2: The O-type population predicted by the PNS model for different parameters.

f	β_{max}	α_{CE}	IMF	$\Phi(q)$	$\langle v_{kick} \rangle$	O_s	O_b	$O_{b,o}$	O_{+CHeB}	O_{cc}+NS	O_{cc}+BH	O_{sb}	$O_{sb,r}$
0.8	1	1	2.7	F	500	0.1	0.51	0.25	0.14	0.03	0.007	0.22	0.05
0.8	1	1	2.7	H	500	0.15	0.59	0.24	0.09	0.03	0.008	0.16	0.05
0.5	1	1	2.7	H	500	0.31	0.31	0.16	0.11	0.02	0.005	0.18	0.03
0.5	1	1	2.7	H	500	0.41	0.41	0.09	0.07	0.02	0.007	0.13	0.05
0.8	1	0.5	2.7	F	500	0.1	0.4	0.18	0.14	0.03	0.007	0.22	0.05
0.8	1	1	2	F	500	0.11	0.59	0.29	0.09	0.02	0.013	0.17	0.04
0.8	1	1	2	H	500	0.15	0.64	0.25	0.06	0.02	0.017	0.13	0.03
0.8	0.5	1	2.7	F	500	0.12	0.64	0.32	0.06	0.03	0.008	0.15	0.08
0.8	0.5	1	2.7	H	500	0.16	0.69	0.28	0.03	0.02	0.010	0.09	0.05
0.8	1	1	2.7	F	150	0.1	0.49	0.18	0.14	0.11	0.007	0.16	0.05
0.8	1	1	2.7	H	150	0.15	0.53	0.13	0.1	0.10	0.009	0.12	0.04

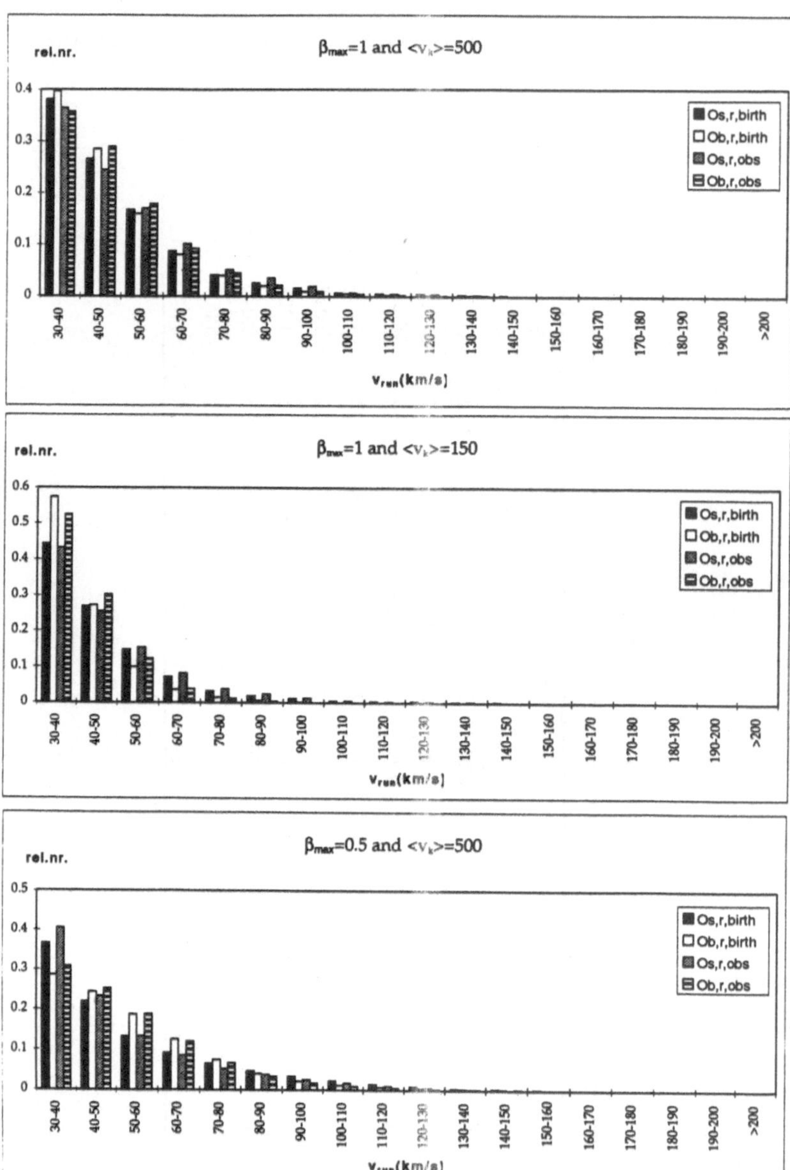

Figure 19.1: The theoretically expected distribution of space velocities of O-type runaways. We separately consider the distributions at birth and the real distribution accounting for the remaining CHB lifetime of the runaway. We furthermore distinct O-type single runaways where the SN explosion disrupted the binary, and the O-type binary runaways where the SN explosion did not disrupt the binary (i.e. the O-type runaway still has a NS companion).

We conclude:

- a comparison between the predicted $O_{b,o}$ numbers and the observed one (33%±13%, section 2.3.2) forces us to conclude that a majority of the O-type stars are formed as primaries in binaries with orbital period between 1 day and 10 years (the binary frequency at birth f > 0.7),
- between 15% and 25% of all O-type stars are post-SN mass gainers of MCBs; if the kick-velocity distribution has an average value of 150 km/s (resp. 450 km/s), ~40% (resp. <12%) of these post-SN O-type mass gainers have a NS companion. It is clear then that if the kick-velocity distribution has an average value of 450 km/s, a very significant fraction of the population of 'single' O-type stars are single post-SN O-type mass gainers of MCBs,
- at most 1-2% of all O-type stars may have a BH companion,
- at most 8% of the O-type stars are runaways due to a previous SN explosion in a binary; if $<v_{kick}>$ = 450 km/s (resp. 150 km/s), < 20% (resp. > 50%) of these runaways have a NS companion. Gies (1987) and Gies and Bolton (1986) investigated the presence of compact companions among O-type runaways and concluded that only a small percentage may hide one. This is predicted by our PNS model provided that the kick velocity during the SN explosion is very large,
- in figure 19.1 we have plotted the theoretically predicted distribution of runaway velocities of the O-type runaways, with ($O_{binary,r}$) and without ($O_{single,r}$) a NS companion. We separately consider the distribution at birth and the distribution where we account for the remaining CHB lifetime of the runaway. It can be concluded that the sample of O-type runaways (with or without a NS companion), who became runaway as a consequence of a previous SN explosion in a binary, have an average space velocity ~50 km/s whereas a runaway with a velocity larger than ~100 km/s is very rare.

19.2.3. THE WR POPULATION

A realistic WR population consists of

- WR_s = number of single WR stars descendent from the O_s class,
- WR_b= number of WR binary components with an OB or low mass component, descendants from the O_b sample; this class can be subdivided into

- $WR_{b,i}$ = number of WR stars with an OB-type or low mass companion, where RLOF and binary interaction played a dominant role,

- $WR_{b,LBV}$ = number of WR stars with an OB-type or low mass companion where, due to very large SW during a LBV phase, RLOF and thus binary interaction may have played only a minor role,

- $WR_{b,merged}$ = number of (single) WR stars, descendent from those interacting binaries that merge,

- $WR_{b,RSG}$ = number of WR binaries with an OB-type or low mass companion orbiting with very large period, where interaction did not happen and where the WR star was formed by SW during the RSG stage,

• WR_{sb} = the number of single post-SN WR stars descendent from the O_{sb} class,

• WR_{cc} = the number of WR+cc binaries that survived the spiral-in phase of the O+cc binary,

• WR_{TZ} = the number of WR stars with a cc in their center, descendent from O+cc binaries where during the spiral-in phase a Thorne-Zytkov object is formed.

The Solar neighbourhood

Tables 19.3a,b list the frequency of each subclass for the Solar neighbourhood (Z = 0.02). Remark that the results depend critically on the value of $M_{WR,min}$. We conclude:

• the observed WR/O number ratio in the Solar neighbourhood cannot be reproduced if it is assumed that $M_{WR,min} = 8\ M_o$; the comparison is better assuming $M_{WR,min} = 5\ M_o$ although the theoretical value is still somewhat smaller than the observed one.
A few remarks are appropriate

- some of the lower luminosity (Log $L/L_o \leq 5$) WR stars may be post-AGB stars, descendants from intermediate mass stars, on their way to become WDs,

- the theoretical O-star lifetime is calculated assuming that a star will be visible as an O-type star from the beginning of CHB (the ZAMS). However, when the theoretical HR diagram of O-type stars is compared to the observed one, there appears to be a lack of stars close to the ZAMS (Garmany et al., 1980; section 5.1.1),

- when the HR diagram of aggregate stars within the Solar neighbourhood is compared to the theoretical evolutionary tracks and one counts the number of stars within time-isochrones of 1 million years apart, it looks as if the incompleteness of massive stars is very large (and not only close to the ZAMS) or that the massive star formation rate was significantly larger a few million years ago,

- when $M_{WR,min}$ = 5 M_o (resp. 8 M_o) it follows that in ~70% (resp. 30%) of the WR+OB binaries (where also the OB star is a massive star) RLOF and mass transfer occurred; the other ~30% (resp. 70%) evolved either according to the LBV scenario either according to the RSG scenario,

- 3-5% of the WR stars may have a normal intermediate mass or low mass companion,

- depending on the adopted efficiency for converting orbital energy into kinetic energy of the envelope of the OB-type star during the spiral-in of the cc, at most 3% (resp. 8%) of the WR stars may have a compact star in their center (WR_{TZ}) if the kick-velocity distribution has an average value of 450 km/s (resp. 150 km/s),

- the PNS model predicts a small percentage of WR stars with a compact companion. Depending on the adopted efficiency for converting orbital energy into kinetic energy of the envelope of the OB-type star during the spiral-in of the cc, at most 2% (resp. 8%) of the WR stars may have a NS companion if the kick-velocity distribution has an average value = 450 km/s (resp. 150 km/s). Between 3% - 5% (resp. ~10%) of the WR stars may have a BH companion when $M_{WR,min}$ = 5 M_o (resp. 8 M_o). The WR+BH numbers depend on the value of the minimum mass of BH formation in binaries. They were calculated assuming $M_{BH,min}^{binary}$ = 40 M_o. Using 50 M_o instead lowers the WR+BH numbers by a factor of 2,

- between 40% and 50% of all WR stars are single WR stars. However, possibly ~70-80% (or more) of these single WR have had a binary history, i.e. are WR stars, descendants from OB type mass gainers in binaries where the SN explosion disrupted the binary, or are WR stars descendants from binaries where during the RLOF both components merged. The latter class is largest when the Hogeveen mass ratio distribution applies and/or when the RLOF in case A/case B_r MCBs is highly non-conservative ($\beta \leq 0.5$),

- at most 5% of all WR stars have a space velocity larger than 30 km/s (WR runaway) caused by the recoil of the SN explosion in a binary.
  The WR space velocity distribution should obviously be very similar to the O-type runaway distribution. Although a binary scenario in order to explain the space velocity of HD 50896 and of 209 BAC (section 13.5.2) cannot be entirely excluded, it can be concluded from figure 19.1 that the probability is very small,

- in figure 19.2a,b we show the theoretically predicted mass ratio and period distribution of WR+OB binaries at the beginning of the WR phase for $M_{WR,min}$ = 5 M_o, for β_{max} = 0.5 in interacting binaries with initial mass smaller than 40 M_o and for a Hogeveen initial mass ratio distribution. When compared to the observed periods and mass ratios of the 17 WR+OB binaries (figures 17.2a and 17.2b), we have to conclude that theory predicts many more WR+OB binaries with periods between 100 days and 1000 days and/or WR+OB binaries with mass ratio larger than 2.2-2.4 or with a small mass ratio ($q < 0.8$).The latter holds also for other values of $M_{WR,min}$ and β_{max}, or for a flat initial mass ratio distribution.
  This discrepancy can obviously be due to still unrecognized physical processes that are not yet included in the evolutionary code. However, also observations may still be highly incomplete. Large period binaries are harder to detect whereas when the mass ratio of the WR+OB binary is smaller (resp. larger) than 0.8 (resp. 2.2-2.4), the OB component (resp. the WR component) may be barely visible,

The inner Milky Way

The metallicity Z in the inner MW ≈ 0.03. The evolution of MCBs hardly depends on Z whereas when compared to Z=0.02, significant differences in single star evolution are expected only if the RSG stellar wind mass loss scales linearly with Z (i.e. the RSG $\dot{M}$ is a factor 1.5 larger compared to Z=0.02).

Our computations reveal that, unless $M_{WR,min}$ is very small, it is hard to explain the observed WR/O number ratio (=0.2) by means of a continuous star formation model.

However, a WR/O number ratio as high as 0.2 is not unusual if, instead of a continuous star formation history, there has been an increased star formation rate a few million years ago.

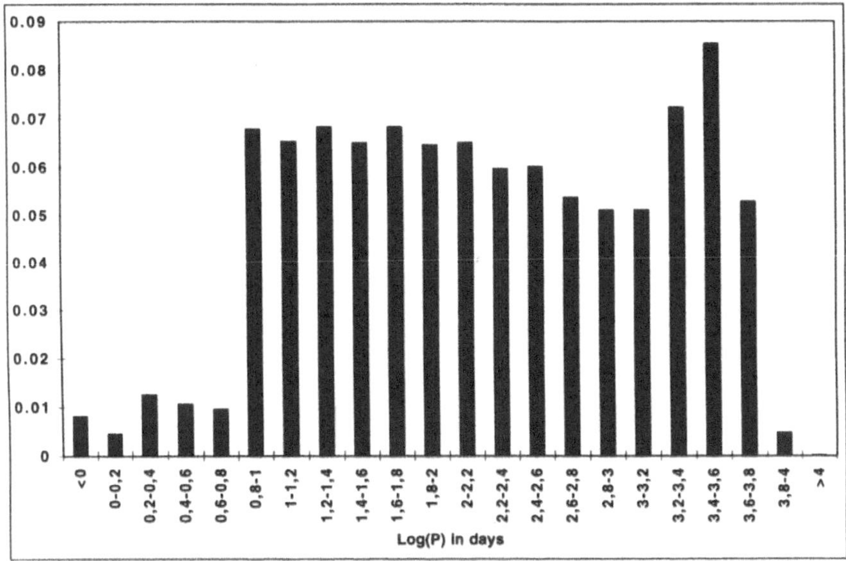

Figure 19.2a: The theoretically predicted period distribution of WR+OB binaries at the beginning of the WR phase for a Hogeveen initial mass ratio distribution, for $M_{WR,min} = 5\ M_{O}$ and for $\beta_{max} = 0.5$ in interacting binaries.

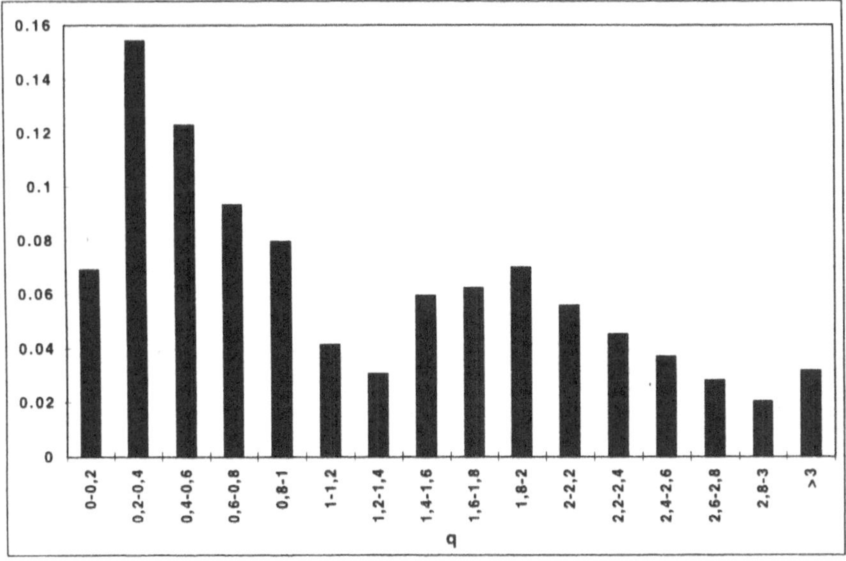

Figure 19.2b: The theoretically predicted mass ratio distribution of WR+OB binaries at the beginning of the WR phase, similarly as figure 19.2a.

TABLE 19.3a: The WR population of the Solar neighbourhood predicted by the PNS model, for different values of the parameters, assuming $M_{WR,min} = 5\ M_o$.

f	β_{max}	α_{CE}	IMF	$\Phi(q)$	$<v_{kick}>$	WR_s	$WR_{b,i}$	$WR_{b,LBV}$	$WR_{b,RSG}$	WR_{merged}	WR+NS	WR+BH	WR_{TZ}	WR_{sb}	WR/O
0.8	1	1	2.7	F	500	0.101	0.319	0.128	0.036	0.084	0.015	0.049	0.03	0.238	0.056
0.8	1	1	2.7	H	500	0.128	0.265	0.162	0.05	0.208	0.008	0.026	0.021	0.132	0.064
0.5	1	1	2.7	F	500	0.309	0.245	0.098	0.028	0.065	0.011	0.038	0.023	0.182	0.056
0.5	1	1	2.7	H	500	0.371	0.191	0.117	0.036	0.15	0.006	0.019	0.015	0.095	0.062
0.8	1	0.5	2.7	F	500	0.101	0.313	0.128	0.036	0.09	0.008	0.05	0.034	0.239	0.053
0.8	1	1	2	F	500	0.111	0.287	0.198	0.032	0.065	0.012	0.081	0.024	0.184	0.063
0.8	1	1	2	H	500	0.138	0.233	0.246	0.044	0.169	0.006	0.043	0.011	0.104	0.069
0.8	0.5	1	2.7	F	500	0.117	0.271	0.148	0.042	0.241	0.006	0.057	0.021	0.097	0.061
0.8	0.5	1	2.7	H	500	0.142	0.2	0.179	0.055	0.339	0.002	0.029	0.01	0.043	0.071
0.8	1	1	2.7	F	150	0.099	0.315	0.126	0.035	0.083	0.066	0.049	0.084	0.144	0.057
0.8	1	1	2.7	H	150	0.127	0.262	0.161	0.049	0.206	0.036	0.026	0.055	0.077	0.065

$M_{WR,min} = 5\ M_o$

TABLE 19.3b: The WR population of the Solar neighbourhood predicted by the PNS model, for different values of the parameters, assuming $M_{WR,min} = 8\ M_o$.

f	β_{max}	α_{CE}	IMF	$\Phi(q)$	$<v_{kick}>$	WR_s	$WR_{b,i}$	$WR_{b,LBV}$	$WR_{b,RSG}$	WR_{merged}	WR+NS	WR+BH	WR_{TZ}	WR_{sb}	WR/O
0.8	1	1	2.7	F	500	0.135	0.118	0.307	0.035	0.233	0.003	0.103	0.011	0.055	0.029
0.8	1	1	2.7	H	500	0.161	0.089	0.363	0.044	0.275	0.001	0.042	0.004	0.02	0.035
0.5	1	1	2.7	F	500	0.368	0.11	0.209	0.024	0.051	0.01	0.07	0.018	0.141	0.026
0.5	1	1	2.7	H	500	0.428	0.085	0.242	0.03	0.112	0.004	0.028	0.01	0.063	0.03
0.8	1	0.5	2.7	F	500	0.127	0.148	0.288	0.033	0.074	0.011	0.097	0.027	0.194	0.024
0.8	1	1	2	F	500	0.135	0.121	0.362	0.026	0.054	0.011	0.013	0.018	0.142	0.035
0.8	1	1	2	H	500	0.164	0.098	0.439	0.035	0.125	0.005	0.056	0.011	0.068	0.038
0.8	0.5	1	2.7	F	500	0.135	0.118	0.307	0.035	0.233	0.003	0.103	0.011	0.055	0.029
0.8	0.5	1	2.7	H	500	0.161	0.089	0.363	0.044	0.275	0.001	0.042	0.004	0.02	0.035
0.8	1	1	2.7	F	150	0.127	0.151	0.287	0.033	0.07	0.056	0.097	0.066	0.114	0.025
0.8	1	1	2.7	H	150	0.157	0.125	0.355	0.043	0.164	0.024	0.041	0.037	0.053	0.029

$M_{WR,min} = 8\ M_o$

The Magellanic Clouds

We focus on the LMC since the SMC numbers are very small. The formation of a WR single star depends primarily on the SW mass loss rate during the LBV phase and/or during the RSG phase, while the duration of the WR phase depends on the WR SW mass loss rate as well. Furthermore the evolution of a MCB hardly depends on the metallicity. Adopting a continuous star formation history in the LMC, the theoretically predicted WR/O number ratio ~ 0.02-0.03 (resp. 0.04-0.05) if $M_{WR,min} = 8\ M_o$ (resp. 5 M_o) which is very close to the observed value.

19.2.4. THE EARLY B-TYPE STAR POPULATION PREDICTED BY THE PNS MODEL

Figure 19.3 shows the fraction, relative to all B-type stars, of the post-SN early B-type mass gainers with and without a NS.

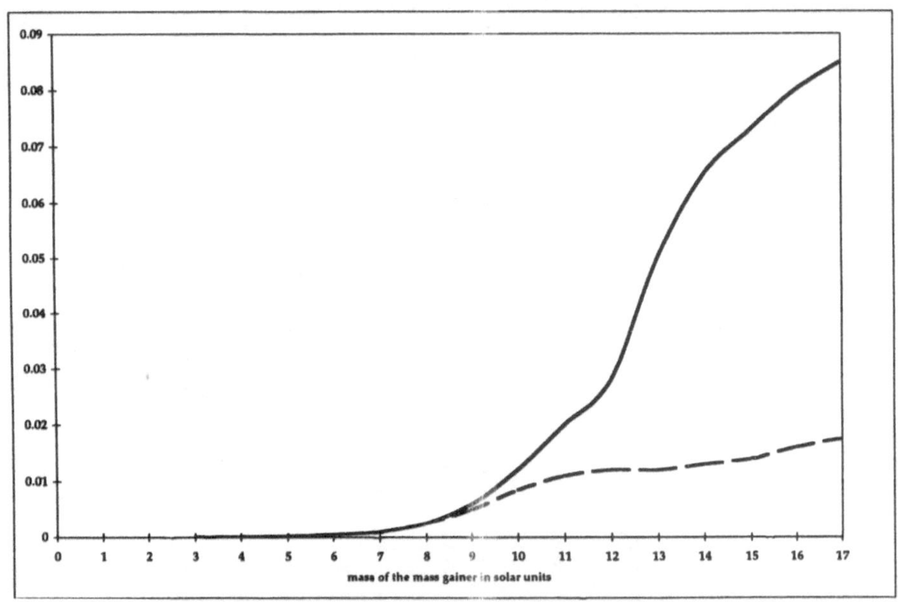

Figure 19.3: The fraction of the post-SN early B-type mass gainers with (dashed line) and without a neutron star (full line) (from Van Bever and Vanbeveren, 1998).

We used

- a total binary frequency at birth f = 0.7; remember that the results scale linearly with this value,
- β_{max} = 1 (section 18.2) for binaries with primary mass ≤ 20 M_o; a smaller value would lower the fractions,
- the v_{kick} distribution predicted by equation (16.20),
- a flat $\Phi(q)$ although the results do depend only marginally on the adopted mass ratio distribution
- a primary in a MCB explodes and forms a NS when its initial mass is larger than ~ 10 M_o.

We conclude

- at most 2% of all B-type stars with spectral type earlier than B3 have a NS companion if $<v_{kick}>$ = 450 km/s,
- at most 10% of all B-type stars with spectral type earlier than B3 may be single binary mass gainers,
- about 5% of the early B-type stars are expected to be runaways (the majority is single if $<v_{kick}>$ = 450 km/s) which is very similar to the O-type runaway frequency. Also the space velocity distribution is very similar to the one for the O-type runaways (figure 19.4).

Since binary mass gainers may be rapid rotators due to the spin-up process during mass transfer (section 9.2), they may be observed as OBe type stars (section 2.4.2). Among all early B-type stars, 15-20% are classified as Be. Furthermore, at least as many normal early B-type stars have rotational velocities similar to the Be stars (section 2.4.2), i.e. 30-40% of all early B-type stars have Be-type rotational velocities. Compared to the binary frequencies quoted above, we conclude

- less than 25% of the rapidly rotating early B-type stars (Be) acquired their rapid rotation as a consequence of binary interaction.

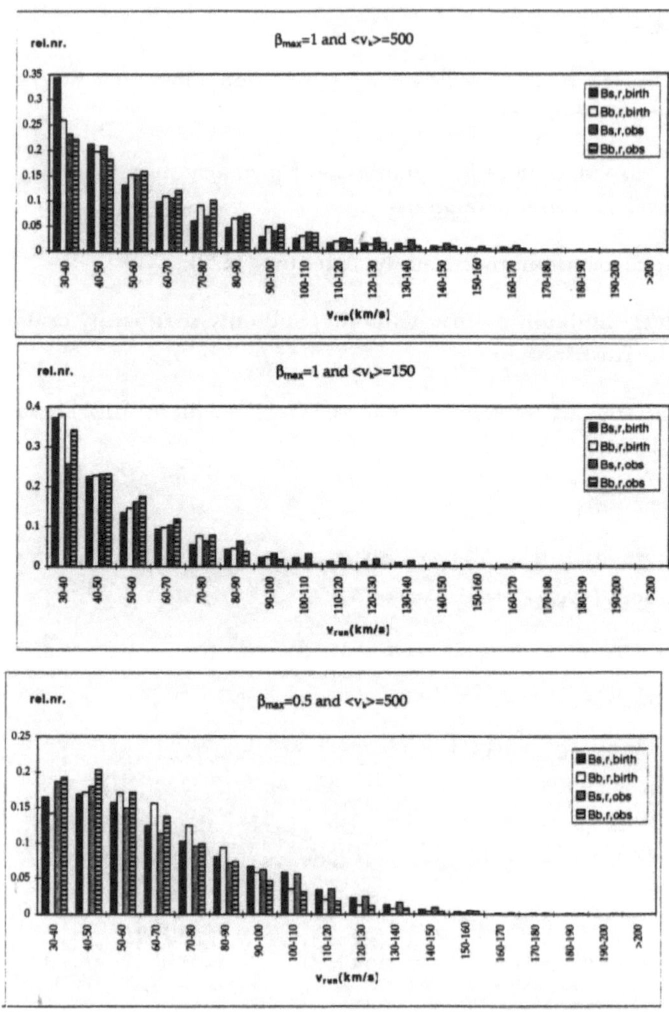

Figure 19.4: Similar to figure 19.1 but for the early B-type runaways.

19.2.5. THE THEORETICALLY EXPECTED NUMBER OF BINARY PULSARS

The PNS model predicts a very small (but non-zero) number of CHeB + NS/BH binaries. Using the same v_{kick} distribution as the one during the first SN explosion, we compute the survival probability of these binaries when the CHeB companion explodes. The predicted frequency of NS+NS binaries then depends critically on the average value of v_{kick} and on the efficiency parameter α for converting orbital energy into potential energy of the envelope during the spiral-in process (section 10.2.1.2). It can be concluded that

- the theoretical birth rate of double NS systems is about 0.0003-0.004 times the formation rate of massive stars. The first (resp. second) number holds when $<v_{kick}> = 450$ km/s and $\alpha = 0.5$ (resp. $<v_{kick}> = 150$ km/s and $\alpha = 1$),

It is generally assumed that our Galaxy is an early type galaxy with $L_B = 2.10^{10}$ L_{B_o}. From table 17.1, one then estimates that the overall SN II+Ib/c rate $\approx$ 1 (2 at most) per century, i.e. the massive star formation rate in our Galaxy $\approx$ 0.01 (0.02) per year and thus the PNS model predicts a NS+NS rate $\approx 3.10^{-6}$-8.10^{-5} per year. Accounting for the crudeness of this estimate, the still existing uncertainties in the PNS model and its input parameters and the uncertainty of the 'observed' value (section 17.5), we conclude

- the PNS model where the asymmetry of the SN explosion is described by the pulsar runaway distribution of equation (16.20), predicts the observed galactic binary pulsar formation rate.

Since our PNS model follows the evolution of MCB periods in detail, we are capable to predict the period distribution at birth of all double compact star binaries. This is shown in figure 19.5 where we have choosen for a binary formation rate f = 0.8, a flat mass ratio distribution, maximum efficiency of the spiral-in/common envelope parameter α. Note however, that the distribution only slightly depends on these adopted PNS input parameters.

Double compact star binaries lose orbital energy and angular momentum by emitting gravitational waves. The binary period shrinks and the system possibly merges. The mathematical formalism has been worked out by Wagoner (1975). The time of complete spiral-in is beside depending on the initial orbital period and on the mass of both components, strongly dependent on the eccentricity e of the system. Assuming an average $<e> = 0.6$ (based on the eccentricities of the observed galactic NS+NS systems), one finds that a double NS binary will merge within 6 million years if its period at birth was smaller than ~2 hours. From figure 19.5 it can be concluded that this will happen in only a tiny fraction (< 5%) of the NS+NS binaries.

The PNS model gives the formation rate of NS+BH and BH+BH binaries as well. It follows that the BH+BH rate may be significantly larger (more than a factor 10) than the NS+NS rate. Figure 19.5 illustrates as well the period distribution at birth of the NS+BH and of the BH+BH binaries. Their periods are much larger than those of the NS+NS binaries. If we repeat the exercise above, but for the BH+BH systems, we have to conclude that complete spiral-in will rarely (not?) happen within evolutionary time (not even within the Hubble time).

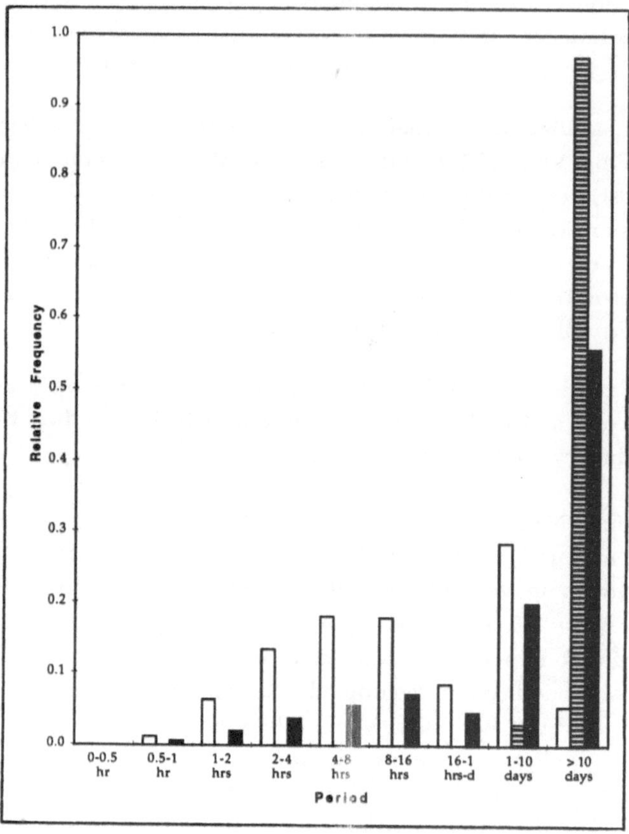

Figure 19.5: The theoretically expected period distribution at birth of NS+NS (white bars), NS+BH (lined bars) and BH+BH (black bars) binaries (from De Donder and Vanbeveren, 1998).

Let us note that owing to the fact that MCB evolution is largely independent of the metallicity

- the binary pulsar formation rate should be independent of the metallicity provided that the binary frequency, the orbital parameter distributions and the IMF are metallicity independent.

19.2.6. THE THEORETICALLY EXPECTED NUMBER OF SINGLE PULSARS WITH A MCB HISTORY

Among the single pulsars with a MCB history, we can distinguish two subclasses:

- the NSs that were members of a CHeB+NS binary in which the SN explosion of the CHeB star disrupted the binary. Accounting for the effect of the asymmetry of the SN explosion, the PNS model predicts that ~80-90% of the CHeB+NS systems will be disrupted. Among the 600 pulsars known in the Galaxy, three are 3 (4) NS+NS binaries. We therefore expect 12 (16) - 27 (36) single pulsars with a MCB history. We conclude that the PNS model that predicts the number of double NS systems in the Galaxy, predicts the existence of ~2-5% pulsars with a MCB history that became single as a consequence of a second SN explosion in a binary,
- the NSs that were members of an OB+cc binary that evolved into a TZO after spiral-in. When the star-matter has been removed by SW, the NS reappears again. The PNS model predicts that up to ~5% of the pulsar sample may be such objects.

19.2.7. THE RUNAWAY VELOCITY OF SINGLE PULSARS

The sample of single pulsars consists of real single pulsars but also of pulsars who originally were binary components and became single as a consequence of the SN explosion and its effect on binary parameters. The observed pulsar velocity distribution (section 16.6) therefore includes the effect of the SN explosion of massive stars and the effect of the pre-SN orbital velocity of disrupted pulsar binary components. To study the consequences, we adopt an initial SN-v_{kick} distribution and we assume the this distribution applies for single stars and for binary components. Since our PNS model follows the evolution of the MCB period in detail, the expected runaway velocity distribution of the pulsars can be determined as function of the PNS input parameters. Figure 19.6 compares the initial v_{kick} distribution with the runaway velocity distribution assuming an overall binary formation rate f = 0.8. We adopt equation 16.20 (or equation 18.2) as initial distribution. As can be noticed, the expected runaway velocity distribution is slightly shifted towards larger velocities compared to the v_{kick} distribution. In the case $<v_{kick}>$ = 450 km/s, the differences are very small, i.e.

if the asymmetry of the SN explosion is such that on the average the NS recieves a kick velocity ≈ 450 km/s, it follows that the runaway velocity distribution of single pulsars nearly equals the v_{kick} distribution.

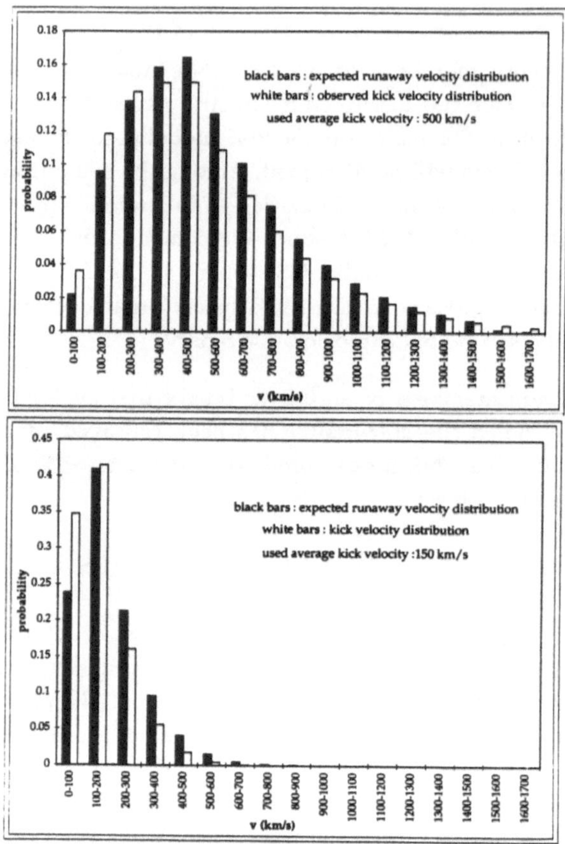

Figure 19.6: The initial vkick velocity distribution compared to the expected runaway velocity of single pulsars; top (resp. bottom) figure corresponds to the case where $<v_{kick}>$ = 150 km/s (resp. 500 km/s).

19.3. The predicted distribution of stars in starburst regions with a realistic binary formation frequency

The Brussels PNS model is capable of following the evolution of a burst of massive stars with a realistic frequency of close binaries. It is possible to predict the time-variation of the number of stars of different spectral subtypes but it is also possible to predict the evolution of starburst in the HR diagram with special emphasis on the evolution of the mass gainers of CBs.

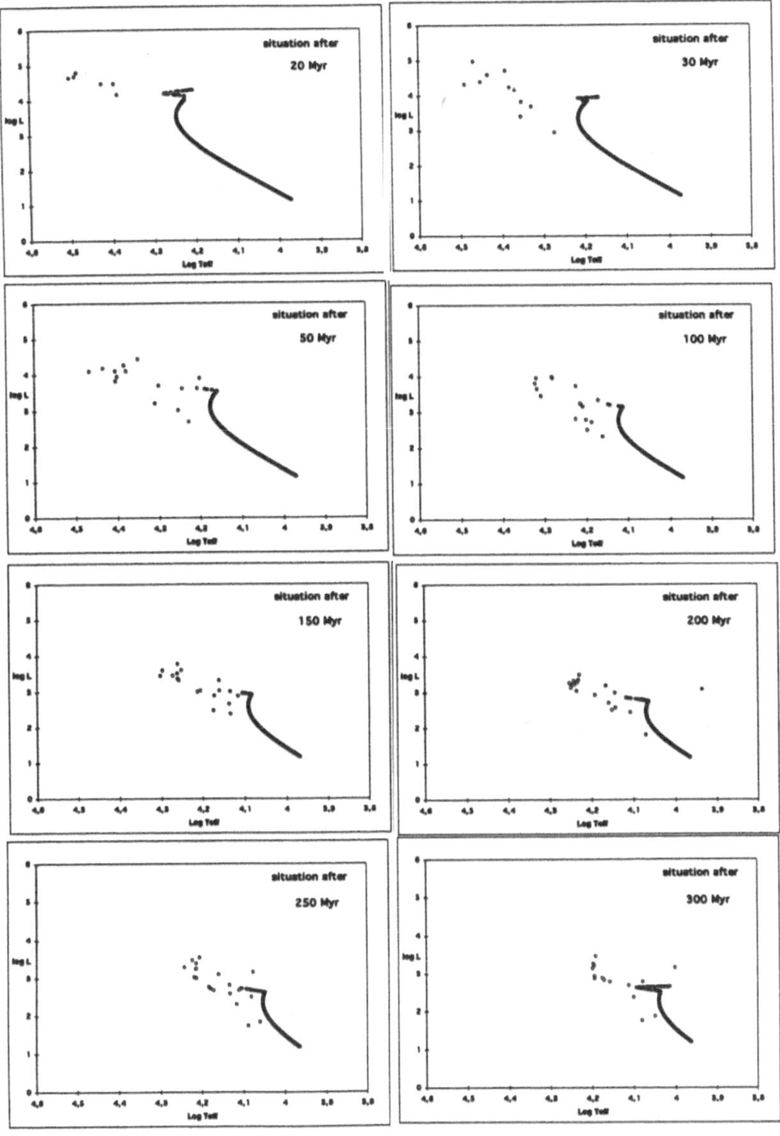

Figure 19.7: An evolutionary simulation in the HR diagram of the time variation of the number of early B-type mass gainers in a starburst region (from Van Bever and Vanbeveren, 1997).

19.3.1. THE POPULATION OF B3-B0 TYPE RAPID ROTATORS IN STARBURSTS

Figure 19.7 shows an evolutionary simulation of a starburst with 4000 stars distributed by means of a Monte Carlo method, accounting for the IMF, early B-type binary formation rate f = 0.7, binary period distribution described by equation (16.17) and a flat mass ratio distribution. We focuss on the early B-type mass gainers which are shown by open circles. As can be noticed, these mass gainers are predominantly the bluest stars of the whole starburst and are situated in a HRD region from which normal B-type stars have disappeared.

When we only consider clusters younger than 30 million years (cluster turn off mass $\geq 10\ M_o$), the number of early B-type mass gainers is smaller than 2% of the total B5-B0-type star sample. Particularly interesting is a comparison of theoretical prediction with the observed number of early Be-type stars in a number of young clusters (table 19.4). Linking the mass gainers to Be stars formed through close binary evolution, the results for NGC 663 and NGC 3760 suggest that binaries only contribute to at most 10% of the Be population. When we also account for the fact that there may be roughly as many 'normal' B-type rapid rotators as Be stars, the percentage above may be as low as 5%.

TABLE 19.4: For a number of young clusters in the Galaxy and in the Magellanic Clouds we compare the observed number of Be-type stars (relative to the total number of B5-B0 type stars) to the theoretically predicted value. The references for the observations: 1 = Mermilliod (1981, 1982), 2 = Slettebak (1985), 3 = Waelkens et al. (1990), 4 = Sanduleak (1990), 5 = Grebel (1995) .

environment	cluster	age (10^6 yrs)	observed Be fraction	predicted Be fraction	ref
Galaxy	h & c Per	<20	0.25-0.5	0.014	1.2.3
	NGC 663	22	0.4	0.015	1,4
	NGC 3760	22	0.33	0.015	1,2
Magellanic	NGC 330	19	0.27	0.014	5
Clouds	NGC 2004	20	0.11	0.015	5
	NGC 1818	25	0.2	0.017	5

19.3.2. THE POPULATION OF O AND WR TYPE STARS IN STARBURSTS

We assume that the overall star formation in starbursts (single stars and binaries) is similar as in regions where star formation was continuous.

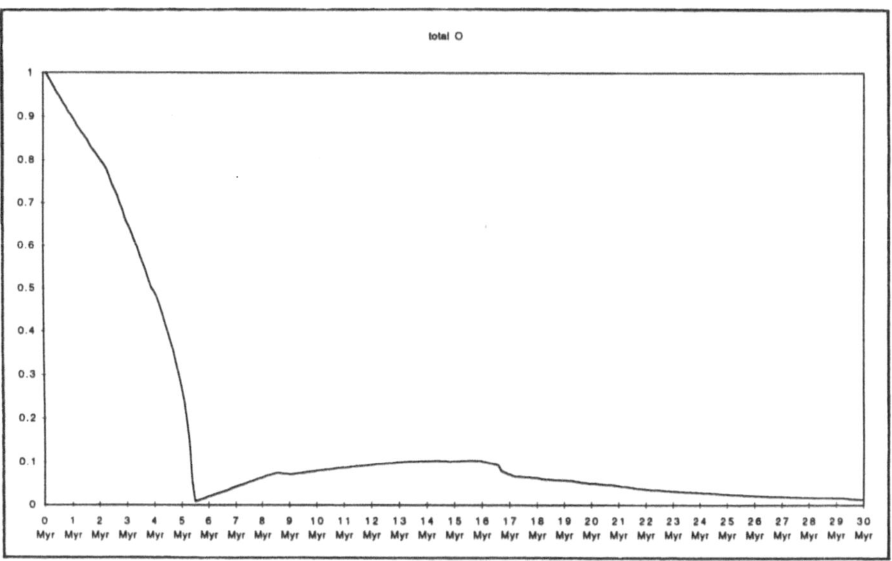

Figure 19.8a: The time evolution of the number of O-type stars (normalized so that initially the number equals one) for an instantaneous starburst with PNS parameters: f = 0.8, IMF $\propto$ $M^{-2.7}$, flat mass ratio distribution, β_{max} = 1 in case A/case B_r binaries.

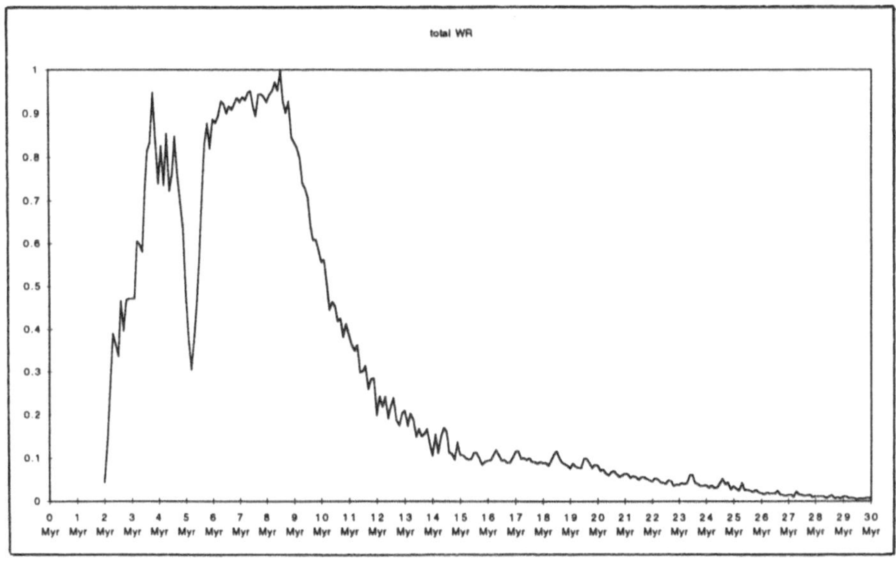

Figure 19.8b: Similar as figure 19.8a but for the WR stars, with $M_{WR,min}$ = 5 M_o; the figure is normalized so that at the moment the WR frequency is maximum, this maximum equals one); the WR/O number ratio is obtained by deviding the numbers of figure 19.8a and 19.8b at the corresponding timesteps and then multiplying the result by 0.033.

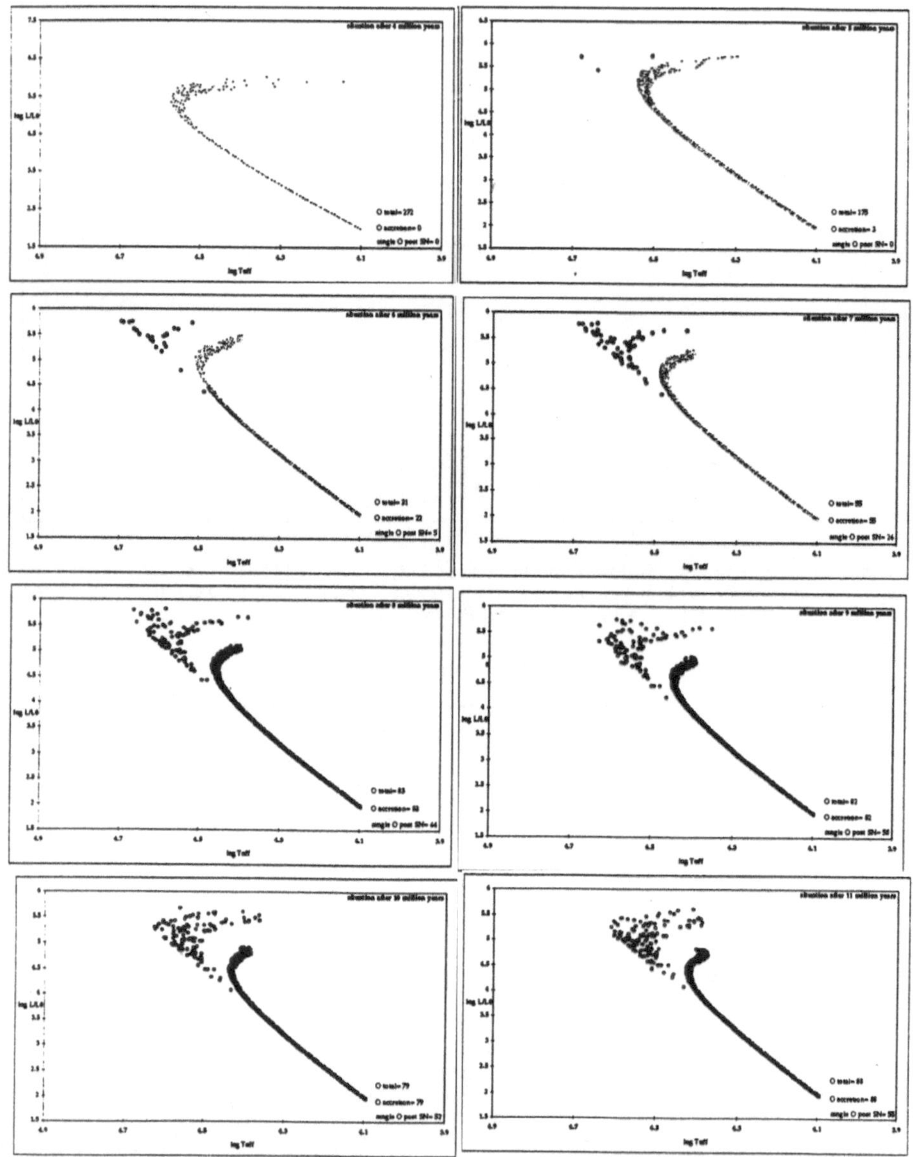

Figure 19.9: The evolution in the HR diagram of a starburst with 1000 massive stars for a starburst which lasted for 1 million years, with the same PNS parameters as in figure 19.5 (from Van Bever and Vanbeveren, 1998).

In figure 19.8, we show the time evolution of the number of O-type stars (normalized so that initially the number equals one) and of WR stars of an instantaneous starburst with PNS parameters: f = 0.8, IMF $\propto M^{-2.7}$, flat mass ratio distribution, $\beta_{max} = 1$ in case A/case Br binaries, $M_{WR,min} = 5\ M_o$.

Figure 19.9 illustrates the evolution in the HR diagram of a starburst with 1000 massive stars where starformation lasted for 1 million years, with the same PNS parameters. We define the O-phase (resp. the WR phase) and the corresponding O-lifetime (resp. WR lifetime) of a starburst as the phase or the time where the starburst shows O-type (resp. WR type) features.

We conclude

- the evolution of a starburst which is older than ~4.10^6 years depends critically on the binary frequency. When binaries are included, the total time that the starburst shows O-type and WR star features is much longer than in starbursts without binaries. This is due to the effect of accretion in MCBs and the appearance of a class of young O-type mass gainers when the starburst is older than ~4.10^6 yrs: we propose to call this the 'rejuvenation of the starburst'. Figure 19.6 shows the situation in which the burst is 7.10^6 years and in which, due to the O-type mass gainers, it looks younger than 4.10^6 years.
  The following simple example illustrates what happens. Suppose one has a starburst with 50 O-type single stars with an average mass of 30 M_o, and 50 MCBs with average mass 30 M_o + 20 M_o (which is a most likely progenitor of the WR binary V444 Cyg, a WNE + O6V, appendix 8). After ~7.10^6 yrs, all single O-type stars have disappeared and only a single WR star class is left. The binaries however, evolved (on average) into a WR+O5/O6 class (just like V444 Cyg), i.e. we will observe a starburst with a significant number of O-type mass gainers, no O-type single stars, and a large WR sample. If one did not account for binary effects, one would conclude that the starburst is at most 3-4.10^6 yrs old (because of the presence of a large number of O5/6 stars) and has a number ratio WR/O $\approx$ 1-2,

- a comparison between theoretical prediction and observations of young starbursts is meaningful only if binaries and the effect of binary evolution are correctly included. Most important is the rejuvenation caused by mass transfer,

- when the observed HR diagram of stellar aggregates in the Galaxy and MCs (figure 17.1) is compared with the predicted HR diagram of starbursts (figure 19.6), although the number of stars is smaller in the former, it looks as if the effects of interaction in binaries give a natural explanation for the presence of a younger class of stars, bluer and more luminous than the cluster turn-off (blue stragglers).

Schaerer and Vacca (1996) concluded that binaries play no role in the interpretation of data of young starburst regions. This may be true for starbursts younger than ~4.10^6 yrs. However, the age determination of a starburst is based on theoretical models where results of stellar evolution are used as input (a.o. Leitherer and Heckman, 1995). Until now, only single star evolution has been included in these models and that is precisely where the shoe pinches, making the conclusion of Schaerer and Vacca doubtful.

19.3.3. THE POPULATION OF RSGs AND WR TYPE STARS IN STARBURSTS

Up to ~4.10^6 years, the starburst shows no signature of RSGs. After ~4.10^6 years RSGs and WR stars are simultaneously present. However, although the lifetime of a starburst where WR stars and RSGs are simultaneously present may be comparable to the one where only WR stars (and no RSGs) will be observed, during most of that lifetime, the number of WR stars is larger than the number of RSGs.

One may wonder whether or not this is compatible with the statement made in section 17.1.2 that only ~27% of the galactic massive stellar aggregates containing WR stars, also contain RSGs.

When compared to starbursts, the number of massive stars in galactic associations or clusters is very small (much smaller than in a typical extra-galactic starburst region). As a matter of fact, inspection of table 3 reveals that a large number of them contain only 1 WR star and no RSG. This is compatible with the statement that when in a starburst (where a large number of massive stars are formed) WR and RSGs are simultaneously present, the number of WR stars is larger than the number of RSGs.

There may be another important difference between extra-galactic starburst regions and galactic massive star aggregates. Due to the fact that many rejuvenated OB-type accretion stars in a starburst are post-SN, they may have a large space velocity (> 30 km/s). It takes a few million years before these OB runaways possibly become WR, i.e. at the moment these OB runaways become WR runaways, they have traveled a few 100 pc from their birth place. The WR runaways will still be visible in the spectrum of an extra-galactic starburst, but as far as galactic aggregates are concerned, they may not any longer be considered as association/cluster member. It is interesting to remark here that this runaway process may be responsible for the fact that the number ratio [aggregate O-type star/field O-type star] is larger than the number ratio [aggregate WR star/field WR star], which is (marginally) observed in the Galaxy within 2.5 kpc from the Sun (Garmany, 1997).

19.3.4. COOKING RECIPE

Accounting for all parameters governing binary evolution and for all binary parameter distributions (IMF, $\Phi(q)$, $\Pi(P)$, $f(v_{kick})$, the direction of the kick velocity expressed by the angles θ and ϕ defined in section 11), the evolution of a starburst with binaries takes significantly more computer time than the evolution of a starburst with single stars only. For population *number* synthesis, this does not appear to be an obstacle. However, at present, to study population *spectral* synthesis of starbursts with a realistic binary frequency at birth, accounting for all spectral features, a cooking recipe seems useful. Based on our PNS computations, the following simplifications are justified.

- Since case B_r binaries are the most frequent class of binaries, do not account for case A, case B_c, case C.
- If the v_{kick}-distribution, predicting the large v_{kick} values discussed in section 16.6, is confirmed, compute the evolution of the starburst assuming that all binaries where the primary explodes, are disrupted. The further evolution of the rejuvenated mass gainer can be simulated by the evolution of a single star with the same mass and chemical composition.
- If the foregoing applies, the detailed evolution of the binary period is not strictly necessary since only a small fraction of the case B_r binaries (with initial mass ratio larger than 0.2) that affect the evolution of a starburst will merge. Since the evolution of case B_r binaries should be largely independent from the initial period, one only needs a case B_r frequency and the integration over period accounting for the distribution $\Pi(P)$ can be omitted.
- Assume that all binaries with initial mass ratio $q \leq 0.2$ merge and simulate the further evolution of these mergers by means of single star evolution with mass equal to the sum of the masses of both components. Whether this is correct or not has to be verified by more evolutionary calculations. We advise to follow this class separately in order to estimate their effect.

19.4. The predicted SN type $I_{b,c}$/type II ratio in regions of continuous star formation with a realistic binary formation frequency

Massive stars which contribute to the SN population should have a mass smaller than the minimum mass of BH formation ($M_{BH\text{-}min}$) which may be different for single stars and for binary components. Furthermore, the two main subclasses are populated by:

$SNI_{b,c}$

- all binary components of interacting binaries with initial mass ≥ 10 M_o,
- binary components or single stars with initial mass larger than M_{min}, which have lost all their hydrogen-rich layers by RSG stellar wind mass loss and/or LBV mass loss. In our scenario, this occurs if the mass of the star was initially larger than ~ 15-20 M_o (Z = 0.02) or larger than 40 M_o (Z = 0.002) if it is a single star, or if the mass of the star became larger than ~ 15-20 M_o (Z = 0.02) or larger than 40 M_o (Z = 0.002) during the mass exchange in a binary,
- all OB-type binary mass gainers which have lost their hydrogen-rich layers by RSG mass loss and/or LBV mass loss and/or spiral-in of the cc where merging is avoided,
- binary mergers which have become more massive than M_{min} after the merging process,

SNII

- single stars with mass between 9±1 M_o and M_{min},
- binary components with mass between 9±1 M_o and M_{min} which have become single after the SN explosion of their companion,
- binary mergers with post-merging masses between 9±1 M_o and M_{min}.

Computations were presented by De Donder and Vanbeveren (1998). It can be concluded that

the number ratio $SNII/SNI_{b,c}$ strongly depends on the adopted massive binary frequency at birth, the binary mass ratio and period distribution.

The latter statement is illustrated by the results of table 19.5.

TABLE 19.5: The theoretical predicted number ratio SNII/SNI$_{b,c}$ for different MCB frequencies at birth f, for different mass ratio distributions. The results hold for an IMF exponent = -2.7, a minimum mass for BH formation = 25 M_o for single stars and 40 M_o for binary components.

f	$\Phi(q)$	II/I$_{b,c}$
0.8	flat	2.01
0.8	Hogeveen	2.77
0.4	flat	3.87
0.08	flat	9.57
0.6	Hogeveen	3.61

The observed number ratio for a certain type of galaxy (section 17.6) has been determined by combining the observations of a large sample of galaxies (of the same type) where possibly the basic parameters that critically affect the population are different (the metallicity and binary frequency). When an average ratio is used in order to estimate the 'observationally expected' SNI/SNII ratio for one galaxy in particular, thus also for our own Galaxy, one assumes implicitely that the basic parameters that affect the ratio are the same everywhere. Therefore,

- using the average SNII/SNI$_{b,c}$ number ratio holding for a large number of galaxies with different metallicities in order to predict the ratio for one galaxy in particular is correct only when all these galaxies have a similar metallicity and a similar massive star distribution (i.e. a similar IMF, a similar binary frequency, a similar binary mass ratio and period distribution). Whether this corresponds to reality is very uncertain.

From our theoretical calculations (with $f \geq 0.75$) we conclude that in the Galaxy, type II's are 2-3 times more frequent than the type I$_{b,c}$'s.

For all the galaxies in their sample, Cappelaro et al. (1992) derive an average ratio of ~4. Since the sample of observed galaxies is large (2461) and since all morphological types are encountered, this value could be considered as some cosmological average. If we try to recover this rate, we have to adopt an average massive binary formation rate between 40 and 60 %, i.e.

the cosmological massive binary frequency at birth may be of the order of 50%.

Cappellaro et al. distinguished early type and late type spiral galaxies. The early types seem to have a ratio of ~2.3 whereas the ratio in the late types ≈ 5.5. If this difference is due to a different massive massive binary formation rate, we may conclude that

the massive binary frequency at birth in late spirals is ~factor 2 smaller than in early type spirals.

Concluding remarks

We have discussed the present state of the evolution of massive single stars and of massive close binaries. Evolution depends on physical processes that contain a number of parameters.

By using these evolutionary computations, starting from initial distribution functions such as binary frequency, mass, binary mass ratio and binary period distribution, accounting properly for the effect of the supernova explosion on binary orbital parameters, it is straightforward to estimate numbers of stars in various evolutionary phases and to predict supernova rates (population number synthesis).

When evolutionary and population number synthesis results are compared with observations of individual stars and observed populations, one may be able to restrict the possible values of the parameters.

Unfortunately, the observations of massive stars are more or less complete only for a very small region in our own Galaxy. Although it is possible to find a massive star evolutionary and population model that explains most of these observations, whether this model applies for the whole Galaxy or for other galaxies is questionable.

Therefore, large efforts must be made in order to improve and increase the number of observations in more distant regions and if possible in other galaxies. Then, it must be clear that

IF ONE COMPARES OBSERVATIONS WITH THEORETICAL EVOLUTIONARY PREDICTION,

THE BRIGHTEST SINGLE STARS

CANNOT BE CONSIDERED SEPARATELY FROM

THE BRIGHTEST BINARIES.

References

Abbott, D.C.: 1982, Ap.J. 259, 282.

Abt,H.A., Levy,S.G., 1985, Ap.J.Suppl.. 59, 229

Allen, C.W.., 1973, Astrophysical Quantities, The Athlone Press, London.

Alongi, M., Bertelli, G., Bressa, A., Chiosi, C.: 1991, A.A. 224, 95.

Anders,E. , Grevesse,N., 1989, Geochimica et Cosmochimica Acta 53, 197.

Annuk, K.: 1991, in 'WR Stars and Interrelations with other Massive Stars in Galaxies', eds. K.A. Van der Hucht & B. Hidayat, IAY Symp. 143, (Kluwer: Dordrecht), p. 245.

Annuk, K.: 1995, in 'WR Stars: Binaries, Colliding Winds, Evolution', eds. K.A. Van der Hucht & P.M. Williams, IAU Symp. 163, (Kluwer: Dordrecht),. p. 231.

Arnett, D.: 1996, 'Supernovae and Nucleosynthesis: an Investigation of the History of Matter, from the Big Bang to the Present', Princeton Univ. Press, Princeton.

Azuma R.E., Buchmann, L., Barker, F.C., Barnes, C.A. et al.: 1994, Nuc. Phys.

Azzopardi, M., Breysacher, J.: 1979, A.A. 75, 120.

Bailes, M.: 1989, Ap. J. 342, 917.

Balona, L., Crampton, D.: 1974, MNRAS 166, 203.

Balona, L.: 1975, MNRAS 173, 449.

Barba, R.H., Niemela, V.S., Baume, G., Vazques, R.A.: 1995, Ap.J. 446, L23.

Barnes,T.G., Evans,D.S., Moffet,T.J.,1976, MNRAS 183, 285

Barnes,T.G., Evans,D.S.,1976, MNRAS 174, 489

Batten,A.H., Fletcher,J.M. , MacCarthy,D.G., 1989, Publications of the Dominion Astrophysical Observatory, Volume XVII: The Eigth Catalogue of the Orbital Elements of Spectroscopic Binary Systems.

Bernasconi, P.A., Maeder, A.: 1996, A.A. 307, 829.

Bethe,H.A. , Brown,G.E. 1995, Ap.J. 445, L129.

Bethe,H.A. , Salpeter,E.E., 1957, Quantum Mechanics of one and Two Electron Atoms, Springer-Verlag, Berlin.

Bhattacharya, D., Van den Heuvel, E.P.J.: 1991, Phys. Rev. 203, 1.

Bhattacharya, D., Srinivasan, G.: 1995, in 'X-ray Binaries', eds. W.H.G. Lewin, J. Van Paradijs, E.P.J. Van den Heuvel, Cambridge University Press, p. 495.

Bhattacharya, D.: 1995, in 'X-ray Binaries', eds. W.H.G. Lewin, J. Van Paradijs, E.P.J. Van den Heuvel, Cambridge University Press, p. 233.

Biehle, G.T.: 1991, Ap.J. 380, 167.

Bjorkman, J.E., Cassinelli, J.P.: 1993, Ap.J. 409, 429.

Bjorkman, J.E.: 1994, in 'Pulsation, Rotation and Mass Loss in Early-type Stars', eds. L.A. Balona, H.K. Henrichs & J.M. Le Contel, IAU 162, p.455.

Blaauw, A.: 1961, Bull. Astr. Inst. Netherlands 15, 265.

Blackwell,D.E., Shallis, M.J.,1977, MNRAS 180, 177

Blaha, C., Humphreys, R.M.: 1989, A.J. 98, 1598.

Böhm-Vitense, E.: 1958, Zeitschrift für Astrophysik 46, 108.

Böhm-Vitense, E.: 1972, A.A. 17, 335.

Bolton, C.T., Rogers, G.L.: 1978, Ap.J. 222, 234.

Bonnet-Bidaud,J.M. , Chardin,G ,1988, Physics Reports 170,325.

Bozic, H., Harmanec, P., Horn,J., Koubsky,P., Scholz,G., McDavid,D., Hubert,A.-M., Hubert,H.,: 1995, A.A. 304, 235.

Bragalia,A., Renzini,A., Bergeron,P., 1993, in "White Dwarfs": Advances in Observation and Theory, 325, M.A. Barstow (ed.), Kluwer.

Bragalia,A., Renzini,A., Bergeron,P., 1995, Ap.J. 443, 735.

Braun, H., Langer, N.: 1995, A.A. 297, 483.

Breysacher, J.: 1981, A.A. Suppl. 43, 203.

Breysacher, J., Perrier, C.: 1991, in IAU Symp. No. 143 'WR Stars and Interrelations with Other Massive Stars in Galaxies', eds. K.A. Van der Hucht and B. Hidayat, (Kluwer: Dordrecht), p. 229.

Brown, G.E., Weingartner, J.C., Wijers, R.A.M.J.: 1996, Ap.J. 463, 297.

Canal, R., Isern, J., Labay, J.: 1990, Ann. Rev. Astron. Astrophys. 28, 183.

Cannon, R., Eggleton, P.P., Zytkow, A.N., Podsiadlowski, Ph.: 1992, Ap.J. 386, 206.

Cappellaro, E., Turatto, M., Benetti, S., Tsvetkov, D.Y., Bartunov, O.S., Makarova, L.N.: 1993, A.A. 273, 383.

Castor, J., Abbott, D.C., Klein, R.: 1975, Ap.J. 195, 157.

Caughlan, G.R., Fowler, W.A.: 1988, Atomic Data and Nuclear Data Tables 40, 238.

Caughlan, G.R., Fowler, W.A., Harris, M.J., Zimmerman, B.A.: 1985, Atomic Data Nuc. Data Tables 32, 197.

Cerruti, M.A.: 1984, IAU Inf. Bull. Var. Stars 2637, 1.

Chaboyer, B., Zahn, J.P.: 1992, A.A. 253, 173.

Chandrasekhar, S., Münch, G.: 1950, Ap.J. 111, 142.

Chandrasekhar, S: 1957, 'An Introduction to the Study of Stellar Structure', Dover, New York

Cherepaschuk, A.M.: 1975, Sov. Astron. A.J. 19, 47.

Cherepashchuk, A.M. , Moffat, A.F.J.: 1994, Ap.J. 424, L53.

Chu, Y.-H., Gull, T.R., Treffers, R.R., Kwitter, K.B., Troland, T.H.: 1982, Ap. J. 254, 562.

Clayton, D.D.: 1968, 'Priciples of Stellar Evolution and Nucleosynthesis', McGraw Hill, N.Y.

Code,A.D., Davis,J.,Bless,R.C.,Hanbury Brown,R.,1976,Ap.J. 203, 417.

Conti, P.S.: 1982, in'Wolf-Rayet Stars: Observations, Physics and Evolution', eds. C. De Loore & A.J. Willis, IAU Symp. 99, Reidel, p. 3.

Conti, P.S., Ebbets, D.: 1977, Ap.J. 213, 438.

Conti, P.S., Niemela, V.S., Walborn, N.R.: 1979, Ap.J. 228, 206.

Conti, P.S., Underhill, A.B., 1988, 'O Stars and Wolf-Rayet Stars', NASA SP- 497.

Cordes, J.M.: 1986, Ap.J. 311, 183.

Counselman, C.C.: 1973, Ap.J. 180, 307.

Cox, A.N., Stewart, J.N.: 1969, Ap.J.Suppl. 19, 243.

Cox, A.N., Tabor, J.E.: 1976, Ap.J. Suppl. 31, 271.

Cox, J.P., Giuli, R.T.: 1968, 'Principles of Stellar Structure I&II', Gordon & Breach.

Cox,A.N., Stewart,J.N.,1969, Publi. of the Astronom. Soc. of the Acad. of the U.S.S.R, number 15.

Crowther, P.A., Hillier, D.J., Smith, L.J.: 1995, A.A. 293, 403.

Dalton, W.W., Sarazin, C.L.: 1995a, Ap.J. 440, 280.

Dalton, W.W., Sarazin, C.L.: 1995b, Ap.J. 448, 369.

Darwin, G.H.: 1908, Scientific Papers, Vol. 2, Cambridge Univ. Press.

Davidson,K., Ostriker,J.P.,1973, Ap.J. 179, 585.

De Donder, E., Vanbeveren, D., Van Bever, J.: 1997, A.A. 318, 812.

De Donder, E., Vanbeveren, D.: 1998, A.A. (in press).

De Jager,C.,1980, The Brightest Stars, D.Reidel Publ. Co.,Dordrecht

De Kool, M., Van den Heuvel, E.P.J., Pylyser, E.: 1987, A.A. 183, 47.

De Loore, C., Doom, C.: 1992, 'Structure and Evolution of Single and Binary Stars', Kluwer Academic Publishers, Dordrecht, The Netherlands.

De Loore, C., Vanbeveren, D.: 1992, AA. 260, 273.

De Loore, C., Vanbeveren, D.: 1994, A.A. 292, 463.

De Loore, C., Vanbeveren, D.: 1995, A.A. 304, 220.

De Loore, C.: 1980, Space Sci. Rev. 26, 113.

De Marchi, G., Clampin, M., Greggio, L., Leitherer, C., Nota, A., Tosi, M.: 1997, Ap.J. 479, L27.

de Vaucouleurs, G., de Vaucouleurs, A., Corwin, H.G., Buta, R.J., Paturel, G., Foque, P.: 1991, Third Reference Catalogue of Bright Galaxies, Springer-Verlag (New-York).

Deutsch, A.J.: 1970, Stellar Rotation, edited by A. Slettebak (D. Reidel, Dordrecht-Holland), p.207.

Dewey, R.J., Cordes, J.M.: 1987, Ap.J. 321, 780.

Dickey,J.M.: 1983, Ap.J. 273, L71.

Drilling, J.S., Jeffery, C.S., Heber, U.: 1985; IAU Circ. No. 4086.

Drissen, L., Lamontagne, R., Moffat, A.F.J., Bastein, P., Seguin, M.: 1986, Ap.J. 343, 426.

Dudley, R.E., Jeffery, C.S.: 1990, MNRAS 247, 400.

Eaton, J.A., Cherepaschuk, A.M., Khaliullin, Kh. F.: 1985, Ap.J. 296, 222.

Eddington, A.S.: 1913, MNRAS 73, 359.

Eggleton, P.P.: 1983, Ap.J. 268, 368.

Esteban, C., Vilchez, J.M., Smith, L.J., Clegg, R.E.S.: 1992, A.A. 259, 629.

Firmani, C., Koenigsberger, G., Bisiacchi, G.F., Moffat, A.F.J., Isserstedt, J.: 1980, Ap.J. 239, 607.

Fitzpatrick, E.L., Bohannan, B.: 1993, Ap.J. 404, 734.

Fitzpatrick, E.L.: 1991, PASP 103, 1123.

Flower, P.J.: 1977, A.A. 54, 31.

Fowler, W.A., Caughlan, G.R., Zimmerman, B.A.: 1975, Ann. Rev. Astron. Astrophys. 13, 69.

Fransson, C. Cassatella,A., Gilmazzi,R., Kirshner,R., Panagia,N., Sonnenborn,G., Warmsteker,W.: 1989, Ap.J. 336, 429.

Fricke,W., Schwan,H., Lederle,T.: 1988,'Fifth Fundamental Catalogue', Astron. Rechen Instit., Heidelberg.

Fryxell, B.A., Arnett, W.D.: 1981, Ap.J. 243, 994.

Gabler, R., Gabler, A., Kudritzki, R.P., Puls, J., Pauldrach, A.W.A.: 1989, A.A. 226, 162.

Garcia-Segura, G., Langer, N., Mac Low, M.-M.: 1996, A.A. 316, 133.

Garmany, C.D.: 1997, in 'Luminous Blue Variables: Massive Stars in Transition', eds. A. Nota, H.J.G. Lamers, ASP. Conf. Ser.

Garmany, C.D., Conti, P.S., Massey, P.: 1980, Ap.J. 242, 1063.

Giacconi, R., Gorenstein, P., Gursky, H., Waters, J.R.: 1967, Ap.J. 148, L118.

Gies, D.R., Bolton, C.T.: 1986, Ap.J. 304, 371.

Gies, D.R., Lambert, D.L.: 1992, Ap.J. 387, 673.

Gies, D.R., Thaller, M.L., Bagnuolo, W.G. Jr., Kaye, A.B., Peeters, G.J., Penny, L.R.: 1996, Bull. American Astron. Soc. 28, 1373.

Gies, D.R.: 1987, Ap.J. Suppl. 64, 545.

Gonzales-Delgado, R.M., Leitherer, C., Heckman, T., Cervino, M.: 1997, Ap.J. 483, 705.

Gosset, E., Vreux, J.M.: 1987, A.A. 178, 153.

Grebel, E.K.: 1995, Ph. D. Thesis, University of Bonn.

Gun, J.E., Ostriker, J.P.: 1970, Ap.J. 160, 979.

Habets, G.M.H.J.: 1985, Ph. D. thesis, University of Amsterdam.

Habets, G.M.H.J.: 1986, A.A. 187, 209.

Hachisu, I., Kato, M., Nomoto, K.: 1996 Ap.J. 470, L97.

Halbwachs, J.L.: 1987, A.A. 183, 234.

Hamann, W.-R., Koesterke, L.: 1996, in 'Wolf-Rayet Stars in the Framework of Stellar Evolution', eds. J.M. Vreux, A. Detal, D. Fraipont-Caro, E. Gosset & G. Rauw, Univ. de Liège, p. 491.

Hamann, W.-R., Koesterke, L., Wessolowski, U.: 1995, A.A. 299, 151.

Hamann, W.-R.,1994, Space Sci. Rev. 66, 237.

Hanbury Brown, R., Davis, J., Herbison-Evans, D., Allan, L.R.: 1970, MNRAS 148, 103.

Hanbury Brown, R., Davis, J., Allen, L.,1974, MNRAS 167, 121

Harries,H.T., Hilditch,R.W. , Hill,G., 1997, MNRAS 285, 277.

Hartman, J.W.: 1997, A.A. 322, 127.

Herbig, G.H., Preston, G.W., Smak, J., Paczynski, B.: 1965, Ap.J. 141, 617.

Herrero, A., Kudritzki, R.P., Gabler, R., Vilchez, J.M., Gabler, A.: 1995, A.A. 297, 556.

Herrero, A., Kudritzki, R.P., Vilchez, J.M., Kunze, D., Butler, K., Haser, S.: 1992, A.A. 261, 209.

Hilditch, R.W.: 1974, MNRAS 169, 323.

Hillier, D.J.: 1996, in 'Wolf-Rayet Stars in the Framework of Stellar Evolution', eds. J.M. Vreux, A. Detal, D. Fraipont-Caro, E. Gosset & G. Rauw, Université de Liège, p. 509.

Hoffleit, D., Warren, Jr. W.H.: 1991, The Bright Star Catalogue, 5th Revised Ed., Yale University Observatory.

Hogeveen, S.: 1991, Ph.D. Thesis, Univ. of Amsterdam.

Hogeveen, S.: 1992, Astrophys. Space Sci. 196, 299.

Howarth, I.D., Prinja, R.K.: 1989, Ap.J. Supp 69, 527.

Humphreys, R.M. , McElroy, D.B.,1984, Ap.J.,284,565.

Humphreys, R.M., Davidson, K.: 1994, PASP 106, 1025.
Hutchings, J.B.: 1975, Ap.J. 200, 122.
Iben, I. Jr., Tutukov, A.V., Yungelson, L.R.: 1995a, Ap.J. Suppl. 100, 233.
Iben, I. Jr., Tutukov, A.V., Yungelson, L.R.: 1995b, Ap. J. Suppl. 100, 217.
Iben, I., Renzini, A.: 1983, Ann. Rev. Astron. Astrophys. 21, 271.
Iglesias, C.A., Rogers, F.J., Wilson, B.G.: 1992, Ap.J. 397, 717.
Israel, F.P.: 1988, A.A. 194, 24.
Jakobsen, P. et al.: 1991, Ap.J. 369, L63.
Jeffery, C.S., Drilling, J.S., Heber, U.: 1987, MNRAS 194, 429.
Jorgensen, H., Lipunov, V.M., Panchenko, I.E., Postnov, K.A., Prokhorov, M.E.: 1997, Ap.J. 486, 110.
Joss, P.C., Rappaport, S.: 1984, Ann. Rev. Astron. Astrophys. 22, 537.
Jura, M.: 1987, Ap.J. 313, 743.
Kaper, L., Lamers; H.J., Ruymaekers, E., Van den Heuvel, E.P., Zuiderwijk, E.J.: 1995, A.A. 300, 446.
Kato, S.: 1966, Publ. Astron. Soc. Japan 18, 374.
Khaliullin, Kh.F., Khaliullina, A.I., Cherepashchuk, A.M.: 1984, Sov. Astron. (Letters) 10, 250.
Kippenhahn, R., Meyer-Hofmeister, E., Thomas, H.C.: 1970, A.A. 5, 155.
Kippenhahn, R., Ruschenplatt, G., Thomas, H.C.: 1980, A.A. 91, 175.
Kippenhahn, R., Weigert, A.: 1967, Zeitschrift für Astrophysik 66, 58.
Kippenhahn, R., Weigert, A.: 1989, 'Stellar Structure and Evolution', Springer, Heidelberg.
Koch, R.H.: 1976, Bull. Am. Astron. Soc. 8, 305.
Koesterke, L., Hamann, W.-R., Wessolowski, U.: 1992, A.A. 261, 535.
Koesterke, L., Hamann, W.-W.: 1995, A.A. 299, 503.
Kopal, Z.: 1972, Astrophys. Space Sci. 16, 3.
Kruszewski, A.: 1963, Acta Astron. 13, 106.
Kudritzki, R.P., Hummer, D.G.: 1990, Ann. Rev. Astron. Astrophys. 28, 303.
Kunth, D., Sargent, W.L.W.: 1981, A.A. 101, L5.
Kurucz, R.L.: 1991, in 'Stellar Atmospheres: Beyond Classical Models', NATO ASI Ser. eds. L. Crivellari, I. Hubeny & D.G. Hummer, p341.
Lamers, H.J.G., Harzevoort, J.M., Schrijver, H., Hoogerwerf, R., Kudritzki, R.P.: 1997, A.A. 325, L25
Lamontagne, R., Moffat, A.F.J.: 1987, A.J. 94, 1008.
Lang, K.R., 1978, Astrophysical Formulae, Springer.
Lang, K.R., 1992, Astrophysical Data, Planets and Stars, Springer.
Langer, N., El Eid, M.F., Fricke, K.J.: 1985, A.A. 145, 179.
Langer, N., Maeder, A.: 1995, A.A. 295, 685.
Langer,N., 1989, A.A., 210, 93.
Langer, N., Heger, A.: 1998, in 'Properties of Hot, Luminous Stars', ed. I. Howarth, ASP. Conf. Proceedings, Vol. 131 (in press).
Lattimer, J.M., Yahil, A.: 1989, Ap.J. 340, 426.
Lauterborn, D., Refsdal, S., Weigert, A.: 1971, A.A. 10, 97.

Lauterborn, D.: 1969, in 'Mass Loss from Stars', ed. M. Hack, D. Reidel. Publ. Com., Dordrecht, Holland, p. 262.

Ledoux, P.: 1947, Ap.J. 105, 305.

Leitherer, C., Heckman, T.M.: 1995, Ap.J. Suppl. 96, 9.

Leitherer, C., Robert, C., Heckman, T.M.: 1995, Ap. J. Suppl. 99, 173.

Lennon, D.J., Dufton, P.L., Fitzsimmons: 1992, A.A. Suppl. 94, 569.

Lennon, D.J., Dufton, P.L., Fitzsimmons: 1993, A.A. Suppl. 97, 559.

Leonard, P.J.T., Duncan, M.J.: 1988, A.J. 96, 222.

Leonard, P.J.T., Duncan, M.J.: 1990, A.J. 99, 608.

Levato, H.: 1976, Ap.J. 203, 680.

Lewin, W.H.G., Van Paradijs, J., Van den Heuvel, E.P.J.: 1995, in 'X-ray Binaries', Cambridge Univ. Press, Cambridge.

Lipunov, V.M., Postnov, K.A., Prokhorov, M.E.: 1996, A.A. 310, 489.

Lipunova, N.A., Cherepaschuk, A.M.: 1982, Sov.Astron. A.J. 26, 45.

Lobel,A., De Jager,C., Nieuwenhuijzen,H., Smolinski,J., Gesicki,K., 1994, A.A. 291, 226.

Lorimer, D.R., Bailes, M., Harrison, P.A.: 1997, MNRAS 289, 592.

Lortet, M.-C.: 1991, A.A. 242, 410.

Lubow, S.H., Shu, F.H.: 1975, Ap.J. 198, 383.

Lucy, L.B., Solomon, P.: 1970, Ap.J. 159, 879.

Lucy, L.B.: 1974, A.J. 79, 745.

Lundstrom, I., Stenholm, B.: 1984, A.A. Suppl. 58, 163.

Lynden-Bell, D., Pringle, J.E.: 1974, MNRAS 168, 603.

Lyne, A.G., Manchester, R.N.: 1988, MNRAS 234, 477.

Lyne, A.G., Lorimer, D.R.: 1994, Nature 369, 127.

Maeder, A., Peytremann, E.: 1970, A.A. 7, 120.

Maeder, A., Peytremann, E.: 1971, A.A. 21, 279.

Maeder, A.: 1983, A.A. 120, 113.

Marsh, M.C., Barstow, M.A., Buckley, et al.: 1997, MNRAS 286, 369.

Marston, A.P., Yocum, D.R., Garcia-Segura, G., Chu, Y.-H.: 1994, Ap.J. Suppl. 95, 151.

Marston, A.P.: 1995, A.J. 109, 2257.

Massey, P.: 1981, Ap.J. 246, 153.

Massey, P., Garmany, C.D., Silkey, M., Degioia-Eastwood, K.: 1989a, A.J. 97, 107.

Massey, P., Parker, J.W., Garmany, C.D.: 1989b, A.J. 98, 1305.

Mazzali, P.A., Lucy, L.B., Butler, K.: 1992, A.A. 258, 399.

McClintock, J.E.: 1992, in 'X-ray Binaries and Recycled Pulsars', eds. E.P.J. Van den Heuvel & S.A. Rappaport, Kluwer Acad. Publish. Dordrecht, p. 27.

McErlean, N.D., Lennon, D.J., Dufton, P.L.: 1998, A.A. 329, 613.

Mermilliod, J.C.: 1981, A.A. 97, 235.

Mermilliod, J.C.: 1982, A.A. 109, 48.

Meurs, E.J., Van den Heuvel, E.P.J.: 1989, A.A.226, 88.

Meynet, G., Maeder, A., Schaller, G., Schaerer, D., Charbonnel, C.: 1994, A.A.Suppl. 103, 97.

Meynet, G., Maeder, A.: 1996, A.A. 313, 140.

Meynet, G., Maeder, A.: 1997, A.A. 321, 465.

Meynet, G., Mermilliod, J.-C., Maeder, A.: 1993, A.A. Suppl. 98, 477.

Mihalas,D,1978, Stellar Atmospheres, 2nd ed.,Freeman, San Fransisco

Mitra, 1996, MNRAS 280, 953.

Moffat, A.F.J., Lamontagne, R., Shara, M.M., McAlister, H.A.: 1986, A.J. 91, 1392.

Moffat, A.F.J., Seggewiss, W.: 1977, A.A. 54, 607.

Moffat, A.F.J., Seggewiss, W.: 1980, A.A. 86, 87.

Moffat, A.F.J., Lamontagne, R., Seggewiss, W.: 1982, A.A. 114, 135.

Moffat, A.F.J.: 1995, in 'WR stars: Binaries, Colliding Winds, Evolution', eds. K.A. Van der Hucht & P.M. Williams, Kluwer: Dordrecht, p.212.

Moffat, A.F.J.: 1996, in 'Wolf-Rayet Stars in the Framework of Stellar Evolution', eds. J.M. Vreux, A. Detal, D. Fraipont-Caro, E. Gosset & G. Rauw, Université de Liège, p. 199.

Morgan,W., Keenan,P.,Kellman,E.:1943, 'An Atlas of Stellar Spectra', Univ. Chicago Press.

Muller, R.A., Newberg, H.J.M., Pennypacker, C.R. et al.: 1992, Ap.J. 384, L9.

Myers, P.C., Fuller, G.A.: 1992, Ap. J. 396, 631.

Neo, S., Miyaji, S., Nomoto, K., Sugimoto, D.: 1977, Publ. Astron. Soc. Japan 29, 249.

Niemela, V.S.: 1995, in 'WR stars: Binaries, Colliding Winds, Evolution', eds. K.A. Van der Hucht & P.M. Williams, Kluwer: Dordrecht, p. 223.

Niemela, V.S., Mandrini, C.H., Mendez, R.H.: 1985, Rev. Mex. Astron. Astrophys. 11, 143.

Niemela, V.S., Mendeze, R.H., Moffat, A.F.J.: 1983, Ap.J. 272, 190.

Niemela, V.S., Moffat, A.F.J.: 1982, Ap.J. 259, 213.

Niemela, V.S., Sahade, J.: 1980, Ap.J. 238, 244.

Niemela, V.S.: 1982, in 'WR Stars: Observations, Physics and Evolution', eds. C. De Loore & A.J. Willis, IAU Symp. 99, Reidel Publ. Co., p. 299.

Nomoto, K., Iwamoto, K., Suzuki, T.: 1995, Physics Reports 256, 173.

Nomoto, K., Suzuki, T., Shigeyama, T., Kumagai, S., Yamaoka, H., Saio, H.: 1993, Nature 364, 507.

Nomoto, K., Yamaoko, H.: 1992, in 'X-ray Binaries and Recycled Pulsars', eds. E.P.J. Van den Heuvel & S.A. Rappaport, Kluwer Acad. Publ. Dordrecht, p. 189.

Nota, A., Livio, M., Clampin, M., Schulte-Ladbeck, R.: 1995, Ap.J. 448, 788.

Ostriker, J.P., Gunn, J.E.: 1969, Ap.J. 157, 1395.

Pacini, F.: 1967, Nature 216, 567.

Packet, W.: 1988, Ph. D. Thesis, Vrije Universiteit Brussels.

Paczynski, B.: 1971, Ann. Rev. Astron. Astrophys. 9, 183.

Paczynski, B.: 1967, Acta Astronomica 17, 355.

Parker, J.W., Garmany, C.D., Massey, P., Walborn, N.R.: 1992, A.J. 103, 1205.

Pauldrach, A.W., Kudritzski, R.P., Puls, J., Butler, K., Hunsinger, J.: 1994, A.A. 283, 525.

Penny, L.R., Bagnuolo, W.G. Jr., Gies, D.R.: 1994, Space Sci. Rev. 66, 323.

Penny, L.R., Gies, D.R., Bagnuolo, W.G., Jr.: 1996, in 'Wolf-Rayet Stars in the Framework of Stellar Evolution', eds. J.M. Vreux, A. Detal, D. Fraipont-Caro, E. Gosset & G. Rauw, Univ. de Liège, p. 289.

Penny, L.R.: 1996a, Ap.J. 463, 737.

Penny,L.R., 1996b, Tomographic Separation of Spectra of O-Type Binary Systems, Ph.D. thesis, Georgia State University.

Petrenz, P., Puls, J.: 1996, A.A. 312, 195.

Petrie, R.M.: 1952, Pub. Dom. Astrophys. Obs. 9, 251.

Petrie, R.M.: 1965, Pub. Dom. Astrophys. Obs. 12, 317.

Phinney, E.S.: 1991, Ap.J. 380, L17.

Pike, C.D., Stickland, D.J., Willis, A.J.: 1983, The Observatory 103, 154.

Plavec, M.: 1970, IAU Coll. on Stellar Rotation, p. 133.

Plavec, M.: 1986, in 'Hydrogen-Deficient Stars & Related Objects', IAU Coll. No. 87, eds. K. Hunger, D. Schonberner, N.K. Rao, Reidel, Dordrecht, p. 231.

Podsiadlowski, Ph., Joss, P.C., Rappaport, S.: 1990, A.A. 227, L9.

Pols, O.R.: 1994, A.A. 290, 119.

Pols, O.R., Coté, J., Waters, L.B.F.M., Heise, J.: 1991, A.A. 241, 419.

Pols, O.R., Marinus, M.: 1994, A.A. 288, 475.

Popova, E.I., Tutukov, A.V., Yungelson, L.R.: 1982, Astron. Space Sci. 88, 55.

Popper, D.M.:1980, Ann. Rev. Astron. Astrophys. 115, 164.

Popper, D.M.:1982, Ap.J. 262, 641.

Portegies Zwart, S.F., Verbunt, F.: 1996, A.A. 300, 179.

Portegies Zwart, S.F.: 1995, A.A. 296, 691.

Portegies Zwart, S.F.: 1996, Ph. D. Thesis, University of Utrecht.

Pringle, J.E.: 1981, Ann. Rev. Astron. Astrophys. 19, 137.

Puls, J, Kudritzki, R.P., Herrer, A. et al.: 1996, A.A. 305, 171.

Rankin, J.M.: 1990, Ap.J. 352, 247.

Rappaport, S., Joss, P.C.: 1983, in 'Accretion Driven Stellar X-ray Sources', eds. W.H.G. Lewin & E.P.J. Van den Heuvel, Cambridge Univ. Press: Cambridge.

Reddish, V.C.: 1978, in 'Stellar Formation', International Series in Natural Philisophy, Vol. 97.

Reid, N., Tinney, C., Mould, J.: 1990, Ap.J. 348, 98.

Reilman,R.F., Manson,S.T., 1979, Astrophys.J.Suppl., 40, 815.

Robert, C., Leitherer, C., Heckman, T.M.: 1993, Ap.J. 418, 749.

Röser,S., Bastian,U.: 1991, 'PPM Catalogue', Astron. Rechen Instit., Heidelberg.

Rousseau, J., Martin, N., Prevot, L., Rebeirot, E., Robin, A., Brunet, J.P.: 1978, A.A. Suppl. Ser. 31, 243.

Roxburgh, I.: 1978, A.A. 65, 281.

Roxburgh, I.: 1989, A.A. 211, 361.

Roxburgh, I.: 1992, A.A. 266, 291.

Ruderman, M., Shaham, J., Tavani, M.: 1989, Ap.J. 336, 507.

Sanduleak, N.: 1990, A.J. 100, 1239.

Sahu, M.: 1992, in 'A study of the ISM in the Puppis-Vela region including the Gum nebula', Ph. D. Thesis, Groningen, The Netherlands.

Sato, N., Nagase, F., Kawai, N. et al.: 1986, Ap.J; 304, 241.

Scalo, J.M.: 1986, Fundamentals of Cosmic Physics 11, 1.

Schaerer, D., Maeder, A.: 1992, A.A. 263, 129.

Schaerer, D., Vacca, W.D.: 1996, in 'Wolf-Rayet Stars in the Framework of Stellar Evolution', eds. J.M. Vreux, A, Detal, D. Fraipont-Caro, E. Gosset & G. Rauw, Université de Liège, p. 641..

Schaller,G., Schaerer,D., Meynet,G. , Maeder,A.: 1992, A&AS,96,268.

Schmutz,W., Geballe,T., Schild,H.: 1996, A.A. 311, L25.

Schmutz, W.: 1996, in 'Wolf-Rayet Stars in the Framework of Stellar Evolution', eds. J.M. Vreux, A. Detal, D. Fraipont-Caro, E. Gosset & G. Rauw, Université de Liège, p. 553.

Schmutz, W.: 1997, A.A. 321, 268.

Schönberner, D., Drilling, J.S.: 1983, Ap.J. 268, 225.

Schönberner, D., Herrero, A., Butler, K., Becker, S., Eber, F., Kudritzki, R.P., Simon, K.P.: 1988, A.A. 197, 209.

Schreier, E., Levinson, R., Gursky, H., Kellog, E., Tannanbaum, H., Giacconi, R.: 1972, Ap.J. 172, L79.

Schwarzschild, M., Härm, R.: 1958, Ap. J. 128, 348.

Schwarzschild, M.: 1958, 'Structure and Evolution of the Stars', Dover Publ., Inc., N.Y.

Shapiro, S.L., Teukolsky, S.A.: 1983, 'Black Holes, White Dwarfs and Neutron Stars', John Wiley & Sons, New York/Chichester.

Shklovskii, I.S.: 1969, Sov. Ast. 13, 562.

Shu, F.H., Lubow, S.H.: 1981, Ann. Rev. Astron. Astrophys. 19, 277.

Slettebak, A.: 1985, Ap.J.Suppl. 59, 769.

Smith, L.F., Shara, M.M., Moffat, A.F.J.: 1996, MNRAS 281, 163.

Smith, K.C., Howarth, I.D., Siebert, K.W.: 1998, in 'Properties of Hot, Luminous Stars', ed. I. Howarth, ASP. Conf. Proceedings, Vol. 131 (in press).

Sobolev, V., 1990, 'Cours d'Astrophysique Théorique', ed. Mir, Moscow.

Sparks, N.M., Stecher, T.P.: 1974, Ap.J. 188, 148.

Srinivasan, G.: 1991, Astron. Astrophys. Rev. 1, 208.

St.-Louis, N., Drissen, L., Moffat, A.F.J., Bastien, P., Tapia, S.: 1987, Ap.J. 322, 870.

St.-Louis, N., Moffat, A.F.J., Drissen, L., Bastien, P., Roberts, C.: 1988, Ap.J. 330, 286.

St.-Louis, N., Howarth, I.D., Willis, A.J.: 1993a, A.A. 267, 447.

St.-Louis, N., Moffat, A.F.J., Lapointe: 1993b, Ap.J. 410, 342.

Stencel, R.E., Pesce, J.E., Bauer, W.H.: 1989, A.J. 97, 1120.

Stevens, I.R., Willis, A.J., 1988, MNRAS 234, 783.

Stone, R.C.: 1979, Ap.J. 232, 520.

Stone, R.C.: 1982, A.J. 87, 90.

Stone, R.C.: 1985, in 'Birth and Evolution of Massive Stars and Stellar groups', Proceedings of a Symposium dedicated to Adriaan Blaauw, D. Reidel Publ. Co., Dordrecht, p.201.

Stone, R.C.: 1991, A.J. 102, 33.

Strittmatter, P.A., Robertson, J.W., Faulkner, D.J.: 1970, A.A. 5, 426.

Sutantyo, W.: 1975, A.A. 41, 47.

Sutantyo, W.: 1978, Ap.&SS 54, 479.

Suzuki, T., Kumagai, S., Shigemayama, T., Nomoto, K., Yamaoka, H., Saio, H.: 1993, Ap.J. 419, L73.

Tammann, G.A.: 1970, A.A. 8, 458.

Tammann, G.A.: 1974, in 'Supernovae and Supernova Remnants', ed. C.B. Cosmovici, Reidel: Dordrecht, p. 155.

Tassoul, J.- L., Tassoul, M.: 1992, Ap.J. 395, 258.

Tassoul, J.- L.: 1987, Ap.J. 322, 856.

Tassoul, J.- L.: 1988, Ap.J. 324, L71.

Tassoul, J.- L.: 1990, Ap.J. 358, 196.

Taylor, J.H., Cordes, J.M.: 1993, Ap.J. 411, 674.

Thomas, H.C.: 1977, Ann. Rev. Astron. Astrophys. 15, 127.

Thorne, K.S., Zytkow, A.N.: 1977, Ap.J. 212, 832.

Torres, G. et al.: 1992, in 'Evolutionary Processes in Close Binary Systems, IAU Symp. 151, eds. Y. Kondo, R. Sistero & R. Polidan, Kluwer: Dordrecht, p. 491.

Turon, et al.: 1992, 'The Hipparcos Input Catalogue', esa SP-1136.

Tutukov, A.V., Yungelson, L.R., Iben, I.: 1992, Ap.J. 386, 197.

Tutukov, A.V., Yungelson, L.R.: 1979, in 'Mass loss and Evolution of O-type Stars', Eds. C. De Loore , C. Conti, Reidel, Dordrecht, p. 401.

Ulrich, R.K., Burger, H.L.: 1976, Ap.J. 206, 509.

Underhill, A.B., 1982, Ap.J., 1982, 263, 741

Underhill, A.B., Divan,L., Prévot- Burnichon, M.L., Doazan, V., 1979, MNRAS 189, 601.

Underhill, A.B., Doazan,V., 1982, "B Stars With and Without Emission Lines", NASA SP- 456.

Underhill, A.B., Greve, G.R., Louth, H.: 1990, PASP 102, 749.

Vacca, W.D., Conti, P.S.: 1992, Ap.J. 401, 543.

Vacca, W.D., Torres-Dodgen, A.V.: 1990, Ap. J. Suppl. 73, 685.

Van Bever, J., Vanbeveren, D.: 1997, A.A. 322, 116.

Van Bever, J., Vanbeveren, D.: 1998, A.A. (in press).

Van den Heuvel, E.P.J., Heise, J.G.: 1972, Nature Phys. Sci. 239, 67.

Van den Heuvel, E.P.J.: 1978, in 'Physics and Astrophysics of Neutron Stars and Black Holes', eds. R. Giacconi & R. Ruffini, North. Holl. Pub. Com., Amsterdam, p. 828.

Van den Heuvel, E.P.J.: 1980, in 'X-ray Astronomy', eds. R. Giacconi & G. Setti, Reidel, Dordrecht, p. 115.

Van den Heuvel, E.P.J.: 1993, in 'Interacting Binaries', Saas-Fee Advanced Course 22, eds. H. Nussbaumer & A. Orr, Pringer-Verlag: p. 263.

Van der Hucht, K.A., Conti, P.S., Lundsrom, I., Stenholm, B.: 1981, Space Sci. Rev. 28, 227.

Van der Hucht, K.A., Schrijver, J., Stenholm, B. et al.: 1997, New Astronomy, Vol. 2, 245.

Van der Hucht,K.A., Hidayat,B., Admiranto,A.G., Supelli, K.R., Doom,C., 1988, A.A. 199, 217.

Van der Hucht,K.A.,1996, Astrophysics and Space Science, 238, 1

Van Kerkwijk, A.H., Van Paradijs, J., et al.: 1995, A.A. 303, 483.

Van Kerkwijk, M.H., Charles, P.A., Geballe, T.R., King, D.L., Miley, G.K., Molnar, L.A., Van den Heuvel, E.P.J., Van der Klis, M., Van Paradijs, J.: 1992, Nature 355, 703.

Van Kerkwijk,M.H.,1993,A.A. ,276, L8.

Van Rensbergen, W., Vanbeveren, D., De Loore, C.: 1996, A.A. 305, 825.

Vanbeveren, D., Conti, P.S.: 1980, A.A. 88, 230.

Vanbeveren, D., Herrero, A., Kunze, D., Van Kerkwijk, M.: 1994, Space Sci. Rev. 66, 395.

Vanbeveren, D., Packet, W.: 1979, A.A. 80, 242.

Vanbeveren, D., Van Bever, J., De Donder, E.: 1997, A.A. 317, 487.

Vanbeveren, D.: 1982, A.A. 115, 65.

Vanbeveren, D.: 1988, Astrophys. Space Sci. 149, 1.

Vanbeveren, D.: 1991, A.A. 252, 159.

Vanbeveren, D.: 1991, Space Sci. Rev. 56, 249.

Vanbeveren, D.: 1994, Space Sci. Rev. 66, 327.

Vanbeveren, D.: 1995, A.A. 294, 107.

Vanbeveren, D.: 1996, in 'Evolutionary Proceses in Binary Stars', eds. R.A.M.J. Wijers, M.B. Davies, C. A. Tout, NATO ASI Series, Kluwer Academic Publishers, p. 155.

Vanbeveren,D., Van Rensbergen,W., De Loore,C., 1982, A.A. 115, 68.

Venn, K.A.: 1994, Space Sci. Rev. 66, 163.

Verbunt, F., Wijers, R.A.M.J., Burn, H.: 1990, A.A. 234, 195.

Verbunt, F.: 1990, in 'Neutron Stars and Their Birth Events', ed. W. Kundt, Kluwer, Dordrecht, p. 179.

Vereshchagin, S., Kraicheva, Z., Popova, E., Tutukov, A., Yungelson, L.: 1987, Pis'ma v Astronom. Zhurnal 13, 63.

Vereshchagin, S., Tutukov, A., Yungelson, L., Kraicheva, Z., Popova, E.: 1988, Astrophys. Space Sci. 144, 245.

Verner,D.A. , Yakovlev, D.G., 1995, A.A.. Suppl. 109, 125

Vreux, J.M, Rauw, G., Bohannan, B., Manfroid, J.: 1997, in'Wolf-Rayet Stars in the Framework of Stellar Evolution', eds. J.M. Vreux, A. Detal, D. Fraipont-Caro, E. Gosset & G. Rauw, Université de Liège, p. 337.

Vreux, J.M.: 1985, PASP 95, 274.

Vreux, J.M.: 1987, in 'Instabilities in Evolved Super- and Hypergiants', eds. C. de Jager, H. Nieuwenhuijzen, North-Holland: Amsterdam, p. 81.

Waelkens, C., Lampens, P., Heynderickx, D. et al.: 1990, A.A. Suppl. 83, 11.

Wagoner, R.V.: 1975, Ap.J. 196, L63.

Walborn, N.R.: 1971, Ap.J. 164, 267.

Walborn, N.R.: 1976, Ap.J. 205, 419.

Waters, L.B.F.M., Pols, O.R., Hogeveen, S.J., Coté, J., Van de Heuvel, E.P.J.: 1989, A.A. 220, L1.

Webbink, R.F.: 1984, Ap.J. 277, 355.

Wheeler, J.C.: 1996, in 'Evolutionary Proceses in Binary Stars', eds. R.A.M.J. Wijers, M.B. Davies, C. A. Tout, NATO ASI Series, Kluwer Academic Publishers, p. 307.

Wheeler, J.C., Benetti: 1997, in 'Astrophysical Quantities', ed. A. Cox, (American Insitute of Physics press): in press.

Wijers, R.A.M.J.: 1996, in 'Evolutionary Proceses in Binary Stars', eds. R.A.M.J. Wijers, M.B. Davies, C. A. Tout, NATO ASI Series, Kluwer Academic Publishers, p. 327.

Williams, P.M., Van der Hucht, K.A.: 1996, in 'Wolf-Rayet Stars in the Framework of Stellar Evolution', eds. J.M. Vreux, A. Detal, D. Fraipont-Caro, E. Gosset & G. Rauw, Université de Liège, p. 353.

Wilson, R.E.: 1953, 'General Catalogue of Radial Velocities', Carnegie Inst.- Washington D.C.

Wilson, R.E.: 1989, Space Sci. Rev. 50, 235.

Wood,M.A., 1992, Astrophys. J., 386, 539.

Woosley, S.E.: 1986, in 'Nucleosynthesis and Chemical Evolution', 16th Saas-Fee Course, eds. B. Hauck et al., Geneva Observatory, p.1.

Wright, A.E., Barlow, M.J.: 1975, MNRAS 170, 41.

Zahn, J.P.: 1975, A.A. 41, 328.

Zahn, J.P.: 1977, A.A. 57, 383.

Zahn, J.P.: 1983, in 'Astrophysical Processes in Upper MS Stars', 13th Saas-Fee Course, eds. B. Hauck & A. Maeder, p.253.

Zahn, J.P.: 1992, A.A. 265, 115.

Zahn, J.P.: 1994, Space Science Rev. 66, 285.

Zeldovich, Y., B., Guseynov, O.H.: 1966, Ap. J. 144, 840.

Index

absolute visual magnitude (M_V): 18-21, 40, 266, 284
absorption coefficient: 67, 237, 238, 240-243
abundances
  CNO-equilibrium: 32, 39, 43, 57, 60, 81, 93, 150, 201, 205, 225, 238, 242-244
  galactic: 31, 155, 238, 243
  SMC-LMC: 60, 84, 155, 256
  solar: 31, 43, 60, 225, 230, 235, 238, 244
accretion: 91-93, 117-119, 159, 163, 170, 182, 189-192, 198, 211, 244, 247, 294, 303, 304, 325, 326
  accretion induced full mixing: 123, 162, 227
  standard accretion: 120, 121, 159, 167, 213, 225, 256
angular diameter: 15, 20
angular momentum: 101, 103, 104
  orbital angular momentum: 106, 111, 117, 136, 137, 136, 317
  spin angular momentum: 101, 103-105, 118, 189
association (cluster):: 15, 18, 40, 90, 225, 227, 231, 233, 281, 326
binaries
  contact: 171, 179
  detached: 171
  double compact, pulsars: (cfr. pulsars)
  post-RLOF: 179
  pre-RLOF: 171
  semi-detached: 171, 179
  Thorne-Zytkow (TZO): 168, 202, 216, 227, 251, 255, 287, 298, 301, 308, 309, 319
  WR+cc: 166, 167, 187, 229, 237, 245, 302, 303, 308, 309
  WR+OB: 43, 106, 137, 146, 153, 207-220, 279, 284-286, 301-303, 309, 310, 311
  X-ray
    HMXB: 184-187, 188
    LMXB: 187, 188
    transient: 185
binary frequency: 170, 262, 263, 265, 274, 275, 289, 299, 307, 318, 325, 327, 329, 330
  early B type stars: 265-269
  O type stars: 265
  pulsars: 196
  WR stars: 284
black hole (BH): 44, 55, 65, 76, 82, 83, 95, 106, 151, 158, 165, 166, 168, 169, 171, 187, 188, 190, 191, 208, 209, 211, 230, 235, 246, 248, 251, 253, 254, 291, 296-298, 300, 302, 303, 307, 308, 309, 316, 328, 329
blue Hertzsprung gap: 73, 96, 97, 162, 166, 303
bolometric correction: 18
chemical abundance; - composition; -species: 31, 44, 59-60, 96,

101, 104, 121, 132, 142, 163, 166, 238, 242, 296
circularization: 107-108, 111, 164, 181
circumstellar shells; -nebulae; -matter; - dust: 42, 46, 47, 86, 94, 261
cluster: see association
common envelope: 113, 129-132, 145, 146, 152, 219, 251, 254, 294
continuous star formation: 200, 279, 303, 314, 328
convection: 60-64, 113, 115, 121-123, 161
core carbon burning (CCB): 74, 76
core helium burning (CHeB): 39, 57, 73, 74, 65, 71, 73, 76-85, 97, 100, 137, 142, 144, 151, 153-159, 167, 200, 205, 222, 236, 248, 249, 251, 261, 290, 291, 293, 295, 301, 302, 304, 316, 319
core hydrogen burning (CHB): 55, 62, 65-66, 71, 72, 74, 77, 79, 80, 85, 90, 91, 100, 101, 103-105, 111, 115, 120, 145, 148, 150, 159, 161-163, 170, 218
diffusion: 61-63, 66, 97, 102, 121, 142, 159, 161, 256-257
eccentricity: 106-108, 136, 172, 173, 179, 185, 229, 288
effective temperature: 15-16, 18
energy generation: 55, 59
energy transport
radiative: 60
convective: 61
entropy: 55, 72, 91-92, 119-120, 123, 141, 164
equation of state:
degenerate plasma: 52, 54, 55, 96, 165
non- degenerate plasma: 51
equipotential surfaces: 109, 110
helium burning: 55, 57, 71, 78
Hertzsprung-Russell diagram: 15, 40, 48, 90, 94, 201, 226-227, 233-234
hydrogen burning: 55-56, 72, 77, 81, 82, 87, 100-101, 103, 105, 115, 149, 150, 154, 160
hydrostatic equilibrium: 28, 50, 54, 67, 69, 110, 113, 119, 223
hypergiants: 38, 42, 199, 229
initial mass function (IMF): 263-265, 280, 297-299, 303, 318, 322, 327, 329
kick velocity: 133, 135, 262-263, 275,276, 278-279, 297-298, 300-302, 307, 309, 315, 317, 319-320, 327
Lagrangian point: 109, 111, 127, 128, 152, 164-165, 191, 294, 303
LBV scenario: 155, 209, 211, 217, 218, 230, 309
limiting mass
white dwarfs: 53-54, 98, 171, 188
neutron stars: 54-55, 99-100, 165, 169, 171, 185, 222, 246

Luminous blue variable (LBV): 36, 42, 66, 80, 83, 84, 85, 113, 125, 132, 150, 209, 254, 291, 294
Magellanic Cloud: 11, 39, 83, 85, 93, 96, 116, 142, 150, 281, 284, 290, 304, 322
mass function (binaries): 205, 246-248, 251, 270-271
mass ratio distribution: 170, 263, 265, 269, 270, 275, 298, 309, 310, 315, 317, 328, 329
    early B type binaries: 272, 273
    O type binaries: 271, 272
    WR binaries: 285, 311
mass transfer: 110, 114, 116-120, 162-166, 191, 200, 201, 204, 209, 211, 218, 251, 260, 294, 325
merger: 131, 132, 151, 255, 257, 274, 294, 297, 300, 303, 304, 327, 328
neutron star (NS): 43, 54, 76, 82, 87, 99, 104-105, 132, 158, 169-171, 184-185, 187-191, 195, 196, 219, 222, 230, 236, 251, 253, 286, 287, 301, 303, 304, 307, 309, 314-320
nuclear reaction: 55, 71
Objects
    Be binary
        Φ Per: 180, 202
        HR 4804: 204
        μ^2 Cru: 204
    Be X-ray binaries
        V 0322+53: 185, 186
    evolved OBN binaries
        HD 163181: 179, 180, 200, 201, 216
        HD 12323: 179, 180, 205
        HD 14633: 179, 180, 205
        HD 193516: 180, 205
    HMXB
        4U1700-37: 229
        Cyg X-1: 185, 186, 229, 236
        Cyg X-3: 186, 187, 236, 245, 260, 286
        Vela X-1: 107, 186, 199, 223, 236
        Wray 977: 186, 228, 229, 235, 236, 254, 296
    LMXB
        Her X-1: 188, 200, 219, 251, 254, 274
    OB+OB post-RLOF binaries
        HD 25638: 180, 206
        HD 209481: 179, 180, 206
        υ Sgr: 180, 206
    Runaways
        HD 50896 (WR):: 43, 65, 187, 199, 220, 221, 236, 238, 245, 286, 310
        HD 197406 (WR): 187, 199, 220, 221, 236, 245, 286
        ρ Cas (hypergiant):: 199
        ζ Oph (O-type):: 199, 230
        ζ Pup (O-type):: 199, 230, 231, 235
semi-convection: 63-64, 70-71, 76, 80, 82-85, 87, 96-97, 105,

121, 123, 142, 150, 162, 166-167, 255-257, 261, 290, 304
Stellar aggregates
LH 9: 281
LH 10: 281
LH 117: 281
LH 118: 281
Vela OB1: 281
Supernova
SN 1987A: 46-48, 94, 166, 167, 255, 261, 304
SN 1993 J: 47, 95
SN 1994 I: 47
WR like binary
V Saggitae: 154, 181, 183, 206, 208, 221, 248, 251
WR+OB binaries
V 444 Cyg: 65, 183, 211
γ^2 Vel: 209, 227, 303
HD 190918: 290, 303
HD 137603: 183, 211
WR 19: 219
WR 104: 219
WR 119: 219
HD 192641: 183, 211
HD 193077: 183, 211
opacity: 60, 70, 81, 117, 164, 238
overshooting: 63, 64, 66, 85, 86, 100, 144, 145, 169, 290
period distribution: 115, 170, 262, 269, 298, 301, 322, 328, 329
double pulsars: 288, 317, 318
early B type binaries: 273
O type binaries: 273
WR binaries: 285, 310, 311
population
binaries: 153, 299
early B-type stars: 314, 322
massive stars: 263, 300
model: 262, 279
O-type stars: 304-307, 322
pulsars: 286, 287
RSG: 326
SN: 44, 45, 255, 328, 329, 330
stars: 265
synthesis: 96, 262, 280, 281, 287, 289, 298, 327
WR stars: 307-314, 322, 326
WR+OB: 216
pulsar
binary: 99, 171, 185, 188, 197, 229, 251, 287, 288, 316-320
millisecond: 196-198
radio: 195-197, 278, 286, 287
single: 105, 137, 262, 263, 275-279, 296, 298, 317, 319, 320
red supergiant (RSG): 38, 42, 77-80, 83, 93, 96-97, 100, 114, 115, 130, 146, 167, 168, 207, 235, 282, 302, 314
rejuvenation: 159-163, 211, 217, 235, 325
Roche lobe overflow (RLOF): 42, 44, 59, 110, 111, 113, 115, 116, 118-119, 125-132, 139-154, 156-158, 160, 163, 164, 169, 170, 179, 191, 217, 294, 299, 300, 303-304
rotation: 28-29, 32, 36, 43, 95, 100-105, 107, 110, 123, 169, 184,

185, 189, 195, 204, 209, 218, 223, 315
RSG scenario: 146, 211, 218-219, 294, 309
runaway: 135, 136, 171, 216, 219, 300
early B-type: 198, 315, 316
O-type: 198, 230-235, 255, 286, 304, 306, 307, 326, 327
pulsar: 137, 262, 263, 275-279, 317, 319, 320
WR: 43, 187, 199, 221, 251, 284, 303, 310
X-ray binary: 199, 200
semi- convection: 63-64, 70, 80, 82-85, 87, 97, 105, 121, 123, 142, 150, 166, 167, 255-257, 290, 304
space distribution (stars): 265-268
spiral in: 113, 129-132, 151, 152, 167-169, 202, 229, 236, 251, 287, 294, 298, 300-302, 308, 309, 317-319
starburst: 100, 200, 264, 279-282, 320-327
stellar structure: 49, 60, 68-69
stellar wind (SW): 15, 26, 36, 38, 40, 42-43, 65, 68, 74-77, 80-83, 94-97, 105, 114, 116, 124, 129, 132, 145, 146, 150, 184, 192, 207, 209, 211, 217, 222, 237, 239, 243, 248, 254, 261, 290, 291, 293, 310
supernova (SN): 43, 48, 59, 75, 76, 82, 94-95, 105, 132, 158, 165-167, 198, 200, 208, 216, 219, 222, 223, 229, 230, 254, 255, 262, 263, 275, 278, 286-288, 291, 300, 302-304, 306, 307, 309, 310, 314, 317, 319, 320, 328
synchronization: 107, 108, 111, 164
thermohaline mixing: 122, 163, 206, 225
Thorne-Zytkow object: see binaries
white dwarf (WD): 44, 54, 74, 75, 98, 99, 171, 188, 196, 197, 204, 206, 308
WN/WC: 65, 182, 279
Wolf-Rayet stars (WR):: 39, 43, 65, 81-82, 93, 106, 150, 153, 155, 168, 181, 187, 199, 206, 208, 218-219, 222, 236, 238, 245, 249, 251, 280, 282, 284, 286, 290-291, 295, 301, 303, 307, 322
see also: binaries
WR/O number ratio: 279, 280, 308-310, 314, 323, 325
X-rays: 119, 184, 188, 191-193, 236-238, 245-250, 286
see also: binaries
Yellow supergiant (YSG):: 38, 42, 290